PHOTOMESIC AND PHOTONUCLEAR REACTIONS
AND INVESTIGATION METHODS WITH SYNCHROTRONS

FOTOMEZONNYE I FOTOYADERNYE REAKTSII
I METODIKA ISSLEDOVANIYA NA SINKHROTRONE

ФОТОМЕЗОННЫЕ И ФОТОЯДЕРНЫЕ РЕАКЦИИ
И МЕТОДИКА ИССЛЕДОВАНИЯ НА СИНХРОТРОНЕ

The Lebedev Physics Institute Series

Editor: Academician D. V. Skobel'tsyn

Director, P. N. Lebedev Physics Institute, Academy of Sciences of the USSR

Volume 25	Optical Methods of Investigating Solid Bodies
Volume 26	Cosmic Rays
Volume 27	Research in Molecular Spectroscopy
Volume 28	Radio Telescopes
Volume 29	Quantum Field Theory and Hydrodynamics
Volume 30	Physical Optics
Volume 31	Quantum Electronics in Lasers and Masers, Part 1
Volume 32	Plasma Physics
Volume 33	Studies of Nuclear Reactions
Volume 34	Photomesic and Photonuclear Processes
Volume 35	Electronic and Vibrational Spectra of Molecules
Volume 36	Photodisintegration of Nuclei in the Giant Resonance Region
Volume 37	Electrical and Optical Properties of Semiconductors
Volume 38	Wideband Cruciform Radio Telescope Research
Volume 39	Optical Studies in Liquids and Solids
Volume 40	Experimental Physics: Methods and Apparatus
Volume 41	The Nucleon Compton Effect at Low and Medium Energies
Volume 42	Electronics in Experimental Physics
Volume 43	Nonlinear Optics
Volume 44	Nuclear Physics and Interaction of Particles with Matter
Volume 45	Programming and Computer Techniques in Experimental Physics
Volume 46	Cosmic Rays and Nuclear Interactions at High Energies
Volume 47	Radio Astronomy: Instruments and Observations
Volume 48	Surface Properties of Semiconductors and Dynamics of Ionic Crystals
Volume 49	Quantum Electronics and Paramagnetic Resonance
Volume 50	Electroluminescence
Volume 51	Physics of Atomic Collisions
Volume 52	Quantum Electronics in Lasers and Masers, Part 2
Volume 53	Studies in Nuclear Physics
Volume 54	Photomesic and Photonuclear Reactions and Investigation Methods with Synchrotrons
Volume 55	Optical Properties of Metals and Intermolecular Interactions
Volume 56	Physical Processes in Lasers
Volume 58	Investigations on Nonlinear Optics and Hyperacoustics
Volume 59	Luminescence and Nonlinear Optics

In preparation

Volume 57	Theory of Interaction of Elementary Particles at High Energies
Volume 60	Spectroscopy of Laser Crystals with Ionic Structure
Volume 61	Theory of Plasmas
Volume 62	Stellar Atmospheres and Interplanetary Plasma: Techniques for Radioastronomical Devices
Volume 63	Nuclear Reactions and Interactions of Neutrons with Matter
Volume 64	Primary Cosmic Radiation
Volume 65	Stellarators
Volume 66	Theory of Coherent Acceleration of Particles and Emission of Relativistic Bunches
Volume 67	Physical Investigations of Strong Magnetic Fields
Volume 68	Emissive Recombination in Semiconducting Crystals
Volume 69	Nuclear Reactions and Accelerators of Charged Particles

Proceedings (Trudy) of the P. N. Lebedev Physics Institute

Volume 54

PHOTOMESIC AND PHOTONUCLEAR REACTIONS AND INVESTIGATION METHODS WITH SYNCHROTRONS

Edited by
Academician D. V. Skobel'tsyn
Director, P. N. Lebedev Physics Institute
Academy of Sciences of the USSR, Moscow

Translated from Russian by
Joachim R. Büchner

SPRINGER SCIENCE+BUSINESS MEDIA, LLC

Library of Congress Cataloging in Publication Data

Main entry under title:

Photomesic and photonuclear reactions and investigation methods with synchrotrons.

(Proceedings (Trudy) of the P. N. Lebedev Physics Institute, v. 54)
Translation of Fotomezonnye i fotoi͡adernye reakt͡sii i metodika issledovanii͡a na sinkhrotrone.
Includes bibliographies.
1. Photonuclear reactions—Addresses, essays, lectures. 2. Mesons—Addresses, essays, lectures. 3. Nuclear physics—Instruments—Addresses, essays, lectures. I. Skobel't͡syn, Dmitriĭ Vladimirovich, 1892- ed. II. Series: Akademii͡a nauk SSSR Fizicheskiĭ Institut. Proceedings, v. 54.
QC1.A4114 vol. 54 [QC794.8.P4] 530'.08s [539.7'56]
ISBN 978-1-4757-6177-1 ISBN 978-1-4757-6175-7 (eBook) 73-79424
DOI 10.1007/978-1-4757-6175-7

The original Russian text was published by Nauka Press in Moscow in 1971 for the Academy of Sciences of the USSR as Volume 54 of the Proceedings of the P. N. Lebedev Institute. The present translation is published under an agreement with the Copyright Agency of the USSR.

PREFACE

This collection of articles contains a systematic outline of original experimental and theoretical research on photoproduction of neutral pions at protons and at a strongly bound system of a few nucleons, i.e., the helium nucleus.

Spark chambers and their use as spectrometers for photons and electrons are described in detail. The articles of the collection include information on a novel method of determining the efficiency of recording apparatus by generating monochromatic photons. The articles describe original theoretical research on the optical anisotropy of nuclei. Problems encountered in experimental studies of operating the synchrotron as a storage-type accelerator of electrons and positrons receive particular attention. The results of this research work are listed, and the problems of oppositely directed electron–positron beams in the 250-MeV synchrotron are considered.

The articles should be of interest to physicists, including research workers, teachers, engineers, graduate students, and students in advanced undergraduate courses.

CONTENTS

PHOTOPRODUCTION OF NEUTRAL PIONS AT NUCLEONS AND NUCLEI

B.V. Govorkov, S.P. Denisov, and E.V. Minarik

The article lists the results of experimental and theoretical research on the photoproduction of π^0 mesons at protons and composite nuclei in the low-energy range. The principal dynamic characteristics of the process were determined for hydrogen. A detailed comparison is made between experimental results and dispersion theory. A discrepancy between theory and experiment was established and the reasons for the discrepancy were investigated. A detailed photoproduction scheme was developed for nuclei. It was shown that elastic coherent formation of π^0 mesons is the dominating process. The experimental results were used to determine structural parameters of nuclei. Several problems are mentioned, whose solution will provide new experimental information required for the development of a theory of pion photoproduction.

Introduction

Research on the interaction between hadrons is one of the most important problems in the physics of high-energy particles. The photoproduction of pions at nucleons is important because in addition to pion–nucleon scattering, pion photoproduction is the simplest reaction in which pions interact with nucleons and from which information on the electromagnetic properties of the pion–nucleon system can be obtained. Experimental work on the photoproduction of pions resulted in the detection of the first resonances in the pion–nucleon system, which were termed Δ (1236), N* (1518), and N* (1688) particle, and in the corresponding quantum numbers. Many of the approximations of strong interactions were checked by comparing the approximations with experimental results obtained in the photoproduction of pions at nucleons. Comparisons of this type involved dispersion theory, the quark particle model, the theory of approximate particle symmetries, and other theories. Experimental results on the photoproduction of pions were extensively used in superconvergent summation rules in order to derive relations between the binding constants of particles and the widths of resonance decays. Interest in photoproduction processes increased recently because it became possible tocheck the charge invariance and the invariance of electromagnetic interactions of hadrons with respect to time reversal.

The photoproduction of pions at composite nuclei is of great interest because this process provides information on both the mechanism of the interaction between photons and a group of nuclei and the density distribution of matter in the nucleus.

The present article describes the results of experimental and theoretical work on the photoproduction of neutral pions at protons and composite nuclei. The research work was done mainly in the near-threshold range of the primary photon energy. The range near the threshold was selected for two reasons in the photoproduction of pions at nucleons. First, the amplitude of the photoproduction is in this range very sensitive to several secondary effects and some parameters of the physics of pions, the parameters being important for dispersion theory, the theory of unitary symmetry, etc. Second, a phenomenological (multipole) analysis, which is the analog to phase analysis in pion–nucleon scattering, can be easily performed in the low-energy range. Accordingly, more or less rigorous quantitative comparisons between the results of theoretical calculations of the dynamic characteristics (energy dependence and angular dependences of the cross section, etc.) and experimental results can be made only in this energy range.

The near-threshold range was also singled out in the photoproduction of neutral pions at composite nuclei. Elastic coherent photoproduction of pions dominates in that range. Pions with low energies are produced so that multiple scattering and absorption are of little importance. The momentum approximation, in which the amplitude of photoproduction at free nucleons is assumed to be known, is used for the interpretation of the photoproduction of pions at nuclei because this approach is more easily taken in the low-energy range.

The photoproduction of pions at nucleons and nuclei was studied in two stages. The general features of the near-threshold photoproduction were determined in the first stage. The details of the processes considered were established in the second stage. Naturally, work done recently received particular attention.

CHAPTER I

The Photoproduction of Neutral Pions at Nucleons

Photoproduction Amplitudes and Differential Cross Sections

General Expressions for Amplitudes, Cross Sections, and Polarization

The amplitude of the photoproduction of pions at nucleons can be stated in the following form [1-4] in the center-of-mass system

$$\mathcal{F} = i\sigma\varepsilon\mathcal{F}_1 + \frac{\sigma q \sigma [k\varepsilon]}{qk}\mathcal{F}_2 + i\frac{\sigma k q\varepsilon}{qk}\mathcal{F}_3 + i\frac{\sigma q q\varepsilon}{q^2}\mathcal{F}_4, \tag{I.1}$$

where k and q denote the momentum of the photon and the pion, respectively, in the center-of-mass system; and σ denotes the vector of the nucleon spin. The amplitudes $\mathcal{F}_i$ are functions of the energy and of the emission angle θ of the pion.

In the center-of-mass system, the differential cross section can be expressed by the amplitude

$$\frac{d\sigma}{d\Omega} = \frac{q}{k}|\chi_f \mathcal{F} \chi_i|^2, \tag{I.2}$$

where χ denotes the Pauli spin matrices.

It follows from (I.1) that the spin independent part of the amplitude of pion photoproduction is entirely determined by the amplitude $\mathcal{F}_2$.

After substituting (I.2) in (I.1), averaging, and summing over the spin states of the nucleon in the initial and final states, the differential cross section of the photoproduction of pions at nucleons by linearly polarized γ quanta (which are incident on nonpolarized nucleons) is

$$\frac{d\sigma}{d\Omega}(\theta,\varphi)=\frac{q}{k}\Big[\Big(|\mathcal{F}_1|^2+|\mathcal{F}_2|^2+\frac{1}{2}|\mathcal{F}_3|^2+\frac{1}{2}|\mathcal{F}_4|^2+\mathrm{Re}\,\mathcal{F}_1\mathcal{F}_4^*+\mathrm{Re}\,\mathcal{F}_2\mathcal{F}_3^*\Big)+$$
$$+\left(\mathrm{Re}\,\mathcal{F}_3\mathcal{F}_4^*-2\,\mathrm{Re}\,\mathcal{F}_1\mathcal{F}_2^*\right)\cos\theta-\Big(\frac{1}{2}|\mathcal{F}_3|^2+\frac{1}{2}|\mathcal{F}_4|^2+\mathrm{Re}\,\mathcal{F}_1\mathcal{F}_4^*+\mathrm{Re}\,\mathcal{F}_2\mathcal{F}_3^*\Big)\times$$
$$\times\cos^2\theta-\mathrm{Re}\,\mathcal{F}_3\mathcal{F}_4^*\cos^3\theta+\Big(\frac{1}{2}|\mathcal{F}_3|^2+\frac{1}{2}|\mathcal{F}_4|^2+\mathrm{Re}\,\mathcal{F}_1\mathcal{F}_4^*+\mathrm{Re}\,\mathcal{F}_2\mathcal{F}_3^*+\mathrm{Re}\,\mathcal{F}_3\mathcal{F}_4^*\cos\theta\Big)\sin^2\theta\cos 2\varphi\Big], \quad (I.3)$$

where φ denotes the angle between the planes of photon polarization and pion formation. After averaging Eq. (I.3) over the direction of photon polarization, the differential cross section of pion photoproduction by unpolarized γ quanta incident on an unpolarized target is

$$\frac{d\sigma}{d\Omega}(\theta)=\frac{q}{k}\Big[\Big(|\mathcal{F}_1|^2+|\mathcal{F}_2|^2+\frac{1}{2}|\mathcal{F}_3|^2+\frac{1}{2}|\mathcal{F}_4|^2+\mathrm{Re}\,\mathcal{F}_1\mathcal{F}_4^*+$$
$$+\mathrm{Re}\,\mathcal{F}_2\mathcal{F}_3\Big)+\left(\mathrm{Re}\,\mathcal{F}_3\mathcal{F}_4^*-2\mathrm{Re}\,\mathcal{F}_1\mathcal{F}_2^*\right)\cos\theta-\Big(\frac{1}{2}|\mathcal{F}_3|^2+\frac{1}{2}|\mathcal{F}_4|^2+$$
$$+\mathrm{Re}\,\mathcal{F}_1\mathcal{F}_4^*+\mathrm{Re}\,\mathcal{F}_2\mathcal{F}_3^*\Big)\cos^2\theta-\mathrm{Re}\,\mathcal{F}_3\mathcal{F}_4^*\cos^2\theta\Big]. \quad (I.4)$$

The polarization P of the recoil nucleon can be expressed by the amplitude $\mathcal{F}_i$

$$P_{\bar{n}}\frac{d\sigma}{d\Omega}=\frac{q}{k}\sin\theta\,\mathrm{Im}\,[(2\mathcal{F}_1\mathcal{F}_2^*+\mathcal{F}_1\mathcal{F}_3^*-\mathcal{F}_2\mathcal{F}_4^*-\mathcal{F}_3\mathcal{F}_4^*)+(\mathcal{F}_1\mathcal{F}_4^*-\mathcal{F}_2\mathcal{F}_3^*)\cos\theta+\mathcal{F}_3\mathcal{F}_4^*\cos^2\theta]. \quad (I.5)$$

Thus, polarization results exclusively from the interference of real and imaginary parts of the various amplitudes $\mathcal{F}_i$. When pions are formed by unpolarized photons, only the polarization component of the recoil nucleon parallel to the unit vector **n** (which is perpendicular to the plane in which the photoproduction takes place) is nonvanishing. Expressions for the polarization have been stated in [4-7] for the case of linearly polarized photons. All expressions for the cross sections and polarizations refer to the center-of-mass system.

The scalar amplitudes $\mathcal{F}_i$ were expanded in [1, 2, 8] in terms of multipole amplitudes which lead to final states with a certain parity:

$$\mathcal{F}_1=\sum_{l=0}^{\infty}\{[lM_{l+}+E_{l+}]\,P'_{l+1}(x)+[(l+1)M_{l-}+E_{l-}]\,P'_{l-1}(x)\},$$
$$\mathcal{F}_2=\sum_{l=1}^{\infty}[(l+1)M_{l+}+lM_{l-}]\,P'_l(x),$$
$$\mathcal{F}_3=\sum_{l=1}^{\infty}\{[E_{l+}-M_{l+}]\,P''_{l+1}(x)+[E_{l-}+M_{l-}]\,P''_{l-1}(x)\}, \quad (I.6)$$
$$\mathcal{F}_4=\sum_{l=1}^{\infty}[M_{l+}-E_{l+}-M_{l-}-E_{l-}]\,P''_l(x),$$

where $P'(x)$ and $P''(x)$ denote derivatives of Legendre polynomials; $x=\cos\theta$, $M_{l\pm}$ and $E_{l\pm}$ denote the amplitudes of the transitions which are caused by magnetic or electric radiation and lead to the final state; πN denotes systems with the total moment $j=l\pm 1/2$ and the parity $(-1)^{l+1}$. The orbital moment of the pion is denoted by l.

According to (I.1) and (I.6), the angular distribution of the pion photoproduction in the center-of-mass system must have the form

$$\frac{d\sigma}{d\Omega}(\theta) = \sum_{n=0}^{N} a_n (\cos\theta)^n, \qquad (I.7)$$

where N = 2j − 1.

Only terms with even powers of $\cos\theta$ are found in the angular distribution of a certain multipole. In the general case, however, the photoproduction depends upon several complex multipole amplitudes so that interference terms with odd powers of $\cos\theta$ appear in the angular dependence of the cross section [see Eq. (I.3)]. For example, when the formation of pions in the s and p wave states is taken into account, the differential cross section assumes the form

$$\frac{d\sigma}{d\Omega}(\theta) = A + B\cos\theta + C\cos^2\theta. \qquad (I.8)$$

We have employed the frequently used notation a_0 = A, a_1 = B, and a_2 = C for the coefficients.

Isospin Structure of the Amplitude

Since it is possible to represent the operator of the electric charge as the sum of the third component I of the isotope spin operator and half-sum of the operators of the barion number N and the strangeness S

$$Q = I_3 + \frac{N+S}{2} \qquad (I.9)$$

the electromagnetic current whose interaction with the photon field results in pion photoproduction can be divided into two parts: an isoscalar part which is related to I_3, and an isovector part which is related to (N + S)/2.

This, in turn, leads to a splitting of the photoproduction amplitude in isotope spin space. We use the following notation for the amplitudes of the various channels of the photoproduction process:

$$\begin{aligned} &\gamma + p \to n + \pi^+ && \mathcal{F}(\pi^+), \\ &\gamma + n \to p + \pi^- && \mathcal{F}(\pi^-), \\ &\gamma + p \to p + \pi^0 && \mathcal{F}(\pi^0), \\ &\gamma + n \to n + \pi^0 && \mathcal{F}(n\pi^0). \end{aligned} \qquad (I.10)$$

The following notation is employed for the currents and amplitudes which correspond to transitions into a final state with certain isotope spin values

$$\begin{aligned} &I_i = \tfrac{1}{2} \to I_f = \tfrac{1}{2} && j^s, \mathcal{F}^s; \\ &I_i = \tfrac{1}{2} \to I_f = \tfrac{1}{2} && j^v = j^{1/2}, \mathcal{F}^{1/2}; \\ &I_i = \tfrac{1}{2} \to I_f = \tfrac{3}{2} && j^v = j^{3/2}, \mathcal{F}^{3/2}. \end{aligned} \qquad (I.11)$$

The quantities

$$\begin{aligned} j^{(+)} &= \tfrac{1}{3}(j^{1/2} + 2j^{3/2}), \\ j^{(-)} &= \tfrac{1}{3}(j^{1/2} - j^{3/2}). \end{aligned} \qquad (I.12)$$

are usually introduced in place of isovector currents. The amplitude of photoproduction in isotope spin space can therefore be written as follows:

$$\mathcal{F} = \mathcal{F}^{(+)}\frac{1}{2}(\tau_\beta\tau_3 + \tau_3\tau_\beta) + \mathcal{F}^{(-)}\frac{1}{2}(\tau_\beta\tau_3 - \tau_3\tau_\beta) + \mathcal{F}^{s}\tau_\beta, \tag{I.13}$$

where τ_β denotes the isotope spin of the pion; the isovector parts of the amplitude ($\mathcal{F}^{(\pm)}$) correspond to the electromagnetic currents (I.12). The amplitudes for certain final charge states of the πN system are

$$\begin{aligned}
\mathcal{F}(\pi^+) &= \sqrt{2}(\mathcal{F}^s + \mathcal{F}^{(-)}) = \sqrt{2}\left(\mathcal{F}^s + \frac{1}{3}\mathcal{F}^{1/2} - \frac{1}{3}\mathcal{F}^{3/2}\right),\\
\mathcal{F}(\pi^-) &= \sqrt{2}(\mathcal{F}^s - \mathcal{F}^{(-)}) = \sqrt{2}\left(\mathcal{F}^s - \frac{1}{3}\mathcal{F}^{1/2} + \frac{1}{3}\mathcal{F}^{3/2}\right),\\
\mathcal{F}(\pi^0) &= \mathcal{F}^{(+)} + \mathcal{F}^{(s)} = \mathcal{F}^s + \frac{1}{3}\mathcal{F}^{1/2} + \frac{2}{3}\mathcal{F}^{3/2},\\
\mathcal{F}(n\pi^0) &= \mathcal{F}^{(+)} - \mathcal{F}^{(s)} = -\mathcal{F}^s + \frac{1}{3}\mathcal{F}^{1/2} + \frac{2}{3}\mathcal{F}^{3/2}.
\end{aligned} \tag{I.14}$$

Thus, the isotope invariance principle can be used to express the photoproduction amplitudes for the four channels by three quantities: $\mathcal{F}^{(\pm)}$ and $\mathcal{F}^s = \mathcal{F}^{(0)}$. The photoproduction amplitudes will be expressed in our ensuing discussion by $\mathcal{F}_i^{(\alpha)}$, where i = 1, 2, 3, 4 in accordance with Eq. (I.1) and $\alpha = \pm, 0$ in accordance with Eq. (I.13). As put into evidence by the expressions for the cross sections and the polarization [(I.1)-(I.5)], the twelve amplitudes $\mathcal{F}_i^{(\alpha)}$ determine all characteristics of the photoproduction process. Since the $\mathcal{F}_i^{(\alpha)}$ are complex, the actual number of parameters describing the reaction is twice as large. The determination of the amplitudes $\mathcal{F}_i^{(\alpha)}$ with the aid of some dynamic principle is the subject of the theory of pion photoproduction. In analogy, the determination of the $\mathcal{F}_i^{(\alpha)}$ from measurements is the object of experiments. The solution is termed a total experiment and requires, in the general case, that results of polarization experiments are available.

Near-Threshold Energy Range

The energy range of primary photons near the threshold (k_{thresh} = 145 MeV for π^0 production at protons) is of particular interest in research on the photoproduction of pions at nucleons. The near-threshold range is usually defined as the energy region of the primary photons for which the momenta of single pions generated in the center-of-mass system is within the limits $0 \le q \le 1$. The range of low energies was selected for the following reasons:

a) The conditions of the Watson theorem [9], which is a consequence of both the unitarity of the S matrix and the invariance of pion processes with respect to time reversal, are satisfied in this range. According to the Watson theorem, the phase of the multipole amplitude $N_{l\pm}^I$, which results in a final state with a certain total moment, parity, and isospin is determined by the phase shift of the pion–nucleon scattering into the $\alpha_{l\pm}^I$ state:

$$N_{l\pm}^I = |N_{l\pm}^I|\, e^{i\alpha_{l\pm}^I}. \tag{I.15}$$

This theorem establishes a relation between the photoproduction and πN scattering and makes it possible to halve the number of unknown parameters which characterize the amplitudes $\mathcal{F}_i^{(\alpha)}$ of photoproduction. The phase shifts are almost vanishing near the threshold and the multipole amplitudes can be assumed as real;

b) Since the πN interaction space is bounded, only some of the lowest multipole amplitudes are significant in the near-threshold range. The amplitude at a particular l is proportional to q^l, and the corresponding cross section is proportional to q^{2l+1}. In other words, the long-wave approximation can be employed. This approximation was the principal method of comparing experimental results with the theoretical conclusions of [10] and of A. M. Baldin and one of the authors of the present article [11]. According to this method, around the point q = 0 the amplitudes $\mathcal{F}_i^{(\alpha)}$ are expanded in powers of the pion momentum:

$$\begin{aligned}\mathcal{F}_1 &= f_1^{(0)} + if_1^{(1)}q + f_1^{(2)} q\cos\theta + f_1^{(3)}q^2 + \ldots,\\ \mathcal{F}_2 &= f_2^{(1)}q + \ldots,\\ \mathcal{F}_3 &= f_3^{(1)}q + \ldots,\\ \mathcal{F}_4 &= f_4^{(2)}q^2 + \ldots\end{aligned} \tag{I.16}$$

and the experimentally determined energy dependence of the cross section is used to determine the multipole amplitudes. In accordance with Eq. (I.16), the differential cross section is given in the form

$$\frac{d\sigma}{d\Omega} = \sum_{i,k} b_{ik} q^i (\cos\theta)^k, \tag{I.17}$$

so that the coefficients at the various powers of the pion momentum provide an additional equation system. Expansion (I.17) greatly facilitates the multipole analysis of the low-energy photoproduction of neutral pions at protons.

Calculation of the Photoproduction Amplitudes with One-Dimensional Dispersion Relations

In the last few years, theoretical considerations of photoproduction processes were based mainly on dispersion relations. One-dimensional dispersion relations are available for the twelve amplitudes $\mathcal{F}_i^{(\alpha)}$, with which the amplitude of pion photoproduction at nucleons is expressed:

$$\mathrm{Re}\,\mathcal{F}_i^{(\alpha)}(W, x) = \mathcal{F}_i^{\mathrm{Born}\,(\alpha)}(W, x) + \frac{1}{\pi}\mathcal{P}\int_{M+1}^{\infty} dW^1 \sum_{j=1}^{4} K_{ij}(W, W', x)\ \mathrm{Im}\,\mathcal{F}_j^{(\alpha)}(W', x), \tag{I.18}$$

where the $K_{ij}(W^l, W, x)$ are unknown functions of kinematic quantities; W denotes the total energy in the center-of-mass system; and $\mathcal{P}$ denotes the principal value of the dispersion integral.

Equation (I.6) was used to derive from Eq. (I.8) dispersion relations for multipole amplitudes. Chew, Goldberger, Low, and Nambu [1] obtained approximate solutions of these equations with several additional assumptions which are not necessary within the framework of dispersion theory:

a) The dispersion relations for the multipole amplitudes were considered integral equations for which the selection of the solutions was governed by the Chew–Low "model with cutoff" [12];

b) the contribution of the angular region without significance in terms of physics ($|\cos\theta| > 1$) was assumed to be negligibly small in the dispersion integrals;

c) 450 MeV was taken as the upper limit of integration;

d) all quantities of the equations were expanded in power series of the parameter 1/M and only the zeroth (static limit) approximation were considered.

It was assumed in the theory of [1] that the most important effects in the range of low-energy photons can be explained when the Born terms and the ($^3/_2$, $^3/_2$) resonance interaction in the final state of the πN system are taken into account. The greatest contribution to the amplitude of photoproduction must be expected from the $M_{1+}^{3/2}$ multipole amplitude which, on the basis of a comparison of dispersion relations for πN scattering and pion photoproduction, was assumed to be proportional to the f_{1+} amplitude of pion scattering. The proportionality coefficient was determined from the ratio of the Born terms of the dispersion relations for f_{1+} and $M_{1+}^{3/2}$ in the static limit. Other multipole amplitudes were calculated under the assumption that the phase shifts of πN scattering can be determined from dispersion relations for πN scattering [13]. The solution for the s-wave amplitude contained the undefined constants $N^{(\pm)}$. The authors of [1] estimated the accuracy of their solutions to the dispersion equations at 5-10% in the energy range of primary photons between the threshold and 500 MeV. Solov'ev [8] obtained independently of the results of [1] solutions to the dispersion relations for multipole amplitudes.

A comparison between theory and experimental results on the photoproduction of charged pions at protons [14] revealed that experimental and theoretical values are basically in agreement. But the differential cross sections calculated for large exit angles of the pions exceeded considerably the measured angles. In order to obtain accurate data on the photoproduction of neutral pions at protons and to compare the data with the results of the theory, R. G. Vasil'kov, V. I. Gol'danskii, and B. B. Govorkov in 1958-1960 measured the differential and total cross sections at the primary photon energies between the threshold and 240 MeV [15].

Cross Section Measurements with the 1γ Method

In order to detect the process

$$\gamma + p \rightarrow p + \pi^0,\ \pi^0 \rightarrow 2\gamma \qquad (I.19)$$

we used methods which are based on detecting γ quanta originating from the decay of neutral pions. When at the beginning the basic features of the photoproduction process were studied, a single decay photon was recorded — hence the name "1γ method." Later on, the details of the process were established and two photons were simultaneously detected ("2γ method"). These methods are more advantageous than methods based upon the detection of recoil protons in the low-energy region, because the differential cross sections of process (I.19) can be measured in a larger angular interval (from about 10° to about 175° in the center-of-mass system) and large-size targets can be used. These conclusions follow from the kinematic "energy-angle" relations in the lab system for particles in the final state of process (I.19), as indicated in Figs. 1 and 2.*

Relation Between the Photon Emission and the Coefficients of the Angular Dependence of the Cross Sections

Let us assume that the pions are formed in a point target by a photon which has a certain energy k and that the differential cross section of the process $\gamma + p \rightarrow p + \pi^0$ can be represented as a series in the center-of-mass system:

$$\frac{d\sigma}{d\Omega}(\theta) = A + B\cos\theta + C\cos^2\theta + \ldots \qquad (I.20)$$

The distributions of the γ quanta of the decay over angles θ_γ and energies E_γ in the lab

* The functions $E_\pi(\theta)_\pi$ and $E_\pi(\theta)_p$ were kindly provided by L. N. Stark. The calculations were made with the following values for the particle masses: $\mu_{\pi^0} = 135$ MeV and $\mu_p = 938.21$ MeV.

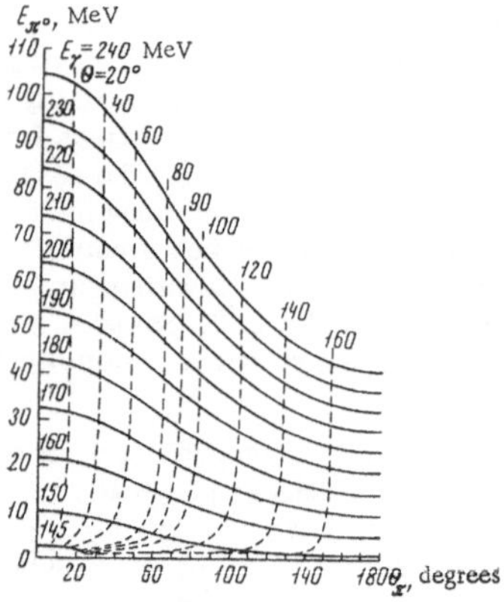

Fig. 1. Kinetic energy of the neutral pion (lab system) as a function of the exit angle (lab system) during photon production by pions. The dashed curves correspond to certain π^0 emission angles in the center-of-mass system. The numbers above the curves denote the energy of the primary photons (lab system).

system can be characterized by the following expressions:

$$\varepsilon(k, E_\gamma, \theta_\gamma) = \frac{2}{q\gamma_c(1-\beta_c\cos\theta_\gamma)}\left\{A + B\left(\frac{\cos\theta_\gamma-\beta_c}{1-\beta_c\cos\theta_\gamma}\right)\left(\frac{1}{\beta}-\frac{1}{2q\gamma_c(1-\beta_c\cos\theta_\gamma)E_\gamma}\right) + C\left[\frac{\sin^2\theta_\gamma}{2\gamma_c^2[1-\beta_c\cos\theta_\gamma]} + \frac{3}{2}\left(\frac{1}{\beta}-\frac{1}{2q\gamma_c(1-\beta_c\cos\theta_\gamma)E_\gamma}\right)^2\left(\left(\frac{\cos\theta_\gamma-\beta_c}{1-\beta_c\cos\theta_\gamma}\right)^2-\frac{1}{3}\right)\right]\right\}+\cdots, \quad \text{(I.21)}$$

where $\beta_c = k/(k + M)$ denotes the velocity of transfer from the center-of-mass system to the lab system; M denotes the mass of the nucleon; q and β denote the momentum and the velocity of the pion in the center-of-mass system; and $\gamma_c = 1/\sqrt{1-\beta_c^2}$. When we are interested only in the angular distribution of the γ quanta, Eq. (I.21) must be integrated over E_γ and the photon-detection efficiency $\varepsilon(E_\gamma)$ of the γ detector as a function of the energy E_γ must be taken into account:

$$\varepsilon(\theta_\gamma) = \int_{E_{\gamma\,\max}}^{E_{\gamma\,\max}} \varepsilon(k, E_\gamma, \theta_\gamma)\,\varepsilon(E_\gamma)\,dE_\gamma, \quad \text{(I.22)}$$

where

$$E_{\gamma\,\min} = (1-\beta)\frac{E}{2\gamma_c(1-\beta_c\cos\theta_\gamma)}; \qquad E_{\gamma\,\max} = (1+\beta)\frac{E}{2\gamma_c(1-\beta_1\cos\theta_\gamma)} \quad \text{(I.23)}$$

denote, respectively, the maximum and minimum energies of the γ quantum spectrum produced in the pion decay (with the energy E referred to the center-of-mass system).

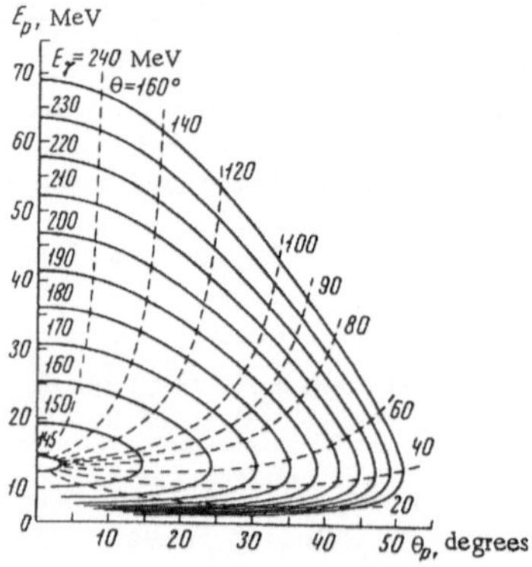

Fig. 2. Kinetic energy of the proton in dependence of the exit angle; photoproduction of neutral pions. The dashed curves correspond to certain π^0 emission angles in the center-of-mass system. The numbers above the curves denote the energy of the primary photons (lab system).

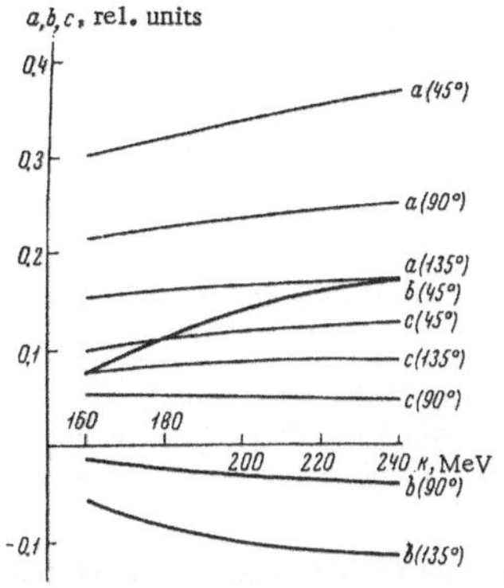

Fig. 3. The coefficients $a(\theta_\gamma)$, $b(\theta_\gamma)$, and $c(\theta_\gamma)$ of Eq. (I.24) as functions of the energy.

With Eq. (I.23) the distribution can be written in the form

$$\varepsilon(\theta_\gamma) = a(\theta_\gamma)\,A + b(\theta_\gamma)\,B + c(\theta_\gamma)\,C + \ldots \tag{I.24}$$

This means that at a particular energy of the primary photons, a unique relation exists between the angular distribution of single γ quanta of the decay in the lab system and the angular distribution of the pions in the center-of-mass system. In order to obtain the general angular dependence of the differential cross section of the process, we measured the emission of single photons at the three angles $\theta_\gamma = 45$, 90, and 135°. Figure 3 shows the coefficients $a(\theta_\gamma)$, $b(\theta_\gamma)$, and $c(\theta_\gamma)$ as functions of the energy k at the angles $\theta_\gamma = 45$, 90, and 135°.

With detectors which do not discriminate the energy of the decay photons, the angular resolution of the 1γ method is characterized by the expression

$$R(\cos\theta_l) = \int_0^\pi \frac{(1-\beta_\pi^2)\,\varepsilon(E_\gamma)\,d\varphi}{[1-\beta_\pi(\cos\theta_l\cos\theta_\gamma + \sin\theta_l\sin\theta_\gamma\cos\varphi]^2}. \tag{I.25}$$

where θ_l and φ denote the pion emission angles in the lab system and β_π denotes the velocity. We note that

$$E_\gamma = \frac{(1-\beta_\pi^2)^{1/2}}{2\,[1-\beta_\pi(\cos\theta_l\cos\theta_\gamma + \sin\theta_\gamma\sin\theta_\gamma\cos\varphi)]}. \tag{I.26}$$

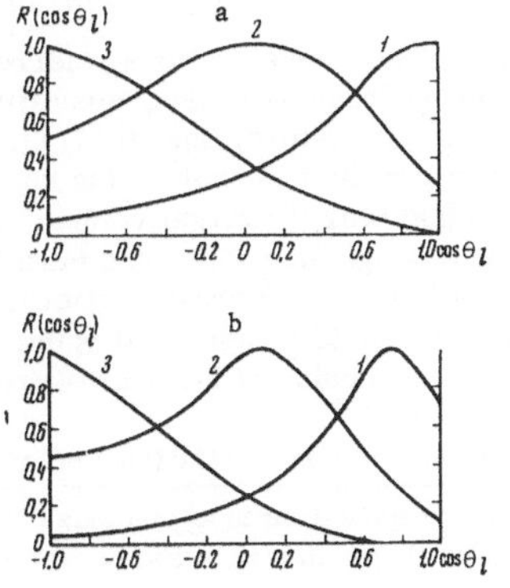

Fig. 4. Angular resolutions R(cos θ_l) which are characteristic of the 1γ method of detecting neutral pion photoproduction. The function R(cos θ_l) is shown for three positions of the γ telescope. Curve 1, 45°; curve 2, 90°; curve 3, 135°; the two energies of the primary photons are: a) 180 MeV and b) 210 MeV.

Figure 4 depicts (in relative units) the calculated angular resolution functions R(cos θ_l) for the three γ-detector positions $\theta_\gamma = 45°$ (curve 1), 90° (curve 2), and 135° (curve 3) at the two primary photon energies k = 180 MeV and k = 210 MeV. The formula

$$\varepsilon(E_\gamma) = \text{const}\left[1 - \exp\left(-\frac{E_\gamma - E_{\text{thresh}}}{43}\right)\right] \quad (I.27)$$

was used to calculate the efficiency of the γ telescope consisting of scintillation counters ($E_{\text{thresh}} = 35$ MeV).

It follows from the figure that the angular resolution is rather complicated in the 1γ method. This is not surprising because the correlation between the directions of motion of the pion and the decay photon is weak at low pion energies and vanishes when the pion energy tends to zero.

The following interesting modification of measuring the total cross section of reaction (I.19) with the 1γ method was employed by Koester and Mills [16] and, later, in our work [17]. When only the first three expansion terms are used in Eq. (I.24), the expression

$$\sigma_t = 4\pi\left(A + \frac{1}{3}C\right), \quad (I.28)$$

is obtained for the total cross section and we have for the emission of γ quanta under angles close to 90° in the lab system

$$\varepsilon(\theta_\gamma \simeq 90°) = a\left(A + \frac{b}{a}B + \frac{c}{a}C\right). \quad (I.29)$$

In the low-energy range k < 300 MeV, the ratio b/a is only slightly energy dependent and amounts to about 0.1 (see Fig. 3). Since coefficient B is small ($B/A \approx 0.3$) in the energy interval under consideration, the term (b/a)B can be ignored at accuracies of 3-5%. Thus

$$\varepsilon(\theta_\gamma = 90°) = a\left(A + \frac{c}{a}C\right). \quad (I.30)$$

This expression differs from Eq. (I.28) by the factor before the coefficient C. In the energy interval under consideration, $c/a \approx 0.2$ to 0.25 and the coefficient ratio is $C/A \approx 0.3$ to 0.6.

We can therefore assume with a rather high accuracy (~5%) that measurements of the emission of single decay γ quanta at angles close to 90° in the lab system are essentially measurements of the total cross section of the process $\gamma + p \to p + \pi^0$.

The above considerations referred to a certain energy k and a pointlike target. In reality, the continuous bremsstrahlung spectrum of the accelerator is present and the target has finite dimensions. Therefore, Eqs. (I.21) and (I.24) must be integrated over the primary photon energies between the threshold value $k_{\text{thresh}} = 145$ MeV and the maximum energy $k_{\max}$ of the spectrum and over the target volume. Additional measurements of the $k_{\max}$ dependence of the emission of single decay photons must be made at certain angles θ_γ in order to obtain the energy dependence of the cross sections. We used in our experiments a measuring technique which enabled us to measure all points of the emission curve at the same time and during a single radiation pulse of the synchrotron [18].

Method of Measuring Excitation Functions

The proposed method of measuring the excitation functions of photomesonic processes is based on the unique relation between the maximum energy $k_{\max}$ of the synchrotron bremsstrahlung spectrum and the magnetic field H ($k_{max} = 300\ H\rho$, where k_{max} is expressed in eV, and H, in gauss; ρ denotes the orbital radius, expressed in cm, of the accelerated electrons). The method makes use of simultaneous sorting of pulses of the "effect" and of the intensity monitor with the aid of the magnetic field scale.

Let us consider the relation between the energy of the accelerated electrons (magnetic field) and time in the case of the 265-MeV synchrotron of the Physics Institute of the Academy of Sciences of the USSR (see Fig. 5a). We assume that the electron bombardment of the accelerator target begins at a magnetic field strength H. The electron energy is then k_1. Naturally, the maximum energy of the energy spectrum of the photon bremsstrahlung has the same value, i.e., $k_{max} = k_1$. We continue the electron bombardment of the target while gradually decreasing the accelerating high-frequency pulse in the resonator until the instant at which the magnetic field reaches the value H_2 (energy k_2); we obtain a radiation pulse in which the maximum energy k_{max} of the bremsstrahlung spectrum runs continuously from the values k_1 to k_2. This process is repeated in the next radiation pulse. The radiation pulses produced in this fashion can be used to study the energy dependence (in the k_1, k_2 interval) of the cross sections of various processes. For this purpose, the pulses from the circuits recording the reaction and from the differential monitor measuring the inhomogeneity and the intensity distribution in the stretched radiation pulse must be sorted in channels corresponding to the various intervals defined by k_{max}.

In the simplest way of practicing this method we made use of conventional time analyzers. The energy spacing of the channels and control of the channel positions were obtained with the circuit measuring the magnetic field strength. A conventional scintillation counter, which was mounted outside the principal accelerator beam and recorded the background during synchrotron operation was used as differential monitor of the intensity. The dependence of the monitor sensitivity upon k_{max} and upon the intensity density in the pulse and the possible excess counting of slow background neutrons were determined in special experiments.

Figure 5 depicts the energy intervals Δk_{max} and the time intervals Δt_i which were selected for measuring the angular dependences of the differential cross sections (Fig. 5a) and the total cross sections (Fig. 5b) of the process $\gamma + p \rightarrow p + \pi^0$. Only part of the intensity of the photon beam was used in the latter case, but the accuracy of k_{max} was in each channel (±1 MeV) considerably increased relative to the case depicted in Fig. 5a. The γ radiation

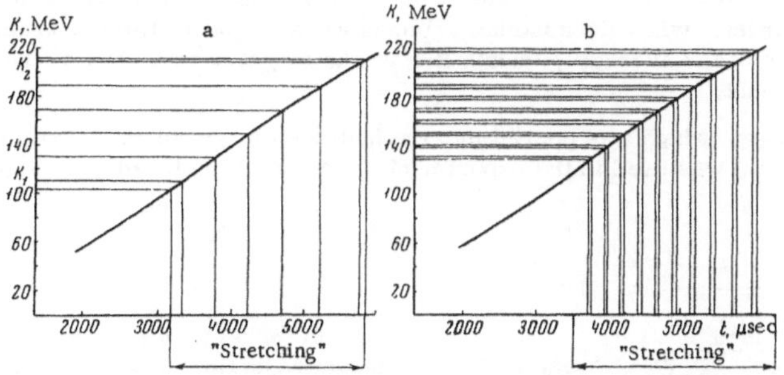

Fig. 5. Time dependence of the energy of the accelerated electrons for the 265-MeV synchrotron of the Physics Institute of the Academy of Sciences of the USSR. Selection of the energy and time intervals in measurements of the angular dependencies of the differential and total cross sections of the reaction $\gamma + p \rightarrow p + \pi^0$.

pulse of the synchrotron (normal duration about 30 μsec) was in both cases stretched to 3000 μsec so that the output of decay photons could be measured in a single experiment in which k_{max} varied from 130 to 220 MeV. Simultaneous determination of numerous points N_γ helped to eliminate several inaccuracies which are related to the instability of accelerator operation, reproducibility of operation conditions, and time-dependent instabilities of the electronic equipment. Moreover, careful adjustments of the intensity distribution in the radiation pulse (which were obtained by proper selection of both the amplitude and the form of the rear front of the accelerating high-frequency pulse) modified the sampling rate to the required statistical accuracy of the Δk_{max} measurements in the various channels and hence substantially reduced the times required for the experiments.

Method of Measurements and Results

1. Coefficients A, B, and C of the Angular Dependences. The angular dependence of the differential cross sections of the reaction $\gamma + p \rightarrow p + \pi^0$ was measured with the setup shown in Fig. 6 on the 265-MeV synchrotron of the Physics Institute of the Academy of Sciences of the USSR. Single γ quanta produced in the decay of neutral pions were detected with the aid of two telescopec comprising four conventional scintillation counters each. A 0.6-cm-thick lead converter, which converted photons into electron–positron pairs, was placed behind the first counter C_1 which was connected in anticoincidence with the three other counters C_2, C_3, C_4, which, in turn, were connected in coincidence. The telescopes were placed at the angles 90° and 135° or 45° and 90° in the lab system. Measurements were made for seven maximum energies of the bremsstrahlung in the various channels, namely 130, 150, 170, 190, 210, and 250 MeV. The experiments rendered directly the ratio of the number of recorded decay γ quanta to the number of readings on the differential monitor for the three angles 45°, 90°, and 135° at the energies listed above. In order to derive from these figures the energy dependence of the photon output at a certain angle, the k_{max} dependence of the differential monitor count per unit γ quanta flux was taken into account. The corrections, which account for this dependence, are the same for measurements made at all angles and have therefore practically no influence upon the measured angular distributions of the decay photons. In order to obtain these distributions from γ quanta counts at various angles, we introduced a correction which characterizes for these angles the difference of the products of effective target volume times the solid "viewing" angle of the telescope. This difference was obtained by comparing the counting rate of γ quanta from the pion decay during irradiation of a polystyrene "point" source with the counting rate obtained with a compact polystyrene foam target having the dimensions of the hydrogen target. When the absolute volumes were introduced, the correction resulting from the difference of the average γ quanta flux on the target surface and on the point source was taken into account.

Liquid hydrogen, which was poured into the hydrogen target of PS-4 foam polystyrene (cooled with nitrogen) was used in the experiments; the target enclosed a 30-cm-long channel

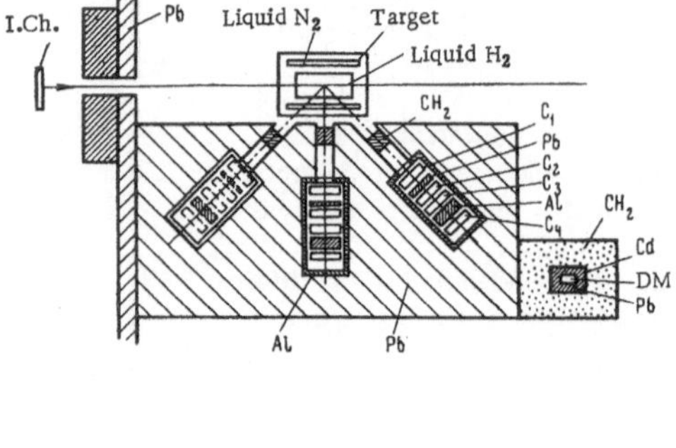

Fig. 6. Scheme of the setup for measuring the photoproduction of neutral pions with the 1γ method. I. Ch., ionization chamber; CH_2 paraffin absorbers; $C_1, \ldots, C_4$, scintillation counters; Al, aluminum filter; DM, differential monitor

TABLE 1. Coefficients A, B, and C (Expressed in 10^{-30} cm^2/sr) and σ_t (Expressed in 10^{-29} cm^2) Measured with the 1γ Method [15]

k, MeV	A	B	C	σ_t
160	0.28±0.03	−0.13±0.03	0.09±0.12	0.31±0.06
180	0.95±0.02	−0.39±0.06	−0.5±0.1	0.98±0.04
200	2.10±0.05	−0.56±0.07	−1.3±0.2	2.08±0.08
220	4.5±0.2	−0.7±0.1	−2.9±0.4	4.5±0.2
240	8.4±0.2	−0.9±0.2	−6.0±0.6	8.1±0.3

which had a diameter of 12 cm and extended in the direction of the beam. The γ quanta beam had a diameter of 6 cm and was incident on the cylinder face. The background counting rate produced by the empty target amounted, on the average, to 30% at 45°, 8% at 90°, and 4% at 135° of the counting rate registered in experiments with hydrogen. The flux of the γ quanta incident on the target was measured in absolute units with the aid of calibrated graphite detectors in which the activation reaction $C^{12}(\gamma, n)C^{11}$ was observed. Absolute measurements with a quantum meter were made in all ensuing experiments. The absolute flux measurements made with the graphite detectors agreed with those made with the quantum meter with an accuracy of 6%.

The curves, which represent the energy dependence of the relative output of decay γ quanta at the angles 45°, 90°, and 135° were used to calculate the emission cross section of the decay γ quanta in the lab system. The cross section figures were compared with the $\varepsilon(\theta_\gamma)$ values obtained with Eq. (I.24) in order to determine the coefficients A, B, and C of the angular distributions of photo-induced pions in the center-of-mass system. The absolute values of A, B, and C were obtained by comparing the γ quanta output observed at 90° in elastic γp scattering at $k_{max} = 130$ MeV with the value calculated with Powell's formula [19]. Table 1 lists the measured values of the coefficients A, B, and C and the total cross section which was calculated with the formula $\sigma_t = 4\pi(A + \frac{1}{3}C)$ for the process $\gamma + p \rightarrow p + \pi^0$.

Measurements of the reaction $\gamma + p \rightarrow p + \pi^0$ established the interference term $B(\cos\theta)$ in Eq. (I.20). This term indicates that pions are produced at low energies in both the p-wave resonance state and in the s-wave state. A comparison of the experimental B/A values with calculations of the photoelectric pion production in the s state in accordance with perturbation theory and with a correction for the finiteness of the nucleon mass [20] revealed that perturbation theory can be used to calculate the amplitudes of pion photoproduction in the s state near the threshold.

Total Cross Sections. The technique of Koester and Mills was used to measure with the "1γ method" the total cross sections of the process $\gamma + p \rightarrow p + \pi^0$. The Δk_{max} en-

TABLE 2. Total Cross Sections Measured with the 1γ Method in the $\gamma + p \rightarrow p + \pi^0$ Reaction

k, MeV	$\sigma_t \cdot 10^{29}$, cm^2	k, MeV	$\sigma_t \cdot 10^{29}$, cm^2	k, MeV	$\sigma_t \cdot 10^{29}$, cm^2
155	0.18±0.07	195	1.73±0.19	225	5.50±0.37
165	0.57±0.09	205	2.84±0.24	235	6.85±0.49
175	0.87±0.012	215	3.40±0.29	245	8.90±0.63
185	1.08±0.13				

TABLE 3. First Terms in the Expansions of the Coefficients A, B, and C (Expressed in Units of $10^4\ \lambda^2_{\text{Compton}}$)

Coefficient	$A^{(0)}$	$A^{(2)}$	$B^{(1)}$	$C^{(2)}$
Experiment	0.08±0.04	1.55±0.03	−0.38±0.04	−0.91±0.14
Theory [1] $N^{(+)}=0$	0.03	0.14	−0.39	0.32

ergy intervals were selected as shown in Fig. 5b. The γ telescope was positioned at 90° with respect to the primary bremsstrahlung beam of the accelerator. Table 2 lists the measured total cross sections.

The energy dependence which we obtained in our work for σ_t agrees strongly with the cross section measured by Koester and Mills [16], though their absolute σ_t values are 30% smaller than the tabulated values.

Comparison Between Experimental Results and Dispersion Theory; Multipole Analysis

Baldin and Govorkov [11] compared the measured coefficients of Table 1 with the predictions of dispersion theory. Under the assumption that the long-wave approximation is valid (the s and p wave amplitudes were considered), an analysis method based on the expansion of the experimental results and dispersion integrals in a series of pion momenta was developed. The coefficients of the angular dependence were accordingly represented as follows:

$$\begin{aligned} A &= q\omega\,[A^{(0)} + A^{(2)}q^2 + A^{(4)}q^4 + \cdots], \\ B &= q\omega\,[B^{(1)}q + B^{(3)}q^3 + B^{(5)}q^{(5)} + \ldots], \\ C &= q\omega\,[C^{(2)}q^2 + C^{(4)}q^4 + C^{(6)}q^6 + \ldots]. \end{aligned} \quad (I.31)$$

The analysis of the experimental values of the coefficients A, B, and C resulted in the first terms in Eqs. (I.31). These terms were compared with the values obtained with the theory of [1] (see Table 3). There, and below, the amplitudes of the photoproduction are stated in units of the Compton wavelength of the pion.

Considerable discrepancies, particularly in the coefficient $C^{(2)}$, exist between the experimental results and the predictions of the theory of [1].

Under the additional assumption that only the multipole amplitudes E_{0+}, M_{1+}, and M_{1-} appear in the energy range near the threshold, an attempt was made to determine the multipole amplitude responsible for the discrepancy. The multipole amplitudes obtained are listed in Table 4 along with the predictions of the theory of [1]. Two different methods of determining the s-wave amplitudes were proposed in [11]: direct determination from the coefficient $A^{(0)} \sim |E_{0+}|^2$, and determination from the interference coefficient $B^{(1)}$. It follows from the table that the two methods render results which agree within the statistical error limits.

TABLE 4. Multipole Amplitudes of the Photoproduction of Neutral Pions (Expressed in Units of $10^4\ \lambda^2_{\text{Compton}}$)

Amplitude	$\lvert E_{0+}\rvert^2$	$\lvert M_{1+}\rvert^2$	$\lvert M_{1-}\rvert^2$
Experiment (in the prediction $E_{1+}=0$)	0.080±0.033 from $A^{(0)}$	0.50±0.03	0±0.09
Theory [1] (for $N(+)=0$)	0.075±0.019 from $B^{(1)}$	0.47	0.23

The resonance amplitude $|M_{1+}|^2$ calculated with dispersion theory agrees very well with the value obtained from an analysis of the experimental results in terms of angular distributions. The same result was obtained in a comparison between measured total cross sections of the $(\gamma p \rightarrow p\pi^0)$ process in the near-threshold energy range (see Table 2) and the theoretical predictions of [1]. Agreement was not obtained for the $|M_{1-}|^2$ amplitude. Experiments rendered a substantially lower value for this amplitude. The experimental cross section values of the photoproduction of charged pions, the results of phase analysis for s-wave phase shifts, and the $|E_{0+}|^2$ values of Table 4 were used to determine the three parameters characterizing the s photoproduction at the threshold [see Eq. (I.14)] and, accordingly, the amplitude of neutral pion production at neutrons: $\sqrt{A^{(0)}} = (0.04 \pm 0.04) \cdot 10^{-2}$. This calculation indicated that the s amplitude of the $n\pi^0$ process is at least one order of magnitude smaller than the amplitude of the π^0 production at protons.

In [11], and in the ensuing work of Baldin [21-23], a new method of comparing experimental results with dispersion relations was developed. The experimental results were inserted on the right and left sides of the dispersion relations. One can therefore in some cases avoid special assumptions which are made to obtain photoproduction amplitudes from the solution of dispersion relations. This is possible, to some extent, when dispersion relations for the photoproduction amplitude of neutral pions are considered at $\theta = 0°$, i.e., only the amplitude part proportional to q is considered. It could be shown that this direct comparison of experimental data and dispersion relations reduces the above-mentioned discrepancies without completely eliminating them. The explanation of the discrepancies seemed to involve a more accurate calculation of the dispersion integrals. It remained unclear whether the discrepancies between experimental and theoretical results originated from neglection of small photoproduction amplitudes or from the disregard for the high-energy range. Further theoretical work can provide answers to these and other questions. Moreover, since the discrepancies between experiments and the solutions of [1] to the dispersion equations were derived basically from a single experiment, new, more accurate measurements using new techniques are highly desirable.

Contrary to our analysis, Dietz, Höhler, and Müllensiefen [24] used the coefficients of Table 1 for an amplitude analysis without any expansion of the experimental cross sections and compared the resulting multipole amplitudes with the solutions of [1]. They assumed that the theoretical expressions of [1] can be used to calculate the imaginary parts of the amplitudes which must be taken into account. The results of their analysis are in almost complete agreement with the conclusions of our work [11]. This is not surprising because the results of the same experimental work were analyzed in both cases.

Short Review of the Improved Solutions of [1] and of Experimental Work

Extensive theoretical work was done in the last few years with the aim to improve the solutions of dispersion theory [1, 8]. The improvements concerned estimates of the possible contribution of the nonphysical region to the dispersion integrals [25, 26], the application of successive approximations to the solution of regular Fredholm equations (to which the linear singular dispersion equations [27, 28] were reduced with the Muskhelishvili–Aumness technique), a relativistic solution for the $M_{1+}^{3/2}$ amplitude [2, 29], the introduction of possible contributions of vector mesons to single pion photoproduction [28, 30], etc. Most extensively used was Ball's method [2] of solving dispersion relations. Ball considered the double spectral representation of the amplitudes and solved the dispersion equations in the "Born + 33" approximation, i.e., in an approximation in which Born terms are taken into account in the equations for the amplitudes, whereas only the contribution of the resonance amplitude is introduced in the calculation of the dispersion integrals. Great expectations were based on the fact that ways to a further refinement of dispersion theory are found in a comparison between calculations and experimental results. Ball's method was further developed in the ensuing work of a group of theoreticians of the Physics Institute of the Academy of Sciences of the USSR [23, 30] and of theoreticians of the University of Bonn [31, 32]. Comparisons between theoretical and experimental work on the reaction $\gamma + p \rightarrow p + \pi^0$ in the low energy range (which was covered in calculations by the majority of authors) could not eliminate the discrepancies between experimental and theoretical values of the differential cross sections at large emission angles of neutral pions, as mentioned in [11] (E_{0+} and M_{1-} multipoles). The asymmetry of the theoretically predicted angular distributions is greater than the asymmetry observed in experiments. Since the approximations in the estimates of the dispersion integrals imply large ambiguities, the contribution of vector mesons to single photoproduction could not be reliably established in a single paper.

A large number of papers dealt with the isobaric model [10, 33, 34]. By selecting a large number of trial parameters, the isobaric model could be adapted to satisfactorily describe photoproduction processes up to energies k = 800 MeV. However, the question whether the solution is unique remains unanswered in this case.

Since the above-mentioned experiments were completed, the situation has changed only insignificantly as far as experiments are concerned. The reason is, in essence, that it is difficult to measure the reaction $\gamma + p \rightarrow p + \pi^0$ near the threshold region of the photons (the differential cross sections are small, i.e., about 10^{-31} cm^2/sr, the reaction products are not easily recorded, the bremsstrahlung spectrum of the accelerator is continuous, and a tremendous electron–photon background interferes). The photoproduction of π^0 mesons at protons was studied in detail by Hitzeroth [35]. In order to identify the process, Hitzeroth measured in his work both the energy and the emission angle of the recoil proton ejected from a gaseous hydrogen target. Nuclear emulsions were used as proton detectors. The accuracy (about 15%) of his results is slightly lower than the accuracy attained in our work [15]. The angular distribution coefficients of [35] agree within the accuracy limits with the results listed in Table 1.

The authors of [36-38] measured the π^0 formation at protons by unpolarized photons in the energy range 220-230 MeV. The results refer to individual values of the parameters θ and k and it is therefore difficult to obtain from their figures an idea of the energy dependence or angular dependence of the differential cross sections and to compare the results with the coefficients of Table 1. The work of [39] plays a particular role in that $\gamma-\gamma$ coincidences were used to measure the outputs from H_2 and He targets in dependence of the maximum energy of the accelerator between the threshold energy and $k_{max} = 170$ MeV. Theoretical and experimental results were compared by integrating the theoretical expressions over the angular resolution and energy resolution of the setup. This comparison, in which the contributions of the small p-wave amplitudes were neglected, led the authors to the following conclusions: (a) the experimental results agree with the calculations of [1] for k > 155 MeV at $N^{(+)} = R$; (b) the experimental results differ from the predictions of [1] for k < 155 MeV. The discrepancy may be explained by the energy dependence of $N^{(+)}$ near the photoproduction threshold (because the ρ-meson diagram renders a contribution, or because the masses of charged and neutral pions differ).

In the last few years (beginning from approximately 1961-1962), polarization experiments were made in addition to measurements of the cross section of pion photoproduction by unpolarized photons (unpolarized target). The polarization of the recoil protons generated in the photoproduction of neutral pions at hydrogen was measured by a group of physicists of the University of Bonn [40]. The measurements were made in the energy range of the primary photons near the first resonance of the πN system (at $k \geq 320$ MeV). The following value of the magnetic dipole amplitude M_{1-} was derived at the energy 360 MeV from an analysis of the angular distribution of the polarization of the recoil protons:

$$\mathrm{Re}\, M_{1-} = (-0.8 \pm 0.2) \cdot 10^{-2}. \quad (I.32)$$

This value agrees with the calculations of dispersion theory according to the "Born + 33" approximation [31]. In the energy range near the threshold, no polarization measurements on recoil nucleons, as they are essential for a full multipole analysis, have been made to date, because the corresponding experiments are difficult (both the energies and emission angles of the recoil nucleons are small and the polarization effect itself is small).

The formation of neutral pions at hydrogen by linearly polarized photons was studied in [41, 42]. The results of that work are listed in Table 5.

The notation of the table is interpreted as follows: θ denotes the emission angle of the pion; P_γ denotes the degree of polarization of the primary photons; and $\sigma_\perp$ and $\sigma_\parallel$ denote the differential cross sections of photoproduction of neutral pions by photons whose electric vector is perpendicular or parallel to the plane of pion production.

The differential cross section of pion production at nonpolarized protons by polarized photons in the s- and p-wave approximation can be expressed through the multipole amplitudes as follows:

$$\begin{aligned}\frac{d\sigma}{d\Omega} = \mathrm{Re}\Big\{\Big[|E_{0+}|^2 + |M_{1-}|^2 + \frac{5}{2}|M_{1+}|^2 + \frac{9}{2}|E_{1+}|^2 + (M_{1-}^{*}M_{1+}) +\\ + 3(M_{1-}^{*}E_{1+}) - 3(M_{1+}^{*}E_{1+})\Big] + \cos\theta\,[-2(E_{0+}^{*}M_{1-}) + 2(E_{0+}^{*}M_{1+}) +\\ + 6(E_{0+}^{*}E_{1+})\Big] + \cos^2\theta\Big[-\frac{3}{2}|M_{1+}|^2 + \frac{9}{2}|E_{1+}|^2 - 3(M_{1-}^{*}M_{1+}) -\\ - 9(M_{1-}^{*}E_{1+}) + 9(M_{1+}^{*}E_{1+})\Big] + \sin^2\theta\cos 2\varphi\Big[-\frac{3}{2}|M_{1+}|^2 + \frac{9}{2}|E_{1+}|^2 -\\ - 3(M_{1-}^{*}M_{1+}) + 3(M_{1-}^{*}E_{1+}) - 3(M_{1+}^{*}E_{1+})\Big]\Big\} = A + B\cos\theta + C\cos^2\theta + \alpha\sin^2\theta\cos 2\varphi. \quad (I.33)\end{aligned}$$

Under the assumption $E_{1+} = 0$ and $\alpha = C$, Eq. (I.33) can be rewritten, since $\sigma_\perp$ corresponds to $\varphi = \pi/2$, and $\sigma_\parallel$, to $\varphi = 0$:

$$\frac{\sigma_\perp - \sigma_\parallel}{\sigma_\perp + \sigma_\parallel} = -\frac{\alpha\sin^2\theta}{\sigma_0}, \quad (I.34)$$

where σ_0 denotes the differential cross section for nonpolarized photons. We obtain the ratio α/A on the right side when $\theta = \pi/2$. This ratio is measured in experiments with polarized photons. Furthermore, when the ratio C/A is deduced from experiments with unpolarized photons, the combination of the results of the two experiments (with polarized and unpolarized photons) renders the ratio α/C. The α/C ratios obtained in this fashion are listed in the penultimate column of Table 5. The fact that the coefficient ratio α/C differs from unity may

TABLE 5. Results of Experiments with Linearly Polarized Photons

k, MeV	θ, degrees	P_γ, %	$\frac{\sigma_\perp - \sigma_\parallel}{\sigma_\perp + \sigma_\parallel}$	$\frac{\alpha}{C}$	Reference
228	90	37	0,246±0.016	0.36±0.07	[42]
235	120	15	0.289±0,047	0.45±0,11	[41[
285	90	14	0.462±0.035	0,74±0.10	[41]
325	90	30.6	0,51±0,01	0.84±0.06	[42]
335	60	16	0.462±0,025	0.90±0.88	[42]
435	90	12	0.529±0.065	0.61±0.09	[42]

indicate that a nonvanishing amplitude E_{1+} exists. The coefficients C and α depend in a different way upon E_{1+}:

$$C \simeq \mathrm{Re}\,[X + 3E^{*}_{1+}(M_1 + M_{1+})],$$
$$\alpha \simeq \mathrm{Re}\,[X - E^{*}_{1+}(M_{-1} - M_{1+})], \quad (I.35)$$

where X stands for the terms which are the same for C and α. An α/C ratio smaller than unity can also be obtained when the contributions of ρ and ω mesons to the photoproduction of single pions are included [41, 42]. Definite conclusions can be drawn once a full experiment has been made in the particular energy range of primary photons. As indicated in the table, experiments with polarized photons were made at energies which are greater than the energy limit of the near-threshold range ($q \leq 1$). Naturally, since the angle and energy intervals considered are bounded (and no experimental results are available in the near-threshold range), the results of the polarization experiments cannot provide answers to questions which originate from discrepancies between experimental results and solutions provided by dispersion theory. The corresponding experiments should be made at lower energies.

Measurements of the Energy Dependence and the Angular Dependence of the Differential Cross Section of the Process $\gamma + p \to p + \pi^0$ with the 2γ Method

A new series of measurements of the neutral pion photoproduction at protons was made in order to determine the reasons for the discrepancies between experimental results and calculations according to dispersion theory.

Measurement Technique

Single photons from the pion decay were measured in all preceding experiments [15-17, 43] in order to derive the energy dependence and the angular dependencies of the differential cross sections of neutral pion photoproduction in the γ-quanta energy range of 145-250 MeV. In order to determine the photoproduction amplitude in the s state, measurements of the energy dependence of $d\sigma/d\Omega(90°)$ must be extended as close as possible to the threshold value of the primary photon energy. The absolute cross section must amount to $(1\text{-}2) \cdot 10^{-31}$ cm^2/sr so that the background produced by elastic photon scattering at protons becomes important. Though the cross section of the Compton effect amounts to $(1\text{-}2) \cdot 10^{-32}$ cm^2/sr, the cross section integrals over the energy (with proper account for the resolutions of the setup) were in most cases of the same order of magnitude of the photoproduction and the Compton effect.

In order to accurately determine the coefficients of the angular distribution, the differential cross sections must be measured in the greatest possible interval of pion emission angles. But due to the tremendous background produced by electrodynamic processes (Compton effect of electrons, etc.), the detection of single decay γ quanta does not allow measurements at angles of less than approximately 30° and more than approximately 150°. Apart from this, the method does not make it possible to obtain good angular resolution.

Thus, in order to discriminate the photoproduction process in a new series of experiments, we recorded simultaneously the two photons produced in the decay of the neutral pion.

The scheme, which is shown in Fig. 7, resembles the experimental scheme with which the photoproduction of neutral pions was observed for the first time [44]. The γ quanta of the bremsstrahlung generated by the 265-MeV synchrotron of the Physics Institute of the Academy of Sciences of the USSR were the primary photon source. The conditions of synchrotron operation were chosen so that the accelerated electrons were ejected onto the target in the region

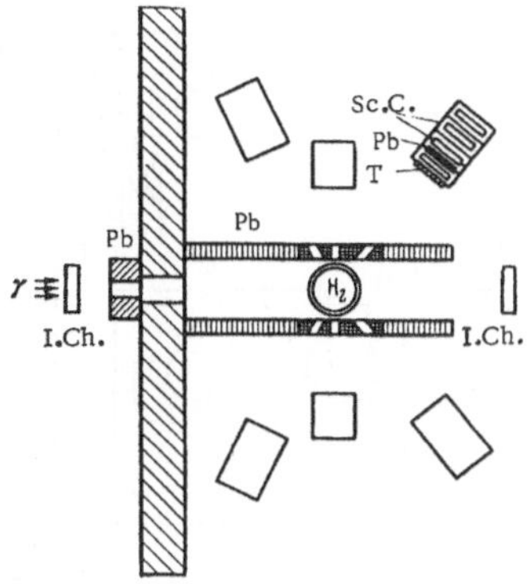

Fig. 7. Setup for the measurement of neutral pion photoproduction with the "2γ method." γ denotes the bremsstrahlung beam; I.Ch. denotes ionization chambers; H_2 denotes liquid hydrogen, Sc.C., scintillation counters, and T, target.

in which the magnetic field passes through the maximum field strength value. The transition from one maximum energy to another was effected by changing the ratio between variable and constant magnetic field components of the accelerator. In order to eliminate random $\gamma-\gamma$ coincidences, the synchrotron radiation pulse was extended to 1000 μsec, which produced a 2% energy spread of the accelerated electrons. The calculations were based on the bremsstrahlung spectra of [45] without taking into account the "stretching." Estimates show that this description of the photon spectra results in errors of about 1% in the final results.

The photon beam from the accelerator was, after collimation, incident on a liquid hydrogen target. Two thin-walled ionization chambers were placed before and behind the hydrogen target to make relative measurements of the primary photon flux. Absolute flux measurements were made with a quantum meter at the center of the target. The accuracy of the energy flux determinations in the photon beam amounted to ±5%.

Liquid hydrogen filled in a foam-plastic container (in the 1962 experiments) or in a dewar vessel of glass (experiments of 1962-1965) was used as the target. In order to reduce the background from the target walls, secondary collimation was introduced, i.e., the counter telescopes did not "see" the target walls and did not register particles which were generated in the target walls hit by the primary photon beam.

The photons emitted in the decay of neutral pions were recorded with scintillation counter telescopes. In the majority of experiments, each telescope comprised four counters: a first counter A separated the photons from the charged particles, while the other counters recorded e^+e^- pairs produced by the photons in a 6-mm-thick lead converter behind anticoincidence counter A. Figure 8 shows the block diagram of the electronic equipment used. The fast fourfold coincidence circuits reduced the number of random counts to negligibly small values. When the energy dependence of the differential cross section was measured at 90°, the γ telescopes were placed in a plane enclosing a 70° angle with the primary photon beam. The emission angle of the photons generated in the decay of the neutral pions was limited by slits in 5-cm-thick lead walls. The walls extended parallel to the beam at a distance of 10 cm from the center of the hydrogen target; the slits defined the 70° plane. The angle of this plane (70°) defined the average exit angle of the recorded pions in the lab system; the correlation angle ψ_k between the telescopes in the 70° plane determined the average energy of the pions. The telescopes were placed symmetrically with respect to the projection of the direction of primary photon motion onto the 70° plane. The measurements were made so that at each maximum energy of the accelerator, the number of $\gamma-\gamma$ coincidences was determined at three different ψ_k values, whereupon the maximum energy of the synchrotron was changed, the number of

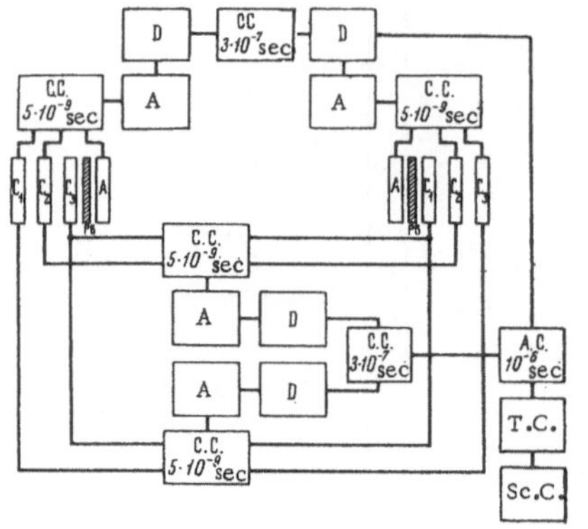

Fig. 8. Block diagram of the setup for coincidence recordings of two photons from π^0 decay. C.C., coincidence circuit; A, amplifier; D, discriminator; A.C., anticoincidence circuit; T.C., trigger circuit; Sc.C., scaling circuit; A, C_1, C_2, C_3, scintillation counters.

coincidences was measured again, and so on. The background generated by the empty target was subtracted from the counts obtained with hydrogen (background 5-10%).

The stability of operation of the equipment was checked by recording $\gamma-\gamma$ coincidences produced by a carbon target whose position was exactly established. The checks indicated that, during the principal measurements, the counting rates of $\gamma-\gamma$ coincidences from the C target (counts referred to a certain number of readings of relative monitors) were maintained with variations of $\pm 2\%$. This means that the operation of the equipment is stable and that the operation conditions of the accelerator and the geometry of the measurements are reproducible.

In the rest of the article, the angle θ_l between the plane of the accelerator orbit and the plane of the target and the two γ telescopes will be termed "position." The average meson emission angle θ_l, which is given by the angular resolution of the setup, differs slightly from the positioning angle θ_k of the telescope plane. This situation results from the fact that θ_k can be set to 0° or 180°, but since the angular resolution of the setup is finite, average angles θ_l of 0° and 180° cannot be obtained at the same time. In the first case there is always a shift toward larger angles, whereas in the second case a shift toward smaller pion exit angles takes place.

Simultaneous measurements in several positions and control measurements during changes of positions could be made with two or three sets of instruments (each set comprising two γ telescopes). The various positions of the γ telescopes were assumed in a cyclic sequence during the measurements. Control measurements were made in a fixed position of the carbon target so that "cross relations" could be established with an accuracy of 1-2%. The measurement procedure was otherwise the same as in the previously considered measurements of the energy dependence of $d\sigma/d\Omega$ for the angle 90°.

Monochromatic photons were used to determine the energy dependence of the γ recording efficiency of the scintillation counter telescope and of the corresponding coincidence circuits [46].

Calculation of Differential Cross Sections from the $\gamma-\gamma$ Coincidence Yields; Angular Resolution and Energy Resolution

The $\gamma-\gamma$ coincidence yields, referred to unit flux intensity of the primary photons, were measured for ten different values of ψ_k and k_{max} in the experiment in which the energy dependence of $d\sigma/d\Omega$ was determined at 90°.

The differential cross section $\frac{d\sigma}{d\Omega}(k, \Omega)$ of the reacting $\gamma + p \rightarrow p + \pi^0$ was calculated with the expression for the yield of decay photons under angles Ω_γ in the lab system and at the maximum energy k_{max} of the bremsstrahlung spectrum of the accelerator:

$$N(k_{max}, \Omega_\gamma) = nC_k \iiint_{\Omega\, V} \frac{d\sigma}{d\Omega}(k, \Omega)\, \eta(k_{max}, k, V)\, \varepsilon(k, \Omega, V)\, dk\, d\Omega\, dV, \tag{I.36}$$

where n denotes the number of nuclei per cm³ target; k denotes the energy of the primary γ quantum causing the reaction; Ω denotes the angular coordinates of the direction of pion emission; $\varepsilon(k, \Omega, V)$ denotes the probability of the equipment recording pions generated in the volume dV; $\eta(k_{max}, k, V)$ denotes the flux density of the primary photons; and C denotes a parameter characterizing the absolute bremsstrahlung flux. The energy resolution and the angular resolution determining the accuracy with which the experimental setup responds to the primary photon energy and the direction of pion emission are given by the following expressions:

$$\frac{dN}{dk} = nC \int_V \eta(k_{max}, k, V)\, dV \int_\Omega \frac{d\sigma}{d\Omega}(k, \Omega)\, \varepsilon(k, \Omega, V)\, d\Omega, \tag{I.37}$$

$$\frac{dN}{d\Omega} = nC \iint_{V\,k} \frac{d\sigma}{d\Omega}(k, \Omega)\, \eta(k_{max}, k, V)\, \varepsilon(k, \Omega, V)\, dV\, dk. \tag{I.38}$$

Under the assumption that the cross section $\frac{d\sigma}{d\Omega}(k, \Omega)$ is constant within the variation intervals of k and Ω and equal to the cross section for the average values $\bar{\theta}_l$ (where θ_l denotes the pion emission angle in the lab system) and $\bar{k}$, and that the process has azimuthal symmetry expressions (I.36)–(I.38) can be rewritten in the form

$$N = 2\pi nC \frac{d\sigma}{d\Omega}(\overline{\cos\theta_l}, \bar{k})\, \varepsilon, \tag{I.39}$$

$$\frac{dN}{dk} = 2\pi nC \frac{d\sigma}{d\Omega}(\overline{\cos\theta_l}, \bar{k})\, R(k), \tag{I.40}$$

$$\frac{dN}{d\cos\theta_l} = 2\pi nC \frac{d\sigma}{d\Omega}(\overline{\cos\theta_l}, \bar{k})\, R(\cos\theta), \tag{I.41}$$

where

$$\varepsilon = \int_k \int_{\cos\theta_l} \int_V \eta(k_{max}\, k, V)\, \varepsilon(k, \cos\theta_l, V)\, dk\, d\cos\theta_l\, dV, \tag{I.42}$$

$$R(k) = \int_V \eta(k_{max}, k, V) \int_{\cos\theta_l} \varepsilon(k, \cos\theta_l, V)\, d\cos\theta_l\, dV, \tag{I.43}$$

$$R(\cos\theta_l) = \int_V \int_k \eta(k_{max}, k, V)\, \varepsilon(k, \cos\theta_l, V)\, dk\, dV, \tag{I.44}$$

and

$$\bar{k} = \left[\int_k kR(k)\, dk\right] \Big/ \left[\int_k R(k)\, dk\right], \tag{I.45}$$

$$\overline{\cos\theta_l} = \left[\int_{\cos\theta_l} \cos\theta_l\, R(\cos\theta_l)\, d\cos\theta_l\right] \Big/ \left[\int_{\cos\theta_l} R(\cos\theta_l)\, d\cos\theta_l\right]. \tag{I.46}$$

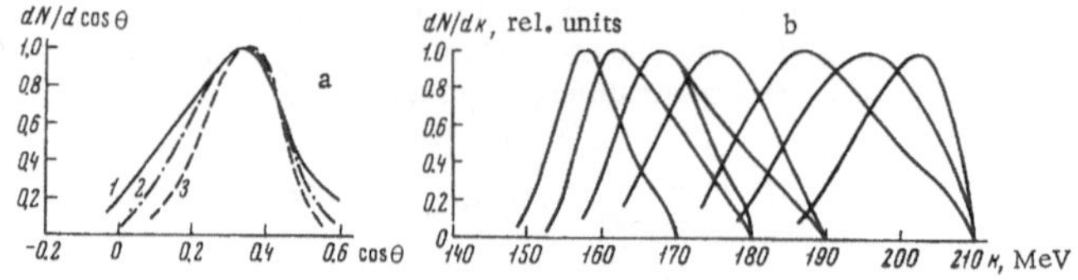

Fig. 9. Angular resolution (a) and energy resolution (b) of the (1) 158.8 MeV; (2) 171.1 MeV; (3) 194.4 MeV.

Thus, in order to obtain the differential cross sections from measured yields and to calculate the energy resolution, angular resolution, and the average $\bar{k}$ and cos θ_l values, one must know the function ε (k, cos θ_l, V) which depends upon the kinematics of the process, the geometry of the experiment, and the parameters of the measuring equipment.

This function and the related quantities were calculated on the M-20 computer of the Physics Institute of the Academy of Sciences of the USSR with a statistical trial (Monte Carlo) method [47]. The statistical method used for calculating the recording probability is based on the fact that all processes occurring during the experiment between the time at which the photon interacts with the proton and the time at which the γ quanta generated in the pion decay are detected by the telescopes can be represented as a sequence of random events which occur with certain probabilities. The sequence of these events is simulated with an appropriate probability scheme and a large number of trials are made with that scheme. When at fixed k_0, Ω_0, and V_0 values l trials out of f independent trials result in a count, the ratio l/f can be used to calculate $\varepsilon(k_0, \Omega_0, V_0)$, and the dispersion of this estimate is l/f^2 for $f \gg l \gg 1$. The function ε (k, Ω, V) was obtained by varying k_0, Ω_0, and V_0.

In the case of the photoproduction of neutral pions the sequence of events determining the recording probability was treated as follows. A pion is formed with the probability ε (V)dV in the volume element dV. Since the average lifetime of pions is very short (about 10^{-16} sec), the decay can be assumed to occur in the same volume element. The function ε(V) was therefore used to select the point of the pion decay. The intensity distribution over the cross section of the bremsstrahlung beam and the absorption of the beam in the target were taken into account when the function ε(V) was determined. The direction in which the γ quanta were emitted in the pion decay was first determined in the pion rest system and then transferred to the lab system. When none of the decay photons was absorbed in the target and both decay photons were incident on, and recorded, by, the telescopes, the trial resulted in a recording of the photoproduction process.

Figures 9a and 9b depict the results of calculations of the angular resolution and the energy resolution of the setup for various ψ_k and k_{max} values.

Results of the Experiments

The differential cross sections of the photoproduction of neutral pions at protons are listed in Table 6 and Fig. 10.

The table includes the calculated average photon energies in the lab system and the average pion emission angles in the center-of-mass system. The errors listed include only the statistical errors [among them the statistical error of the calculation of the efficiency $\varepsilon(\theta, k)$] but exclude errors made in the calibration of the photon energy of the accelerator beam. Figure 10 includes the data of other work for the purpose of comparison. Obviously, all sets of differential cross section data for the photoproduction of neutral pions agree for the angle $\theta \sim \pi/2$. As far as absolute values are concerned, the points obtained in [43] deviate to some extent.

TABLE 6. Results of Measurements of the Energy Dependence of $\frac{d\sigma}{d\Omega}(\theta^\circ, k)$

k, MeV	θ, degrees	$\frac{d\sigma}{d\Omega}$, μbarn/sr	k, MeV	θ, degrees	$\frac{d\sigma}{d\Omega}$, μbarn/sr
158.8	94.4	0.27±0.03	175.5	88.6	0.83±0.05
161.4	94.1	0.31±0.03	181.6	87.0	1.46±0.13
164.9	91.5	0.46±0.03	188.5	87.2	1.53±0.10
168.4	89.9	0.58±0.03	194.4	86.5	1.87±0.13
171.1	90.0	0.68±0.04	198.7	86.1	2.38±0.16

The $\gamma-\gamma$ coincidence yields were measured for 6-7 pion emission angles θ_l in the experiments for establishing the angular dependence of the differential cross sections at the maximum energies k_{max} = 180, 200, 230, and 264 MeV. The efficiency $\varepsilon(k, \theta)$ of pion detection, the average energy of the primary photon in the lab system, and the average pion emission angle in the center-of-mass system were also in this case determined with statistical trials. The latter two quantities and the differential cross section figures calculated from the yields are listed in Tables 7 and 8. As above, the errors of the differential cross sections include only the statistical errors of the yield measurements and the efficiency calculations. The errors of the absolute values amount to 10% and originate from insufficient information on the bremsstrahlung spectrum, ambiguities in the absolute measurements of the energy flux through the target, and systematic errors in the calculation of the pion-recording efficiency by means of the Monte Carlo method. Figures 11 and 12 depict the curves representing the energy resolution and angular resolution, respectively, of our experimental setup.

Angular Dependences of the Differential Cross Sections

As can be inferred from Tables 7 and 8, the average energies $\bar{k}$ differ slightly for the various positions at a given maximum energy k_{max} of the spectrum. In order to determine the angular dependences of the differential cross sections, all results obtained at a fixed k_{max} value were reduced to a single average energy. For example, for k_{max} = 180 MeV, the differential cross sections were reduced to the average energy 163 MeV, for k_{max} = 200 MeV, to k = 181 MeV, and for k_{max} = 230 MeV, to k = 212 MeV. The method was not applied to the case k_{max} = 264 MeV because average energies very close to the value 234 MeV had been obtained for all positions from calculations of the energy resolution.

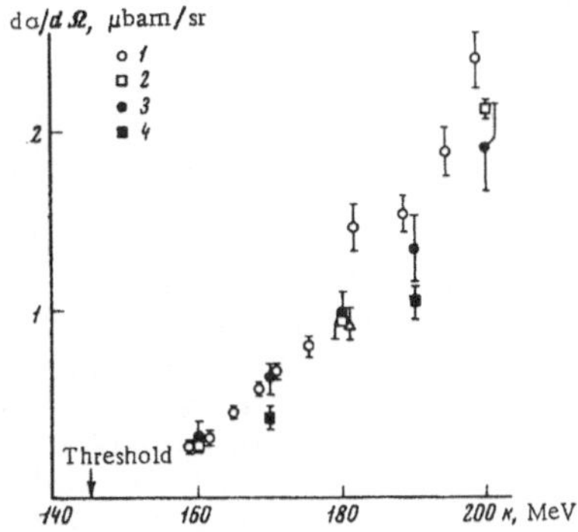

Fig. 10. Differential cross section of the reaction $\gamma + p \rightarrow p + \pi^0$ for the pion emission angle $\theta \approx 90°$ in the center-of-mass system. (1) Results of our measurements [50]; (2) [15]; (3) [35]; (4) [43].

TABLE 7. Differential Cross Sections of the Photoproduction of Neutral Pions at Protons

κ, MeV	θ, degrees	$\frac{d\sigma}{d\Omega}$, μbarn/sr	κ, MeV	θ, degrees	$\frac{\partial\sigma}{\partial\Omega}$, μbarn/sr	κ, MeV	θ, degrees	$\frac{\partial\sigma}{\partial\Omega}$, μbarn/sr
164.4	25.8	0.18±0.02	181.0	25.8	0.45±0.08	211.5	13.8	1.38±0.23
165.6	80.8	0.42±0.05	183.5	45.9	0.53±0.07	212	40.2	1.76±0.30
164.7	111.3	0.42±0.04	186.0	77	1.01±0.22	212.5	79.8	2.57±0.21
164.0	126.6	0.53±0.06	186.0	106.7	1.21±0.11	212.5	106.3	2.27±0.21
162,3	145.9	0,55±0.06	181.3	135.8	1.02±0.08	213	128.6	3.22±0.24
161.1	157.4	0,66±0.05	181.0	161	0.91±0.08	214	147.2	2.89±0.17
						212	168.3	2.13±0.19

TABLE 8. Differential Cross Sections of the Photoproduction of Neutral Pions at Protons

κ, MeV	θ, degrees	$\frac{d\sigma}{d\Omega}$, μbarn/sr	κ, MeV	θ, degrees	$\frac{d\sigma}{d\Omega}$, μbarn/sr
234.4	13.8	4.16±0.42	236.2	124	5.92±0.74
234.1	41.1	5.94±0.69	234.1	168.2	3.72±0.57
234.6	74.5	6.21±0.78	233.9	147.3	4.43±0.51
234.2	105.8	6.75±0.73			

Both the energy dependence and the angular coefficients of our previous work [15] were used to determine correction factors. Calculations of similar factors with the aid of theoretical coefficients A, B, and C [1, 8] modify only slightly (1–2%) the correction coefficients. It is therefore justified to reduce the differential cross sections to a single average energy with the above procedure.

Table 9 and Fig. 13 are compilations of the differential cross sections of neutral pion photoproduction at protons; the figures were obtained for certain energies.

The angular dependences of the cross sections were approximated by polynomials of powers of the cosine of the meson emission angle in the center-of-mass system:

$$\frac{d\sigma}{d\Omega} = A + B\cos\theta + C\cos^2\theta + D\cos^3\theta + \ldots,$$

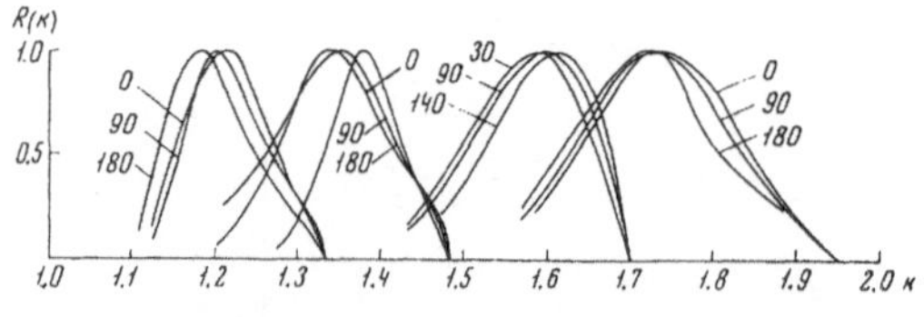

Fig. 11. Energy resolution of the setup for measurements of the angular dependence $(d\sigma/d\Omega)(\theta)$ of the reaction $\gamma + p \rightarrow p + \pi^0$. The figures at the curves indicate the angle (in degrees) between the orbital plane of the accelerator and the plane defined by the γ telescopes and the hydrogen target. The energy k of the primary photons is expressed in multiples of $m_\pi c^2$, where m_π denotes the mass of the π meson.

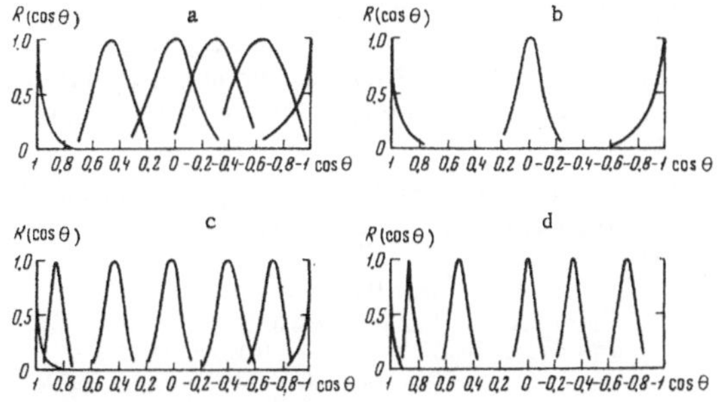

Fig. 12. Angular relations R(cos θ) of the setup at the energies (a) k = 163 MeV; (b) 181 MeV; (c) 212 MeV; and (d) 234 MeV.

TABLE 9. Differential Cross Sections of the Reaction $\gamma + p \rightarrow p + \pi^0$

$\bar{k}$ = 163 MeV		$\bar{k}$ = 181 MeV		$\bar{k}$ = 212 MeV		$\bar{k}$ = 234 MeV	
θ, deg	$\frac{d\sigma}{d\Omega}$, μbarn/sr	θ, deg	$\frac{d\sigma}{d\Omega}$, μbarn/sr	θ, deg	$\frac{d\sigma}{d\Omega}$, μbarn/sr	θ, deg	$\frac{d\sigma}{d\Omega}$, μbarn/sr
25.9	0.17±0.02	25.8	0.45±0.08	13.8	1.44±0.24	13.8	4.16±0.42
80.8	0.35±0.04	45.9	0.50±0.07	40.2	1.76±0.30	41.4	5.94±0.69
110.3	0.37±0.04	77	0.83±6.18	79.8	2.57±0.21	74.5	6.21±0.78
126.6	0.49±0.06	106.7	1.01±0.09	106.3	3.27±0.21	105.8	6.75±0.73
145.9	0.58±0.06	135.8	1.02±0.08	128.6	3.06±0.23	124	5.92±0.74
157.4	0.58±0.04	161	0.91±0.08	147.2	2.73±0.16	147.3	4.43±0.51
				168.3	2.13±0.19	168.2	3.72±0.57

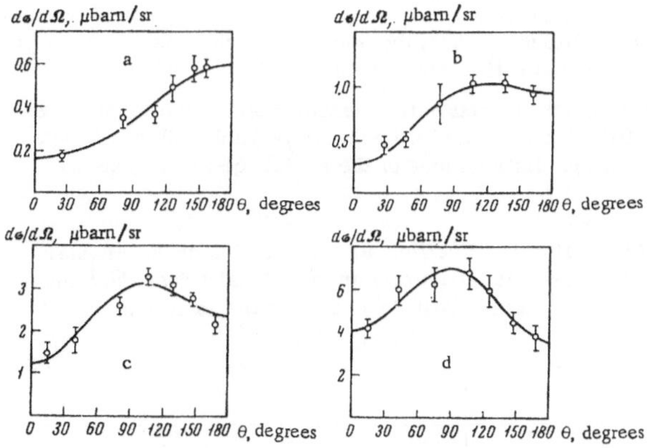

Fig. 13. Differential cross section of the reaction $\gamma + p \rightarrow p + \pi^0$ for four primary photon energies: (a) 163 MeV; (b) 181 MeV; (c) 212 MeV; (d) 234 MeV.

The coefficients of these polynomials are functions of the kinematic quantities and multipole amplitudes. The following condition was imposed on the magnitude of the coefficient at the highest power of cos θ: the error of the coefficient must not exceed its magnitude. This means that the last coefficient which can be calculated for the approximating polynomial must be substantially different from zero.

The following final approximation formulas (in which the coefficients are expressed in μbarn/sr) were obtained for the angular distributions of the pions:

$$k = 163 \text{ MeV}$$

$$\frac{d\sigma}{d\Omega} = \begin{cases} (0.36 \pm 0.02) - (0.22 \pm 0.02) \cos\theta; \\ (0.34 \pm 0.03) - (0.22 \pm 0.02) \cos\theta + (0.04 \pm 0.05) \cos^2\theta; \end{cases}$$

$$k = 181 \text{ MeV}$$

$$\frac{d\sigma}{d\Omega} = (0.93 \pm 0.08) - (0.32 \pm 0.05) \cos\theta - (0.33 \pm 0.13) \cos^2\theta;$$

$$k = 212 \text{ MeV}$$

$$\frac{d\sigma}{d\Omega} = \begin{cases} (3.02 \pm 0.15) - (0.57 \pm 0.12) \cos\theta - (1.25 \pm 0.23) \cos^2\theta; \\ (2.92 \pm 0.15) - (1.53 \pm 0.42) \cos\theta - (1.15 \pm 0.24) \cos^2\theta + (1.23 \pm 0.52) \cos^3\theta; \end{cases}$$

$$k = 234 \text{ MeV}$$

$$\frac{d\sigma}{d\Omega} = (6.9 \pm 0.5) + (0.27 \pm 0.28) \cos\theta - (3.06 \pm 0.68) \cos^2\theta.$$

The χ^2 criterion does not point to a preferable expression among the two possible expressions for $d\sigma/d\Omega$ at k = 212 MeV. However, the expression which is quadratic in cos θ was chosen as initial data for the ensuing analysis, because the coefficient D, which assumes the value 1.23 ± 0.52 at k = 212 MeV, is the result of a unique measurement insofar as even at higher energies no values exceeding the experimental errors were found for this coefficient. Actually, the present results at k = 234 MeV and attempts to approximate the angular dependences of the reaction $\gamma + p \rightarrow p + \pi^0$ (which were measured in the γ-quanta energy interval 270-400 MeV [48]) by a third-power polynomial in cos θ indicate that the coefficient D must be positive in that energy interval. So far, one can only say that the coefficient D differs from zero at energies exceeding 360 MeV.

The two possible forms of the angular distribution are practically identical at k = 163 MeV and, in the multipole analysis, lead to results which are identical within the accuracy limits of the unknown amplitudes. Thus, the polynomial which is quadratic in cos θ was also in this case chosen as approximating expression.

At the four energies considered in the measurements, the formula $d\sigma/d\Omega = A + B\cos\theta + C\cos^2\theta$ (with the coefficients A, B, and C as stated in Table 10) was adopted as best description of the angular π^0 meson distributions of the center-of-mass system.

TABLE 10. The Coefficients A, B, and C of the Angular Distributions (in μbarn/sr) and the Total Cross Sections $\sigma_t = 4\pi(A + \frac{1}{3}C)$ (in μbarn) of the Reaction $\gamma + p \rightarrow p + \pi^0$

k, MeV	A	B	C	σ_t
163	0.34±0.03	−0.22±0.02	0.04±0.05	4.44±0.23
181	0.93±0.08	−0.32±0.05	−0.33±0.13	10.30±0.62
212	3.02±0.15	−0.57±0.12	−1.25±0.23	32.66±1.26
234	6.90±0.50	+0.27±0.28	−3.06±0.68	73.85±3.60

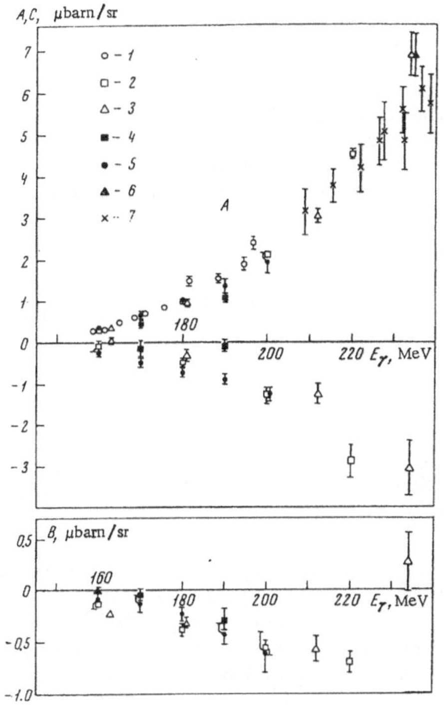

Fig. 14. Coefficients of the angular dependencies of the differential cross sections of the reaction $\gamma + p \rightarrow p + \pi^0$. 1) Cross sections obtained in our work [50]; 2) [15]; 3) result of the regression analysis of measurements made with the 2γ method [49]; 4) [43]; 5) [35]; 6) [38]; 7) [37].

Figure 14 depicts the coefficients A, B, C obtained in our work [49, 50] and in [15, 43, 35, 37]. It follows from the experimental data of Fig. 14 that the values of the coefficient A agree in all experimental results with an accuracy of 10-15%. New measurements of the angular distributions confirm our previous measurements of the coefficient A and agree very satisfactorily with an individual measurement of $\frac{d\sigma}{d\Omega}\left(\theta \sim \frac{\Omega}{2}\right)$ [50]. Good agreement between new data and our previous measurements or the results of others is also observed in the case of coefficient B. The new absolute values of coefficient C are smaller than the C obtained in our earlier work [15] and in [35], probably because the angular resolution was greatly improved in the setup used in the last experiments. Moreover, pion recordings through two γ quanta facilitate experiments in a much greater interval of pion-emission angles and, more particularly, at angles close to zero.

Comparison between Measured Angular Dependences of the Differential Cross Sections and Calculations Based on One-Dimensional Dispersion Relations

Comparisons of theoretical and experimental angular dependences of the differential cross sections with the aid of the χ^2 criterion were the first step in the analysis of calculations based on dispersion theory. The following form of the functional was used:

$$\chi^2 = \sum_i \left(\frac{\sigma_i^e - \sigma_i^t R}{\Delta\sigma_i^e} \right)^2 + \left(\frac{R-1}{\delta} \right)^2, \tag{I.47}$$

where σ_i^e and σ_i^t denote the experimental and theoretical differential cross sections, respectively; $\Delta\sigma_i^e$ denotes the error of the cross section; R denotes a normalizing factor; and δ denotes the relative error which is made when the experimental results are converted to absolute values. The first term characterizes the agreement between the form of the experimental and theoretical angular dependences. The second term expresses the deviation of the normalization factor from unity as a consequence of inaccurate conversion to absolute values. Accordingly, the number of degrees of freedom is $f = m - 1 + 1 = m$, where m denotes the number of experimentally measured cross sections. The normalization factor is defined by the condition that χ^2 be a minimum:

$$R = \frac{\sum_i \frac{\sigma_i^e \sigma_i^t}{(\Delta\sigma_i^e)^2} + \frac{1}{\delta^2}}{\sum \left(\frac{\sigma_i^t}{\Delta\sigma_i^e}\right) + \frac{1}{\delta^2}}. \tag{I.48}$$

When the second term in Eq. (I.47) is not taken into account, we have

$$R = \frac{\sum_i \frac{\sigma_i^e \sigma_i^t}{(\Delta\sigma_i^e)^2}}{\sum_i \left(\frac{\sigma_i^t}{\sigma_i^e}\right)^2}. \tag{I.49}$$

When the conversion to absolute values is made with poor accuracy ($\delta \to \infty$), then $(R-1)/\delta \to 0$. With a sufficiently large number of experimental points (10-20 points) obtained with an accuracy of 5-10% at $\delta \sim 10\%$ (as in our case), the contribution of the second term to χ^2 is small. Since the conversion to absolute values was made when angular dependences were measured for each energy k, various R values are obtained for the angular distributions measured at various energies.

Table 11 lists the results obtained from a comparison of angular distributions measured at k = 163, 181, and 212 MeV and calculated distributions [8, 28]. The angular distributions were calculated with the aid of the multipole amplitudes stated in those papers. The solutions of the dispersion relations of [1] are defined as the quantities calculated with the formulas of [1] for multipole amplitudes, with the experimentally determined phases of πN scattering used in the calculations. The value −0.020, which renders good agreement with our results for this amplitude, was used for the parameter $N^{(+)}$ in the calculation of $\mathrm{Re}F_{0+}$. We note that the solutions of [1] were previously defined as solutions of dispersion relations for photoproduction, in which phase shifts calculated with the effective radius formulas of [13] were inserted.

TABLE 11. Application of the χ^2 Criterion to the Comparison of Theoretical and Experimental Angular Dependences of the Differential Cross Sections

Reference	Without $\left(\frac{R-1}{\sigma}\right)^2$			With $\left(\frac{R-1}{\sigma}\right)^2$		
	χ^2	f	P	χ^2	f	P
Solov'ev [8]	22.3	16	0.13	45.4	19	0.0007
Schmidt [31]	106.7	16	0.00000	131.1	19	0.00000
Chew et al. [1]	17.8	16	0.33	33.2	19	0.024
Donnochie and Shaw [28]	101.6	16	0.00000	124.44	19	0.00000

Note: P denotes the probability of χ^2 exceeding the value of the second column.

When the angular distributions were calculated with the multipole amplitudes stated in the work of L. D. Solov'ev, $\mathrm{Im}\,E_{0+} \approx \mathrm{Im}\,E_{0+}^{(+)}$ was also taken into account, in accordance with the solution of [1].

As can be inferred from the table, the solutions to the dispersion relations of [28, 31] do not describe the form of the angular distributions since they render excessive asymmetry with respect to the angle $\theta = 90°$. On the other hand, the solutions of [1] and [8] appropriately describe the form of the angular dependences of the cross sections at all primary photon energies considered. But none of the solutions of dispersion theory can adequately describe the absolute values and the energy dependence of the cross sections.

Determination of Multipole Amplitudes from Experimental Results

The multipole amplitudes were calculated from the measurements in order to expand the analysis of neutral pion photoproduction at protons and to obtain a better idea of the reason for the discrepancies between experimental data and the solutions to the dispersion equations. We had in mind that knowledge of the multipole amplitudes in the near-threshold energy range of the primary photons may be useful in the development of new theoretical methods of calculating the reaction under consideration (e.g., when the summation rules are employed, the unitary symmetry of particles is considered, etc.).

Calculation of the E_{0+} Amplitude from the Energy Dependence of Coefficient A

In order to determine the amplitude of neutral pion photoproduction at photons in the s state at the energy threshold, the energy dependence of the differential cross sections of Table 6 was analyzed for the pion emission angle ~90° in the center-of-mass system. In addition, the values of the coefficient A of our previous work [15] were used.

In accordance with Eq. (I.17), the differential cross section was represented as a series

$$\frac{d\sigma_0}{d\Omega}\frac{k}{q} = \left[\frac{d\sigma}{d\Omega} - \frac{d\sigma}{d\Omega}(E_{0i})\right]\frac{k}{q} = \sum_m A_m (qk)^m, \qquad m = 0, 2, \ldots, \tag{I.50}$$

where $d\sigma/d\Omega$ denotes the differential cross section from Table 6; and $(d\sigma/d\Omega)(E_{0i})$ denotes the contribution of the imaginary part of the s-wave amplitude to the differential cross section. The amplitude was calculated with the formula

$$\frac{d\sigma}{d\Omega}(E_{0i})\frac{q}{k}\left[\frac{2}{3}ef(a_1 - a_3)\,q_+F_s\right]^2. \tag{I.51}$$

where $ef = 2.416 \cdot 10^{-2}$; $a_1 - a_3 = 0.25$; q_+ denotes the π^+ meson momentum corresponding to the generation of a π^0 meson with momentum q; $F_s = 1 - \frac{1}{3}\frac{q^2}{\omega^2}$. The contribution of $\frac{d\sigma}{d\Omega}(E_{0i})$ to all measured differential cross sections was less than 10%. The least square method yields the following parameter values in calculations of the regression coefficients:

two-term analysis

$$\frac{d\sigma_0}{d\Omega} = \frac{q}{k}\,[(0.06 \pm 0.06) + (3.16 \pm 0.17)(kq)^2]\;\frac{\mu\text{barn}}{\text{sr}}\;; \tag{I.52}$$

three-term analysis

$$\frac{d\sigma_0}{d\Omega} = \frac{q}{k}\,[(0.06 \pm 0.14) + (3.19 \pm 0.66)\,(kq)^2 + (0.23 \pm 0.62)(kq)^4]\;\frac{\mu\text{barn}}{\text{sr}} \tag{I.53}$$

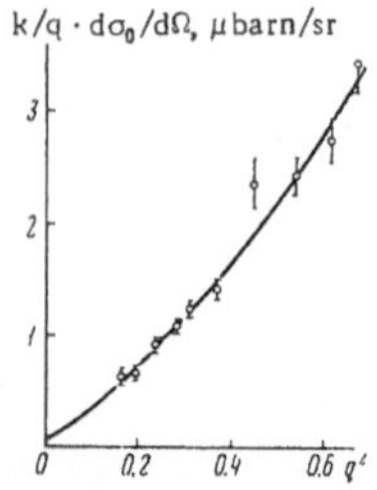

Fig. 15. Energy dependence of $d\sigma_0/d\Omega$ for $\theta \approx 90°$. The curve is the approximation of the measured points by the two-term polynomial (I.52).

According to the Gauss criterion, optimum description of the experimental results is obtained for a quadratic form. Figure 15 shows the curve corresponding to the approximation (I.52) and, in addition, the experimentally determined differential cross sections.

We obtain with the coefficients of our paper [15] for the three energies k = 160, 180, and 200 MeV:

$$\frac{d\sigma_0}{d\Omega} = \frac{q}{k}\,[(0.09 \pm 0.08) + (2.70 \pm 0.12)(kq)^2]\;\frac{\mu\text{barn}}{\text{sr}}\,. \tag{I.54}$$

When the dependence of coefficient A is restricted to two terms and the average of expressions (I.52) and (I.54) is taken, we obtain

$$A^{(0)} = (0.074 \pm 0.048)\;\frac{\mu\text{barn}}{\text{sr}} \tag{I.55}$$

and

$$E_{0+} = (-\,0.27 \pm 0.09)\,10^{-16}\ \text{cm} \tag{I.56}$$

or $E_{0+} = (0.18 \pm 0.06) \cdot 10^{-2}$ in units $\lambda = h/m_\pi c$. The sign of E_{0+} is taken in agreement with the sign of the interference term $B\cos\theta$ in the angular dependence in the near-threshold energy range of the photons. We have not included in our analysis a possible detailed structure of the differential cross section at small momenta, which may be caused by the threshold of the reaction $\gamma + p \rightarrow p + \pi^+$. Estimates of this effect were given in [52] with the Lapidus–Chou Huan-chao method outlined in [51]. It could be shown that the threshold anomaly results in a maximum addition of 20% of the amplitude value to the amplitude. Moreover, when the s-wave amplitude is determined from the energy dependence of the coefficient A and the total cross section σ_t as in [52], the following error is made. As has been mentioned above, the expansion

$$A = A^{(0)} + q^2A^{(2)} + q^4A^{(4)} + \ldots,$$

$$\frac{\sigma_+}{kq} = \sigma^{(0)} + q^2\sigma^{(2)} + q^4\sigma^{(4)} + \ldots$$

can be used at $q < 1$, where $|E_{0+}|^2 = A^{(0)} = \sigma^{(0)}/4\pi$.

It was assumed in the analysis of the experimental data that the p-wave ($A^{(2)}$ and $\sigma^{(2)}$) and s-wave ($A^{(0)}$ and $\sigma^{(0)}$) portions of the cross sections were essentially determined from experiments in which small differences (~10%) occur in the normalization. This assumption was a consequence of the relations for the statistical accuracy of the points measured at small and large momenta q in the various experiments. But the small difference in the normalization of the experimental data of [15] and [17, 48, 53] resulted in an inaccurate determination of the E_{0+} amplitude when the data of all those papers were combined for the analysis. The sensitivity of $|E_{0+}|$ is a consequence of the ratio of the quantities $A^{(0)}$ and $A^{(2)}$ at low energies: $A^{(0)} \sim 0.1$, but $A^{(2)} \sim 1.6$.

Calculation of Multipole Amplitudes from the Angular Dependences of Differential Cross Sections

It has been shown above that the experimentally obtained angular dependence in the center-of-mass system can be optimally approximated by a second-order polynomial in cos θ. This means that the s- and p-wave approximations suffice in the energy range considered (as far as the accuracy with which our results were obtained is concerned). The coefficients A, B, and C can be expressed in this approximation by the following multipole amplitudes of photoproduction:

$$
\begin{aligned}
A &= \frac{q}{2k}\left[2\,|E_{0+}|^2 + |2M_{1+} + M_{1-}|^2 + |M_{1+} - M_{1-} - 3E_{1+}|^2\right],\\
B &= \frac{2q}{k}\,\mathrm{Re}\left[E_{0+}^{*}(M_{1+} - M_{1-} + 3E_{1+})\right],\\
C &= \frac{q}{2k}\left[2\,|M_{1+} - M_{1-} + 3E_{1+}|^2 - |M_{1+} - M_{1-} - 3E_{1+}|^2 - |2M_{1+} + M_{1-}|^2\right].
\end{aligned}
\qquad (I.57)
$$

In other words, a total of three equations are available for determining the four complex quantities E_{0+}, M_{1+}, M_{1-}, and E_{1+}. When only the experimental data on angular dependences are used, a full amplitude analysis is not feasible and all pertinent multipoles cannot be determined, because the number of unknowns is in this case always greater than the number of equations for the unknowns. Only three equations, which define the experimental coefficients A, B, and C, are available for the determination of the four amplitudes E_{0+}, M_{1+}, M_{1-}, and E_{1+} in the p and s wave approximation. When the d-wave amplitudes $E_{2\pm}$ and $M_{2\pm}$ are included in the analysis (these amplitudes are important once the third and fourth powers of cos θ with the corresponding coefficients $\mathcal{D}$ and $\mathcal{E}$ are considered in the angular dependences), the number of equations is increased to five, but the number of unknowns increases to eight. The coefficient $\mathcal{D}$ depends upon the interference of p and d waves; only the amplitudes $E_{2\pm}$ and $M_{2\pm}$ appear in the expression for the coefficient $\mathcal{E}$. To date no information is available on these amplitudes except for, maybe, the amplitude E_{2-}.

Thus, when multipole amplitudes are to be determined from experimental angular distributions, certain additional assumptions must be introduced. These assumptions are more or less plausible within the framework of existing concepts and make the phenomenological analysis feasible. The traditional assumption, that the electric quadrupole amplitude E_{1+} is small is one of these simplifying assumptions in the s,p-wave approximation. When we assume that $E_{1+} = 0$ in the energy interval under consideration and introduce in some way the imaginary parts of the other s- and p-wave amplitudes, we obtain for the three unknowns Re E_{0+}, Re M_{1+}, and Re M_{1-} a system of three equations which are determined by expressions (I.57).

As far as the phenomenological analysis of the photoproduction of neutral pions at protons is concerned, there exists a quantity which is highly sensitive to the multipole E_{1+}. As has been mentioned above, the cross section of neutral pion photoproduction by plane-polarized photons has the form

$$
\frac{d\sigma}{d\Omega}(\theta, \varphi) = \frac{d\sigma}{d\Omega}(\theta) + \alpha \sin^2\theta \cos 2\varphi, \qquad (I.58)
$$

where $\frac{d\sigma}{d\Omega}(\theta)$ denotes the cross section for nonpolarized photons, and the coefficient α accounts for the azimuthal symmetry of the process (φ denotes the angle between the plane of meson generation and the plane of photon polarization). In the s,p-wave approximation

$$
\frac{\alpha}{C} = 1 - 12\frac{q}{k}\frac{\mathrm{Re}\left[(M_{1+} - M_{1-})^{*}E_{1+}\right]}{C}
$$

can differ substantially from unity, if $E_{1+} \neq 0$, since the numerical factor 12 is important. The experimental values of the ratio α/C can then be used as additional condition to Eq. (I.57). A completely determined system of four equations for the four unknowns Re E_{0+}, Re M_{1+}, Re M_{1+}, and Re E_{1+} is obtained in this case.

The data of [41, 42] for α/C refer to the energies 228 and 235 MeV and are used in one of the following sections for a total amplitude analysis at the energy 234 MeV and for the determination of the s-wave and all p-wave multipoles Re E_{0+}, Re $M_{1\pm}$, and Re M_{1-} of the reaction $\gamma + p \to p + \pi^0$. Since the corresponding experimental results for α/C are not available, a similar analysis cannot be performed at the energies 163, 181, and 212 MeV and we consider only the dependence of the experimental amplitudes Re M_{1+} and Re M_{1-} (obtained at these energies under the assumption $E_{1+} = 0$) upon the magnitude of the electric quadrupole amplitude Re_{1+}, which is taken as a free parameter.

Amplitude Analysis under the Assumption $E_{1+} = 0$

When we solve the system of Eqs. (I.57) in this section, we use an approximation which is stated in the form of the following additional assumptions:

1) Re $E_{1+} = 0$.

2) The imaginary parts Im M_{1-} and Im E_{1+} of the p-wave amplitudes are also zero. In order to assess the importance of this assumption for the final results, the imaginary parts of these amplitudes were included in the analysis. The following theoretical expressions for [1]

$$\text{Im}\, M_{1-} = \frac{2}{9} ef\, F_M kq \left(1 + \frac{\omega}{M}\right)^{-1} (\sin 2\alpha_{11} - \sin \alpha_{31}) + \frac{\mu_v}{f} \frac{k}{q} \left(\frac{1}{3} \frac{\sin^2 \alpha_{11}}{q} + \frac{2}{3} \frac{\sin^2 \alpha_{31}}{q}\right),$$

$$\text{Im}\, E_{1+} = \frac{11}{9} ef F_Q kq \left(1 + \frac{\omega}{M}\right)^{-1} (\sin 2\alpha_{13} - \sin 2\alpha_{33})$$

and various sets of experimentally determined scattering phases [29, 54, 55] were used. The observed changes in the results were within the error limits of the unknown amplitudes.

3) Im E_{0+} and Im M_{1+} have significant values in the energy range under consideration. These quantities were taken into account with the aid of theoretical expressions of [1]:

$$\text{Im}\, E_{0+} \approx \text{Im}\, E_{0+}^{(+)} = \frac{2}{3} ef (\alpha_1 - \alpha_3) F_s \left(1 + \frac{\omega}{M}\right)^{-1},$$

$$\text{Im}\, M_{1+} \approx \text{Im}\, M_{1+}^{(+)} = -\frac{1}{9} ef F_M kq \left(1 + \frac{\omega}{M}\right)^{-1} (\sin 2\alpha_{13} - \sin 2\alpha_{33}) + \frac{\mu_v}{f} \frac{k}{q} \left(\frac{1}{3} \frac{\sin^2 \alpha_{13}}{q} + \frac{2}{3} \frac{\sin^2 \alpha_{33}}{q}\right).$$

Since $\sin 2\alpha_{13}$ and $\sin^2 \alpha_{13}$ can be ignored, the last expression simplifies to

$$\text{Im}\, M_{1+} \approx \frac{1}{9} ef F_M kq \left(1 + \frac{\omega}{M}\right)^{-1} \sin 2\alpha_{33} + \frac{2}{3} \frac{\mu_v}{f} \frac{k}{q} \frac{\sin^2 \alpha_{33}}{q}, \qquad \text{(I.59)}$$

which, in the near-threshold range, differs by approximately 25% from the Im M_{1+} approximation with only the μ part of the resonance amplitude $M_{1+}^{3/2}$:

$$\text{Im}\, M_{1+} \approx \text{Im}\, M_{1+,\,\mu}^{3/2} = \frac{2}{3} \frac{\mu_v}{f} \frac{k}{q} \frac{\sin^2 \alpha_{33}}{q} \qquad \text{(I.59a)}$$

Im M_{1+} was also estimated from the experimentally determined total cross sections of the reaction $\gamma + p \to p + \pi^0$, as proposed in [15, 21-23]:

$$\text{Im}\, M_{1+} = \sqrt{\frac{\sigma_t}{4\pi} \frac{k}{q}} \sin \alpha_{33}. \qquad \text{(1.59b)}$$

The values

$$e^2 = \frac{1}{137}; \quad f^2 = 0.081; \quad \mu_v = 1.446 \cdot 10^{-2}; \quad \frac{\alpha_1 - \alpha_3}{q} = 0.291.$$

were used for the parameters in the formulas for Im E_{0+} and Im M_{1+}. In the energy range extending to the first resonance, the phase α_{33} is known with an accuracy of several per cent; the experimental value of the phase can be appropriately described with the effective-radius formula

$$\frac{\sin^2 \alpha_{33}}{q^3} = \frac{q^3}{q^6 + \omega^2 \left(\frac{0.75}{f_c}\right)^2 \left(1 - \frac{\omega}{\omega_r}\right)^2}$$

and the parameters $f_c^2 = 0.087$, $\omega_r = 2.24$ [31].

The calculated Im E_{0+} and Im M_{1+} values are listed in Table 12. The Im M_{1+} values calculated with Eqs. (I.59) and (I.59a) were used in the analysis. It was therefore possible to estimate the influence of various forms of Im M_{1+} upon the final results. The absolute value of Re M_{1-} is independent of the expression [Eq. (I.59) or (I.59a)] used to approximate Im M_{1+}. When Eq. (I.59a) is used for Re E_{0+} and Re M_{1+} version (I.59) results in an Re E_{0+} value which at 234 MeV is about 10% greater than that obtained with Eq. (I.59a), whereas the Re M_{1+} value is about 4% greater. The difference between the results obtained with Eqs. (I.59) and (I.59a) is insignificant at 163 MeV. The results of the analysis are now outlined for the case of Eq. (I.59).

The system of equations (I.57) was solved under the above assumptions. The combination of coefficients A + C with B results in

$$[\mathrm{Re}\, E_{0+}]^2_{\pm} = \frac{a \pm \sqrt{a^2 - 4b}}{2}, \tag{I.60}$$

$$[\mathrm{Re}\,(M_{1+} - M_{1-}]^2_{\pm} = \frac{a \mp \sqrt{a^2 - 4b}}{2} = [\mathrm{Re}\, E_{0+}]^2_{\mp}; \tag{I.61}$$

whereas with the combination A − C

$$[\mathrm{Re}\,(2M_{1+} + M_{1-})]^2_{\pm} = c^2_{\pm}, \tag{I.62}$$

where

$$\begin{aligned} a &= (A + C)\frac{k}{q} - (\mathrm{Im}\, E_{0+})^2 - (\mathrm{Im}\, M_{1+})^2 = [\mathrm{Re}\, E_{0+}]^2 + [\mathrm{Re}\,(M_{1+} - M_{1-})]^2, \\ 4b &= \left(B\,\frac{k}{q} - 2\,\mathrm{Im}\, E_{0+}\,\mathrm{Im}\, M_{1+}\right)^2 = 4\,[\mathrm{Re}\, E_{0+}]^2\,[\mathrm{Re}\,(M_{1+} - M_{1-})]^2, \\ c^2_{\pm} &= (A - C)\frac{k}{q} - (\mathrm{Im}\, E_{0+})^2 - 4\,(\mathrm{Im}\, M_{1+})^2 - [\mathrm{Re}\, E_{0+}]^2_{\pm}. \end{aligned} \tag{I.63}$$

TABLE 12. Im E_{0+} and Im M_{1+} for the Photoproduction of π^0 Mesons

k, MeV	Im$E_{0+}\cdot 10^2$	Im$M_{1+}\cdot 10^2$		
		a	b	c
163	0.172	0.009	0.007	0.009
181	0.236	0.045	0.034	0.035
212	0.300	0.211	0.159	0.162
234	0.330	0.450		0.360

We obtain from Eqs. (I.60) and (I.61)

$$[\mathrm{Re}\,E_{0+}]_{\pm\pm} = \pm\sqrt{\frac{a \pm \sqrt{a^2-4b}}{2}};$$

$$[\mathrm{Re}\,(M_{1+} - M_{1-})]_{\pm\pm} = \pm\sqrt{\frac{a \mp \sqrt{a^2-4b}}{2}} = [\mathrm{Re}\,E_{0+}]_{\pm\mp}. \tag{I.64}$$

Since in the energy interval under consideration

$$\left(B\,\frac{k}{q} - 2\,\mathrm{Im}\,E_{0+}\,\mathrm{Im}\,M_{1+}\right) < 0,$$

we obtain from Eq. (I.57)

$$\mathrm{Re}\,E_{0+}\,\mathrm{Re}\,(M_{1+} - M_{1-}) < 0,$$

i.e., the signs of $\mathrm{Re}\,E_{0+}$ and $\mathrm{Re}\,(M_{1+} - M_{1-})$ differ. Since according to all theories of photoproduction, $|\mathrm{Re}\,M_{1+}| > |\mathrm{Re}\,M_{1-}|$ must hold in the energy range under consideration and the sign of $\mathrm{Re}\,M_{1+}$ must be positive, both negative solutions must be used:

$$[\mathrm{Re}\,E_{0+}]_{\pm} = -\sqrt{\frac{a \pm \sqrt{a^2-4b}}{2}},$$

$$[\mathrm{Re}\,(M_{1+} - M_{1-})]_{\pm} = +\sqrt{\frac{a \mp \sqrt{a^2-4b}}{2}} = -[\mathrm{Re}\,E_{0+}]_{\mp}, \tag{I.65}$$

$$[\mathrm{Re}\,(2M_{1+} + M_{1-})]_{\pm\pm} = \pm c_{\pm}.$$

Simultaneous solution of the last two equations results in

$$[\mathrm{Re}\,M_{1+}]_{\pm\pm} = \frac{1}{3}\{\pm c_{\pm} - [\mathrm{Re}\,E_{0+}]_{\mp}\},$$

$$[\mathrm{Re}\,M_{1-}]_{\pm\pm} = \frac{1}{3}\{\pm c_{\pm} + 2\,[\mathrm{Re}\,E_{0+}]_{\mp}\}. \tag{I.66}$$

In other words, two solutions exist for the s-wave amplitude E_{0+}, and four solutions for the p-wave amplitudes M_{1+} and M_{1-}. However, the two solutions corresponding to the minus sign before $c_{\pm}$ can be omitted for $\mathrm{Re}\,M_{1+}$ and $\mathrm{Re}\,M_{1-}$, because $\mathrm{Re}\,M_{1+} < 0$ in this case. Consequently, there remain two signs which correspond to plus and minus before the term $\sqrt{a^2-4b}$ in Eq. (I.64) for each amplitude. Our task is to select a proper set of multipole amplitudes from the two possible expressions (I.65) and (I.66). Moreover, transitions from plus to minus solutions are possible when we switch from one energy of the primary photons to another. The momentum dependence of the matrix element is proportional to q^l in the long-wave approximation. Accordingly, $\mathrm{Re}\,E_{0+}$ ($l = 0$) must be almost constant near the photoproduction threshold, whereas $\mathrm{Re}\,(M_{1+} - M_{1-})$ ($l = 1$) must increase linearly with increasing q. This means that the relation between $\mathrm{Re}\,E_{0+}$ and $\mathrm{Re}\,(M_{1+} - M_{1-})$ must change with increasing energy. First, $|\mathrm{Re}\,E_{0+}| > |\mathrm{Re}\,(M_{1+} - M_{1-})|$, then the amplitudes become comparable at a certain k_0 value, whereupon

$$|\mathrm{Re}\,E_{0+}| < |\mathrm{Re}\,(M_{1+} - M_{1-}|$$

when the energy increases further on to $k > k_0$.

This means that the quantity

$$\sqrt{a^2-4b} = \pm\{[\mathrm{Re}\,E_{0+}]^2 - [\mathrm{Re}\,(M_{1+} - M_{1-})]^2\} = \pm\,\mathrm{Re}\,(M_{1+} - M_{1-} - E_{0+})\,\mathrm{Re}\,(M_{1+} - M_{1-} + E_{0+}) \tag{I.67}$$

must change its sign at increasing energies. The quantity $\sqrt{a^2-4b}$ is expressed by the experimental values A, B, and C and the imaginary parts of the amplitudes E_{0+}, M_{1+}, and M_{1-}:

$$a^2 - 4b = \left\{(A + B + C)\frac{k}{q} - [\mathrm{Im}\,(M_{1+} - M_{1-} + E_{0+})]^2\right\}\left\{(A - B + C)\frac{k}{q} - [\mathrm{Im}\,(M_{1+} - M_{1-} - E_{0+})]^2\right\}. \tag{I.68}$$

TABLE 13. Multipole Amplitudes Re E_{0+}, Re M_{1+}, and Re M_{1-} of the Reaction $\gamma + p \to p + \pi^0$ (under the Assumption $E_{1+} = 0$)

κ, MeV	Re $E_{0+}\cdot 10^2$	Re $M_{1+}\cdot 10^2$	Re $M_{1-}\cdot 10^2$	Disregarded solution for Re $E_{0+}\cdot 10^2$
163	−0.204±0.028	0.352±0.025	−0.216±0.046	−0.568
181	−0.220±0.041	0.483±0.019	−0.122±0.064	−0.606
212	−0.255±0.042	0.864±0.019	−0.131±0.050	−0.984
234	−0.048±0.058	1.228±0.055	−0.192±0.080	−1.421

In view of the large errors, no definite conclusions can be drawn from our work [15], but it can be assumed that the quantity $\sqrt{a^2 - 4b}$ vanishes in the energy range of about 180 MeV. Since our latest, very accurate measurements [47-50] do not indicate a change in the sign of $\sqrt{a^2 - 4b}$ in that energy range, we disregarded the "crossover" solutions for the multipole amplitudes.

We have to make a choice between plus and minus solutions. To this end, we use the results of the previous analysis of the energy dependence of coefficient A. The following E_{0+} amplitude was obtained at the threshold of the γ quanta

$$(E_{0+})_{\text{thresh}} = (-0.18 \pm 0.06)\cdot 10^{-2}.$$

Extrapolation to the Re E_{0+} threshold values obtained in our analysis at the energies 163, 181, and 212 MeV gave the value

$$(\text{Re } E_{0+})_{\text{thresh}} = (-0.22 \pm 0.02)\cdot 10^{-2}.$$

for one of the solutions. This solution, which corresponds to the minus sign before $\sqrt{a^2 - 4b}$ in Eq. (I.65), i.e., to small solutions for the s-wave amplitude, was chosen as the final result in the amplitude analysis of the angular distributions. Table 13 and Figs. 16 and 17 (open circles) indicate the Re E_{0+}, Re M_{1+}, and Re M_{1-} values corresponding to the solution adopted. The last column of Table 13 lists the s-wave amplitudes corresponding to the second, ignored solution.

The errors indicated at the amplitudes result from errors in the differential cross section measurements and were calculated with a formula for the dispersion D in which the cor-

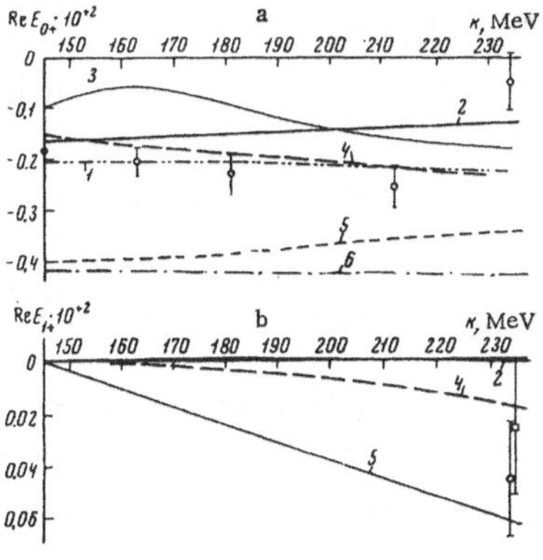

Fig. 16. Electric multipole amplitudes of neutral pion photoproduction. (a) Result of the analysis for the amplitude E_{0+} of the energy dependence of the coefficient A (●) and the angular dependence of the differential cross section (○); b) result of the multipole analysis for the E_{1+} amplitude using the α values from [42] (○) and [41] (□); 1) calculations with the formulas of [1]; 2) [8, 57]; 3) [90]; 4) [30]; 5) [28]; 6) [31]. The theoretical curves, including those for the E_{0+} amplitude, are indicated.

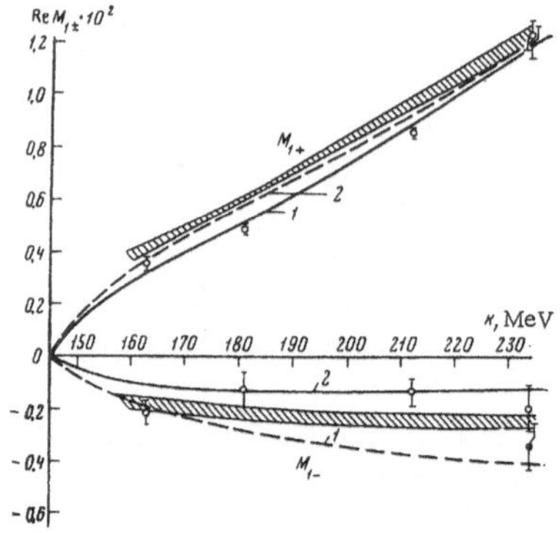

Fig. 17. Magnetic multipole amplitudes of neutral pion photoproduction at protons. (○) Results of multipole analysis under the assumption $E_{1+} = 0$; (●) results of an analysis in which E_{1+} was included (α from [42]). The shaded areas correspond to the amplitude spread which is obtained when various sets of phase shifts of πN scattering are inserted in the expressions of [1]. 1) For the M_{1+} amplitude, results of [8, 30, 31, 57]; 2) [29]; 1) for the M_{1-} amplitude of [28, 30, 31]; 2) [57].

relation of A, B, and C was taken into account:

$$D[\mathrm{Re}\,Y] = \sigma_0^2 \sum_{i,k} \frac{\partial[\mathrm{Re}\,Y]}{\partial A_i} G_{ik}^{-1} \frac{\partial[\mathrm{Re}\,Y]}{\partial A_k}, \tag{I.69}$$

where Y denotes the E_{0+}, M_{1+}, or M_{1-} amplitudes; $A_1 = A$, $A_2 = B$, $A_3 = C$; G_{ik}^{-1} denotes the element of the error matrix calculated in the regression analysis; and σ_0^2 denotes the dispersion per unit weight.

Amplitude Analysis in Which E_{1+} Is Included

The present section outlines the results of the full amplitude analysis of π^0 meson photoproduction at protons in the s- and p-wave approximation for the energy 234 MeV. The analysis is based on the solution of the system of four equations (I.57) and (I.58) for the four unknowns Re E_{0+}, Re M_{1+}, Re M_{1-}, and Re E_{1+}. The solution method and the assumptions made for the solution are as described above. Since Re E_{0+} is independent of Re E_{1+}, the set of p-wave amplitudes corresponds to the solution adopted in the preceding section.

The following expressions were obtained for Re M_{1+}, Re M_{1-}, and Re E_{1+}:

$$[\mathrm{Re}\,M_{1+}]_- = \frac{1}{6}\{[\mathrm{Re}\,E_{0+}]_+ + d_+ + 2f_-\}, \tag{I.70}$$

$$[\mathrm{Re}\,M_{1-}]_- = -\frac{1}{3}\{[\mathrm{Re}\,E_{0+}]_+ + d_+ - f_-\}, \tag{I.71}$$

$$[\mathrm{Re}\,E_{1+}] = \frac{1}{6}\{[\mathrm{Re}\,E_{0+}]_+ - d_+\}, \tag{I.72}$$

where

$$d_+ = \sqrt{[\mathrm{Re}\,E_{0+}]_+^2 - (C-\alpha)\frac{k}{q}},$$
$$f_- = \sqrt{(A-\alpha)\frac{k}{q} - (\mathrm{Im}\,E_{0+})^2 - 4(\mathrm{Im}\,M_{1+})^2 - [\mathrm{Re}\,E_{0+}]_-^2}. \tag{I.73}$$

The experimental values of [41, 42] were used for the coefficient α when d and f were calculated. The resulting multipole amplitudes are listed in Table 14. The error stated for α in the table does not include the error of $(d\sigma/d\Omega)(\theta)$ in the expression $(\sigma_\perp - \sigma_\parallel/\sigma_\perp + \sigma_\parallel) = -(\alpha \sin^2\theta) : [(d\sigma/d\Omega)(\theta)]$. The latter error was taken into account when the errors of the mul-

TABLE 14. Multipole Amplitudes of the Reaction $\gamma + p \to p + \pi^0$ at the Energy 234 MeV

	$E_{1+} = 0$	α	
		-2.30 ± 0.39 [41]	-1.70 ± 0.11 [42]
$ReE_{0+} \cdot 10^2$	-0.048 ± 0.058	-0.048 ± 0.058	-0.048 ± 0.058
$ReM_{1+} \cdot 10^2$	-1.228 ± 0.055	1.219 ± 0.063	1.208 ± 0.062
$ReM_{1-} \cdot 10^2$	-0.192 ± 0.088	-0.271 ± 0.101	0.342 ± 0.092
$ReE_{1+} \cdot 10^2$		-0.026 ± 0.026	0.046 ± 0.022

tipole amplitudes listed in the table were calculated. The dark dots of Fig. 17 indicate the quantities Re M_{1+} and Re M_{1-} for $\alpha = -1.70$. The Re E_{1+} values corresponding to the two α values are depicted in Fig. 16. This amplitude is in both cases negative and nonvanishing within two standard deviations for the α measured with the smallest error in [42].

We restricted our considerations to the influence of Re E_{1+} upon the quantities Re M_{1+} and Re M_{1-} at the energies 163, 181, and 212 MeV, since Re E_{1+} was used as a free parameter in the analysis. We obtain in this case in place of Eq. (I.66)

$$\{\mathrm{Re}\,M_{1+}]_- = \frac{\sqrt{c_-^2 - 12\,\mathrm{Re}\,E_{1+}[\mathrm{Re}\,E_{0+}]_+ - 36(\mathrm{Re}\,E_{1+})^2} - [\mathrm{Re}\,E_{0+}]_+}{3} - \mathrm{Re}\,E_{1+}, \quad (I.74)$$

$$\{\mathrm{Re}\,M_{1-}]_- = \frac{\sqrt{c_-^2 - 12\,\mathrm{Re}\,E_{1+}[\mathrm{Re}\,E_{0+}]_+ - 36(\mathrm{Re}\,E_{1+})^2} + 2[\mathrm{Re}\,E_{0+}]_+}{3} + 2\,\mathrm{Re}\,E_{1+}. \quad (I.75)$$

Figure 18 is a plot of the experimentally determined Re M_{1+} and Re M_{1-} values versus the parameter Re E_{1+}. Since the results of polarization experiments seem to indicate a negative Re E_{1+} in the region of the first resonance, one can assume that the Re M_{1+} values obtained under the assumption $E_{1+} = 0$ are the upper limit of the Re M_{1+} amplitude, whereas the Re M_{1-} values are the lower limit of the absolute value. Evidently Re M_{1+} is very little sensitive to Re E_{1+}, whereas Re M_{1-} is strongly affected when the electric quadrupole is included in the analysis. It follows from both the table and the curves that for Re $E_{1+} = -0.04 \cdot 10^{-2}$, the change in Re M_{1-} amounts approximately to a factor of 2 (increasing negative value), compared to the value obtained with $E_{1+} = 0$.

Stability of the Solutions for the Multipole Amplitudes

In order to obtain conclusions of greater validity, we considered the stability of the multipole amplitudes with respect to several disregarded effects.

Let us consider the influence of the previously ignored d-wave amplitudes upon the results of our analysis. Experimental results concerning the Re E_{2+} and Re $M_{2\pm}$ amplitudes are

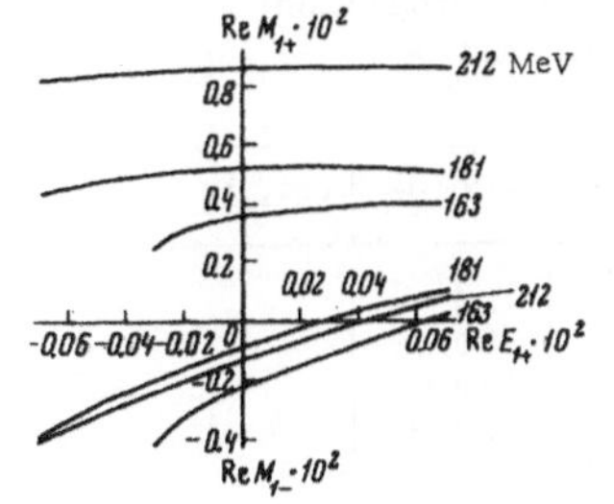

Fig. 18. Dependence of Re M_{1+} and Re M_{1-} upon the parameter Re E_{1+}.

not available at the present time. Calculations based on dispersion relations have shown that the E_{2-} amplitude is the strongest of the d-wave amplitudes considered. The other multipole amplitudes can be ignored in comparison to the d-wave amplitudes. Accordingly, we considered only the E_{2-} amplitude when we estimated the importance of the d-wave amplitudes. We write the $\mathcal{F}_i$-amplitudes with the aid of E_{0+}, E_{1+}, E_{2-}, M_{1+}, and M_{1-} as follows:

$$\begin{aligned} \mathcal{F}_1 &= E_{0+} + E_{2-} + 3(M_{1+} + E_{1+})\cos\theta, \\ \mathcal{F}_2 &= 2M_{1+} + M_{1-}, \\ \mathcal{F}_3 &= 3(E_{1+} - M_{1+}), \\ \mathcal{F}_4 &= -3E_{2-}. \end{aligned} \tag{I.76}$$

The E_{2-} amplitude leads to the following additions to the angular distribution coefficients

$$\begin{aligned} \Delta A &= \frac{q}{2k}\operatorname{Re}[5|E_{2-}|^2 - 2E_{2-}E^*_{0+}], \\ \Delta B &= \frac{2q}{k}\operatorname{Re}[\operatorname{Re}E^*_{2-}(M_{1+} - M_{1-} - 6E_{1+})], \\ \Delta C &= \frac{q}{2k}\operatorname{Re}[3(2E_{2-}E^*_{0+} - |E_{2-}|^2], \\ \Delta\alpha &= \frac{q}{2k}\operatorname{Re}[-3(2E_{2-}E^*_{0+} - |E_{2-}|^2]. \end{aligned} \tag{I.77}$$

In the energy interval considered, a substantial change can occur only in coefficient B, because a term accounting for the interference from the M_{1+} resonance amplitude in the E_2 amplitude appears only in ΔB. Estimates show that ΔA, ΔC, and $\Delta\alpha$ are negligibly small compared to A, C, and α obtained from experiments. The E_2 values calculated by Lebedev and Kharlamov [30] were used to determine the influence of the E_2 amplitude. Only ΔB was included in the estimates. The amplitudes calculated under these assumptions are listed in Table 15.

The second column of the table lists the theoretical Re E_{2-} amplitude values [30] used in the analysis.

It can be inferred from the results listed in Tables 13 and 15 that inclusion of the Re E_{2-} amplitude in the analysis changes the Re M_{1+} values by less than 2% at all photon energies. The changes in Re M_{1-} and Re E_{1+} are similarly small (less than 4%). The fact that the changes in Re M_{1+}, Re M_{1-}, and Re E_{1+} are small when Re E_{2-} is included in the analysis can be easily explained. The point is that the Re E_{0+} s-wave amplitude is essentially determined by coefficient B to which only the increment ΔB was added. The other amplitudes depend only insignificantly upon this coefficient. On the other hand, Re E_{2-} greatly affects the real part of the E_{0+} amplitude in that at the energies 181 and 212 MeV, the absolute value of the real part de-

TABLE 15. Multipole Amplitudes of π^0 Photoproduction Calculated with Theoretical Estimates of Re E_{2-}; $\alpha = -1.70$

k, MeV	Re $E_{2-}\cdot 10^2$ (theoret.)	Re $E_{0+}\cdot 10^2$	Re $M_{1+}\cdot 10^2$	Re $M_{1-}\cdot 10^2$	Re $M_{1+}\cdot 10^2$
163	−0.02	−0.183	0.358	−0.218	—
181	−0.037	−0.188	0.492	−0.128	—
212	−0.058	−0.195	0.861	−0.137	—
234	−0.070	−0.042	1.209	−0.346	−0.C44

TABLE 16. Multipole Amplitudes for k = 234 MeV When the Absolute Cross Section Figures Are Varied (α = −1.70)

R	1.1	0.9	R	1.1	0.9
$\mathrm{Re}\,E_{0+}\cdot 10^2$	−0.046	−0.051	$\mathrm{Re}\,M_{1-}\cdot 10^2$	−0.363	−0.308
$\mathrm{Re}\,M_{1+}\cdot 10^2$	1.290	1.139	$\mathrm{Re}\,E_{1+}\cdot 10^2$	−0.051	−0.038

creases by approximately 20%, and by 10% at 163 MeV. But we emphasize that this change in $\mathrm{Re}\,E_{0+}$ does not affect the arguments for the selection of the negative solution as the meaningful solution in multipole analysis.

Finally, let us consider the influence of the undetermined absolute values of the measured differential cross sections upon the results of the multipole analysis. The resulting ambiguity was estimated at ±10%. In order to observe possible amplitude changes produced by systematic errors, the data were analyzed when for $\alpha = -1.70$ all differential cross sections measured at the energy k = 234 MeV were multiplied by the factors R = 1.1 and R = 0.9. The corresponding results are listed in Table 16.

It follows from the comparison of the results listed in Tables 14 and 16 that a 10% change in the differential cross section changes the multipole amplitudes by approximately 5-7%. The statistical errors in the analysis of the multipole amplitudes $\mathrm{Re}\,E_{0+}$, $\mathrm{Re}\,M_{1-}$, and $\mathrm{Re}\,E_{1+}$ are much greater than the possible shifts produced by the ambiguities in the conversion of experimental results to absolute values. It is therefore not sensible to include these shifts in the calculations of these three amplitudes.

However, the statistical errors of $\mathrm{Re}\,M_{1+}$ are smaller than the possible shifts produced by the ambiguous parameter R. Since calibration tests were made at each primary photon energy, shifts to different sides can occur at the various energies. We are inclined to explain the influence of calibration errors by the observed spread of the γ values around the theoretical curve of [8, 30, 31].

For the purpose of obtaining additional information on the multipole amplitudes of π^0 meson photoproduction at protons, we analyzed in the above fashion results from experiments in which single photons from the decay of neutral pions were recorded (our work [15], [43]). Table 17 lists the values calculated from the data.

A three-term analysis was performed at all energies of the primary photons. The resulting multipole amplitudes agreed, within the statistical error limits, with the amplitude values obtained from the analysis of our more recent data (Figs. 16 and 17). It can be inferred

TABLE 17. Results of the Multipole Analysis of the Experimental Data of [15] and [43]

k, MeV	$\mathrm{Re}\,E_{0+}\cdot 10^2$	$\mathrm{Re}\,M_{1+}\cdot 10^2$	$\mathrm{Re}\,M_{1-}\cdot 10^2$	Reference
160	0.192±0.139	0.323±0.062	−0.056±0.190	[15]
170	−0.191	0.286	−0.113	[43]
190	−0.162	0.508	−0.268	[43]
200	−0.443±0.487	0.644±0.252	0.131±0.326	[15]
220	−0.353±0.090	0.977±0.043	+0.080±0.132	[15]

from the table that the errors of the multipole amplitudes obtained are rather large and the results can therefore not provide additional information on the data listed in Tables 13 and 14.

Discussion of the Results of Multipole Analysis

1. The Re E_{0+} Amplitude. Isospin amplitudes of π^0 meson photoproduction in the s state. The values obtained for the Re E_{0+} amplitude are listed in Fig. 16 which includes the results of theoretical calculations. Curve 1 refers to a calculation with a formula of Chew et al. [1]: $\mathrm{Re}\,E_{0+} = -\omega\,(2f\mu_s - efN^{(+)} - ef\,(q^2/M\omega))$ with $N^{(+1)} = -0.02$. By appropriate selection of the parameter of Chew's theory, agreement between the theoretical calculations and experimental data can be obtained. Chew et al. [1] estimated $|N^{(+)}| < 0.1$. Solov'ev mentioned that this parameter can assume smaller values [8]. Curve 2 of the figure represents the results of [8], curve 3, the results of the initial form of the calculations made by Lebedev and Kharlamov [30]. After our measurements of the energy dependence of the differential cross section at the angle 90° and after comparing the experimental and theoretical values of the $E_{0+}^{(+)}$ amplitude at the energy threshold [50], Lebedev and Kharlamov made new calculations in which the Im E_{0+}, Im M_{1+}, and E_{1+} amplitudes were introduced in the dispersion integrals in a form slightly different from the previous calculations. Curve 4 of Fig. 16 represents their new values of the E_{0+} amplitude. Obviously, slight modifications of Solov'ev's solution and of the last calculations of Lebedev and Kharlamov make it possible to match the Re E_{0+} results obtained with one-dimensional dispersion relations and the experimental amplitudes. The absolute values of the work of Schmidt [31] (curve 6) and of Donnoshie and Shaw [28] (curve 5) are two times greater than the absolute values of the experimental data. This means that the solution obtained for this amplitude in the "Born + 33" approximation (this solution was used in [31]) does not correspond to the real conditions and that higher resonances in the πN channel, contributions of nonresonance amplitudes to dispersion integrals, etc., must be taken into account. The discrepancy between experimental and theoretical Re E_{0+} amplitude values is the main reason for the above described disagreement between measured and calculated angular dependencies of the differential cross section of neutral pion photoproduction at protons.

In analogy to the analysis of pion photoproduction at nucleons in the s state (see our work [11]), we determined the $E_{0+}^{(+)}$ amplitude at the photon-energy threshold ($q \to 0$). Very accurate values of the $E_{0+}^{(-)}$ and $E_{0+}^{(s)}$ amplitudes were recently obtained from the analysis of the photoproduction of charged pions [56]. In experimental determinations of the isotope structure of the s-wave amplitude, the greatest difficulties arise in measurements of the $E_{0+}^{(+)}$ amplitude because this amplitude is determined from the $E_{0+}(\pi^0)$ amplitude which, in turn, has a – by at least one order of magnitude – smaller absolute value than the $E_{0+}(\pi^+)$ and $E_0(\pi^-)$ amplitudes. With Eq. (I.15), the threshold value $E_{0+}(\pi^0) = (0.22 \pm 0.02) \cdot 10^{-2}$ obtained from the angular dependence of the cross section, and the threshold value of $E_{0+}^{(s)}$ from [56], the amplitude

$$E_{0+}^{(+)} = (-0.12 \pm 0.03)\cdot 10^{-2} \tag{I.78}$$

was obtained. We add for the sake of completeness the $E_{0+}^{(-)}$ and $E_{0+}^{(s)}$ amplitudes from [56]:

$$\begin{aligned} E_{0+}^{(-)} &= (2.12 \pm 0.01)\cdot 10^{-2}, \\ E_{0+}^{(s)} &= (-0.10 \pm 0.01)\cdot 10^{-2}. \end{aligned} \tag{I.79}$$

These three quantities determine completely the pion photoproduction at nucleons in the threshold range. The $E_{0+}^{(+)}$ and $E_{0+}^{(s)}$ amplitude values can be used to estimate the amplitude of neutral pion photoproduction in the s state:

$$E_{0+}(n\pi^0) = E_{0+}^{(+)} - E_{0+}^{(0)} = (-0.02 \pm 0.03)\cdot 10^{-2}. \tag{I.80}$$

The calculations confirm the conclusion of our previous work [11], i.e., the probability of neutral pion photoproduction at neutrons in the s state is much smaller than the probability of pion photoproduction at protons.

We obtain from $E_{0+}^{(+)}$ and $E_{0+}^{(-)}$ the amplitudes for certain isospin states of the pion–nucleon system:

$$E_{0+}^{1/2} = E_{0+}^{(+)} + 2E_{0+}^{(-)} = 4.12 \pm 0.04,$$
$$E_{0+}^{3/2} = E_{0+}^{(+)} - E_{0+}^{(-)} = 2.24 \pm 0.03. \tag{I.81}$$

It is generally accepted that the $E_{0+}(\pi^+)$ and $E_{0+}(\pi^0)$ amplitudes are

$$\frac{E_{0+}(\pi^+)}{\sqrt{2}} = E_{0+}^{(0)} + \frac{1}{3}(E_{0+}^{1/2} - E_{0+}^{3/2}),$$
$$E_{0+}(\pi^0) = E_{0+}^{(0)} + \frac{1}{3}(E_{0+}^{1/2} + 2E_{0+}^{3/2}). \tag{I.82}$$

These expressions and the above-determined $E_{0+}^{1/2}$ and $E_{0+}^{3/2}$ values prove that the $E_{0+}(\pi^0)$ amplitude is much smaller than the $E_{0+}(\pi^+)$ amplitude, because $2|E_{0+}^{3/2}| \simeq 1|E_{0+}^{1/2}|$, that the sign of these amplitudes are opposite, and that $|E_{0+}^{(0)}| \ll |E_{0+}^{1/2}|; |E_{0+}^{3/2}|$. The s-wave amplitudes which have been recently determined agree with one of the first simple predictions about their magnitude. It was predicted that in the E1 absorption of the photon, the amplitude of pseudoscalar pion photoproduction in the s state is proportional to the dipole moment of the pion–nucleon system in the final state. Thus,

$$|E_{0+}(\pi^-)| : |E_{0+}(\pi^+)| : |E_{0+}(\pi^0)| : |E_{0+}(n\pi^0)| = \left(1 + \frac{1}{M}\right) : 1 : \frac{1}{\sqrt{2M}} : 0 \simeq 1.15 : 1 : 0.1 : 0. \tag{I.83}$$

According to experience, the first three amplitudes have the following relation (1.10 ± 0.03): 1:(0.07 ± 0.01). A relation similar to Eq. (I.83) is obtained with perturbation theory.

2. The Re M_{1+} Amplitude. The shaded area in Fig. 17 indicates the spread of the Re M_{1+} values due to the ambiguity of the experimentally determined πN scattering phases which were used in calculations of the amplitude with a theoretical expression of [1]:

$$\mathrm{Re}\, M_{1+} = \left\{-\frac{2}{9} ef F_M (\cos^2 \alpha_{13} - \cos^2 \alpha_{33}) + \frac{2}{3}\frac{\mu_s f}{\omega}\right\} kq \left(1 + \frac{\omega}{M}\right)^{-1} + \frac{\mu_v}{f}\frac{k}{q}\left(\frac{1}{3}\frac{\sin \alpha_{13} \cos \alpha_{13}}{q} + \frac{2}{3}\frac{\sin \alpha_{33} \cos \alpha_{33}}{q}\right). \tag{I.84}$$

The solid curve represents the results of [30, 31, 57] (the three calculations result in amplitude values which differ by 1-2%); the dashed curve represents the results of [28]. Very satisfactory agreement of the various theoretical calculations is observed for this amplitude. The difference amounts only to several per cent at 230 MeV; the difference increases slightly with decreasing energy but does not exceed 15% in the most unfavorable case. Optimum agreement between theoretical results and the experimental value of the present work is observed in the case of [30, 31, 57]. The δ function was used in [57] as approximation of the imaginary part of M_{1+}, whereas the "Born + 33" approximation was employed to calculate Re M_{1+}. The same approximation was used in [31] to calculate the quantity Re M_{1+}, whereas the phase analysis of α_{33} was used to calculate Im M_{1+}. When Lebedev and Kharlamov calculated Re M_{1+}, they considered many of the possible contributions (second πN resonance, nonresonance p-wave amplitudes, etc.) to dispersion integrals. The fact that the various calculations do not render substantially different Re M_{1+} means that this amplitude is not affected by the majority of possible small effects.

3. The Re M_{1-} Amplitude. The other shaded area of Fig. 17 is the result of calculations of the Re M_{1-} amplitude with the following expression of [1]:

$$\mathrm{Re}\, M_{1-} = \left\{\frac{4}{9} ef F_M (\cos^2 \alpha_{11} - \cos^2 \alpha_{31}) - \frac{4}{3}\frac{\mu_s f}{\omega}\right\} kq \left(1 + \frac{\omega}{M}\right)^{-1} + \frac{\mu_v}{f}\frac{k}{q}\left\{\frac{1}{3}\frac{\sin \alpha_{11} \cos \alpha_{11}}{q} + \frac{2}{3}\frac{\sin \alpha_{31} \cos \alpha_{31}}{q}\right\}. \tag{I.85}$$

The solid curve is from [57], and the dashed curve from [28, 30, 31]. The latter three papers give very similar Re M_{1-} values. The spread of the experimental Re M_{1-} values, which are characterized by rather large (up to 40–50%) errors, make it impossible to give preference to any one of the theoretical amplitude calculations. However, the amplitude analysis for the energy 234 MeV under inclusion of the electric quadrupole amplitude (black dots of Fig. 17) and consideration of the Re E_{1+} dependence of Re M_{1-} at 163, 181, and 212 MeV (Fig. 18) indicate that the calculations of [28, 30, 31] satisfactorily describe the M_{1-} amplitude. Should this conclusion be corroborated by future work, the failure of [28, 30, 31] to describe the experimental values of the $\gamma + p \rightarrow p + \pi^0$ reaction near the threshold must basically result from incorrect calculations of the s-wave Re E_{0+} amplitude.

4. The Re E_{1+} Amplitude. In addition to the Re E_{1+} values which were experimentally obtained at 234 MeV for the α values, Fig. 16 shows the results of theoretical Re E_{1+} calculations of [8, 30, 31]. So far, the theory does not allow definite predictions of the electric quadrupole amplitude or its sign. The experimental determination of this amplitude is therefore of great importance. The results of the present analysis, which are based on experiments with nonpolarized and plane-polarized photons, establish clearly both the sign and the order of magnitude of the quantity Re E_{1+}. But the poor accuracy of the available experimental data does at the present time not allow definite conclusions concerning the absolute value of the amplitude.

Let us outline the basic conclusions of the determination of multipole amplitudes from experimental results and compare them with calculations made with one-dimensional dispersion relations.

1. Extrapolation of the energy dependence of the differential cross section at 90° renders the following E_{0+} amplitude of the photoproduction of neutral pions at protons for the threshold of the photon energy: $E_{0+} = (-0.18 \pm 0.06) \cdot 10^{-2}$.

2. The real parts of the multipole amplitudes E_{0+}, M_{1+}, and M_{1-} were obtained from an analysis of our measured coefficients of the angular dependence of the differential cross sections of the reaction $\gamma + p \rightarrow p + \pi^0$ at the energies 163, 181, 212, and 234 MeV. A unique solution was selected with an experimental criterion: it was stipulated that the Re E_{0+} values obtained in two different experiments agree at the threshold energy of the γ quanta; one experiment consisted of measurements of the angular dependence of the differential cross sections at 90°, the other, of measurements of the interference term $B \cos \theta$ in the angular dependence of the differential cross sections.

3. The analysis of the photoproduction of charged pions was used to estimate both the isovector $E_{0+}^{(+)}$ part of the s amplitude and the s-wave amplitudes of certain isospin states at the energy threshold. The $E_{0+}(n\pi^0)$ amplitude of the photoproduction of pions at neutrons was estimated: $E_{0+}(n\pi^0) = (-0.02 \pm 0.03) \cdot 10^{-2}$.

A certain discrepancy is found between our $E_{0+}^{(+)}$ amplitude and the corresponding calculations based on dispersion theory. This discrepancy may be ascribed to improper introduction of the imaginary part of the s-wave amplitude, of nonresonance p-wave amplitudes, and of the high-energy range into the dispersion integrals.

4. With the experimental results gathered at 234 MeV with linearly polarized photons, the following estimate is obtained for the real part of the quadrupole amplitude: Re $E_{1+} = (-0.04 \pm 0.02) \cdot 10^{-2}$. The quantity Re E_{1+} has practically no influence upon the results of the analysis which was made under the assumption $E_{1+} = 0$ for Re E_{0+} and Re M_{1+}, but greatly affects Re M_{1-}. Accordingly, the Re M_{1-} values obtained under the assumption $E_{1+} = 0$ represent only the upper limit of Re M_{1-}. A large number of the previously observed discrepancies between experimental and theoretical values of the Re M_{1-} amplitude and, hence, of the differ-

ential cross sections, may be caused by neglecting the E_{1+} amplitude in the analyses and calculations.

5. The multipole amplitudes of the phenomenological analysis are, within the statistical error limits, stable with respect to both the possible contribution of $\mathrm{Im}\, M_{1-}$ and the various forms in which $\mathrm{Im}\, M_{1+}$ can be taken into account. Estimates of the possible contributions of d-wave amplitudes (particularly of $\mathrm{Re}\, E_{2-}$) indicate practically negligible changes in the real parts of the M_{1+}, M_{1-}, and E_{1+} amplitudes. The influence upon $\mathrm{Re}\, E_{0+}$ is rather noticeable and amounts to 10-20%. The absolute value of the $\mathrm{Re}\, E_{0+}$ amplitude decreases and is practically independent of the photon energy. The ambiguity in the absolute normalization of the measured differential cross sections modifies $\mathrm{Re}\, E_{0+}$, $\mathrm{Re}\, M_{1-}$, and $\mathrm{Re}\, E_{1+}$, but the changes are small compared to the statistical errors of these amplitudes. The change is comparable to the statistical error in the case of $\mathrm{Re}\, M_{1+}$.

6. The values of the $\mathrm{Re}\, M_{1+}$ amplitude are in excellent agreement (to within 5%) with the values calculated by Solov'ev [8, 57], Schmidt [31], and Lebedev and Kharlamov [30] with the aid of dispersion theory and in slightly poorer agreement with the calculations based on the work of [1] and the work of Donnoshie and Shaw [28].

7. Our $\mathrm{Re}\, E_{0+}$ amplitude figures agree roughly with Solov'ev's calculations based on dispersion theory and with the work of Lebedev and Kharlamov who used one version of [1]. The theory of [1] can be matched with the experimental $\mathrm{Re}\, E_{0+}$ values by introducing the additional parameter $N^{(+)} = -0.02$. The $\mathrm{Re}\, E_{0+}$ from the analysis of experimental data have about half the absolute value of the results of Schmidt and Donnoshie and Shaw.

8. Our upper limits of the $\mathrm{Re}\, M_{1-}$ can precisely establish the sign (−) of that amplitude. As far as the absolute value is concerned, the best agreement with our values is obtained with the calculations of Solov'ev. But when the $\mathrm{Re}\, E_{1+}$ amplitude is taken into account, the agreement with the calculations of other researchers is improved. The question whether inclusion of the $\mathrm{Re}\, E_{1+}$ amplitude suffices for obtaining full agreement between experimental and theoretical $\mathrm{Re}\, M_{1-}$ values of [28, 30, 31] must be answered by experiments, i.e., experiments with linearly polarized photons must be performed at the energies 181 and 212 MeV.

9. The calculated multipole amplitudes agree with the results of polarization experiments (measurements of the polarization of the recoil nucleon) and of the less accurate measurements of the neutral pion photoproduction at protons in the work of [43] and in our work [15].

10. This work has furthered our understanding of the photoproduction of neutral pions at protons. It follows from a comparison of calculations, based on dispersion theory and various assumptions, and experimental results, that the p-wave amplitudes are rather well determined by the "Born + 33" approximation. The contribution of the real part of the E_{1+} amplitude and the E_{1+} amplitude must be taken into account, the resonances of the πN system must be considered in detail, etc. The determination of the multipole amplitudes of photoproduction implies the determination of characteristic quantities of the reaction, which are only slightly influenced by ambiguities in the dispersion integral calculations. This will make it possible to determine the contributions of the vector mesons to the pion photoproduction. An increased future knowledge about the reaction $\gamma + p \rightarrow p + \pi^0$ can be expected mainly from polarization experiments in the near-threshold energy range.

Estimates of the Upper Limit of a Possible Violation of C and T Invariance in the Photoproduction of Neutral Pions

In connection with the observation of the K_2^0 meson decay (with forbidden CP parity) into two charged pions [58], the possible noninvariance of the electromagnetic hadron interaction with respect to C and T transformations, i.e., charge conjugation and time reversal, has been extensively discussed in the recent past [59-61]. In order to estimate the limits of a possible violation of C and T invariance, the use of experimental data on the reaction $\gamma + p \rightarrow p + \pi^0$ was proposed in [61]. The upper limit of a possible T invariance of the interaction was determined from a comparison of the theoretical expression for the coefficient B_{theor} of [1] with the experimental B_{exp} value obtained in the energy interval $\Delta(1236)$ by interpolation of the closest measurements of the angular dependence. However, the theoretical expression for the asymmetry coefficient of the angular dependence can hardly be evaluated in this case.

In order to obtain an estimate of the upper limit at which T invariance is violated in pion photoproduction, one can use another method which depends only weakly upon theoretical ambiguities [62]. This method is based on the possible determination of the E_{0+} amplitude by two experimental techniques in the near-threshold energy range: The E_{0+} amplitude can be determined from measurements of the energy dependence of the coefficient A or from measurements of the angular dependence of the differential cross section. The amplitude part, which is independent of the momentum q but depends upon the interference of s and p waves, can be derived from measurements of the angular dependences at several photon energies. A comparison of the E_{0+} amplitude values obtained from two experiments at the threshold energy ($\dot{q} \rightarrow 0$) enables estimates of the upper limit at which T invariance can be violated in the reaction $\gamma + p \rightarrow p + \pi^0$.

When we assume that the S matrix of photoproduction is split into a T-invariant part T_1 and a T-noninvariant part T_2

$$S = 1 - i\,(T_1 + iT_2), \tag{I.86}$$

the momentum-independent term forms the following sum in the first method of determining the s-wave amplitude (from the energy dependence of A):

$$A^0 = |E_{0+,1}|^2 + |E_{0+,2}|^2, \tag{I.87}$$

where $E_{0+,1}$ and $E_{0+,2}$ denote the T-invariant and T-noninvariant parts of the E_{0+} amplitude, respectively.

In the second method, the E_{0+} amplitude can be basically determined from the coefficient B, which can also be split into two parts

$$B = B_1 + B_2. \tag{I.88}$$

We have

$$B_1 \simeq \frac{2q}{k} \operatorname{Re}[E^*_{0+,1}(M_{1+,1} - M_{1-,1} + 3E_{1+,1})], \tag{I.89}$$

$$B_2 \simeq \frac{2q}{k} [\operatorname{Im}(E^*_{0+,2} M_{1+,1}) + \operatorname{Im}(M^*_{1+,2} E_{0+,1})]. \tag{I.90}$$

Expressions (I.89) and (I.90) depend in various ways upon the pion momentum q. In the near-threshold region, we have $B_2 \ll B$, and we therefore obtain with good accuracy $E_{0+} \simeq E_{0+,1}$ from an analysis of the angular dependences. For example, when we use the expressions of [51] for estimating the T-invariance violating $E_{0+,2}$ and $M_{1+,2}$ amplitudes, we obtain $B_2 < 0.05B_1$.

By using the coefficient A obtained in the analysis of the energy dependence

$$A^{(0)} = (0.035 \pm 0.022) \cdot 10^{-4}$$

and the value $\operatorname{Re} E_{0+} = \operatorname{Re} E_{0+,1}$ at $q = 0$ from the analysis of the measured angular dependences at k = 163, 181, and 212 MeV, we obtain with 95% probability

$$|E_{0+,2}|^2 = (-0.014 \pm 0.024)\cdot 10^{-4}, \tag{I.91}$$

$$|E_{0+,2}|^2 < 0.03\cdot 10^{-4}. \tag{I.92}$$

When we use for $E_{0+,2}$ the amplitudes of the Born expression stated in [61]

$$E_{0+,2} \simeq E_{0+,2}^{\text{Born}} \simeq \frac{l_r f_r}{4\pi} (\lambda_A \omega + \lambda_C k) \frac{\omega}{M^2},$$

we can estimate the sum of the phenomenological constants as $|\lambda_A + \lambda_C| < 3.5$ with 95% probability. This limit agrees with the value obtained from a comparison of B_{exp} and B_{theor} in the energy interval $\Delta(1236)$, provided that two standard deviations of B_{exp} are used for the estimate. But our limit can be accepted with greater confidence, since our value is based on experimental concepts. In essence, we have made only one rather general assumption, namely that the s-wave amplitude of the reaction $\gamma + p \to p + \pi^0$ has the form $E_{0+} = E_{0+,1} + iE_{0+,2}$.

Indications of a somewhat abnormal behavior of the amplitude at photon energies below 155 MeV were obtained in [39], as mentioned above. A possible anomaly was not taken into consideration when the differential cross section was approximated by Eq. (I.50). For the purpose of accurate determinations of the s-wave amplitude at $q \to 0$, the energy dependence and the angular dependence of the differential cross sections must be measured in the energy interval between the threshold (145 MeV) and 155-160 MeV. Once these measurements have been made, it will be possible to indicate the exact limit at which T invariance can be violated in pion photoproduction.

CHAPTER II

Photoproduction of Neutral Pions at Composite Nuclei

The Pion Photoproduction Mechanism in the Near-Threshold Energy Region

First Experiments

We started our research on the photoproduction of neutral pions at composite nuclei in 1956 with measurements of the simplest characteristic of the reaction, namely the dependence of the pion yield upon the mass number N(A) of the nuclei [63]. It had been shown in research work done by that time [44, 64, 65] that the ratio of pion yield in the photoproduction at a composite nucleus with mass number A to the pion yield at the carbon nucleus varies in proportion to $A^{2/3}$. But the range of light nuclei was poorly studied and the ratio of cross sections of composite nuclei to hydrogen was either completely ignored or determined with low statistical accuracy from difference measurements involving paraffin and graphite targets. The goal of our first investigations was to establish the A dependence of neutral pion photoproduction, particularly in the range of low mass numbers.

We compared in our experiments the single γ quantum yields of the decay pions from liquid hydrogen, liquid nitrogen, and liquid oxygen which had been poured into a cylindrical 5-liter target of foam polystyrene. The measurements were made with scintillation telescopes. Two telescopes situated at the angles 90° and 135° were used at the same time.

TABLE 18. Relative Yields of Neutral Pion Photoproduction at Nuclei

Nucleus	k_{max} = 180 MeV 90°	k_{max} = 200 MeV 135°	k_{max} = 256 MeV	
			90°	135°
H^1	1	1	1	1
C^{12}	23.9±2.6	22.1±1.4	12.7±0.1	12.0±0.4
N^{14}	31.4±3.4	23.2±1.5	15.1±0.1	12.5±0.5
O^{16}		29.5±1.9	17.1±0.2	16.5±0.5
Al^{27}	53.8±6.2	40.5±2.9	22.1±0.2	26.0±0.9
Fe^{56}	71.1±8.7	60.4±5.2	38.3±0.4	39.8±1.4
Cu^{64}	73.6±9.2	82.0±6.8	44.7±0.5	41.0±0.5
Cd^{112}	100±16.0	109±13	62.3±0.5	67.6±2.6
Pb^{207}	152±28	176±31	105.2±1.2	91±4.5

We used also a plane graphite target with a thickness of 3.3 g/cm^2 which was placed at an angle of 45° relative to the bremsstrahlung beam in the same geometrical configuration as targets of aluminum (1.79 g/cm^2), iron (1.30 g/cm^2), copper (1.27 g/cm^2), cadmium (0.928 g/cm^2), and lead (0.541 g/cm^2). Plane C, Al, Fe, Cu, Cd, and Pb targets were mounted on a special device which could be used for remote-control exchange of targets in the bremsstrahlung beam of the accelerator.

The results of the tests are listed in Table 18. The table indicates that at the maximum energies (180 and 200 MeV) of the bremsstrahlung spectra, the yields obtained with light nuclei (C, N, O; yields referred to that obtained with hydrogen) exceed considerably the values calculated with the $A^{2/3}$ dependence and go beyond the A dependence proper. Only the yields at k_{max} = 256 MeV correspond to the A dependence. When the measured yields are referred to the pion yield from the carbon target, the yields at all k_{max} (beginning from nitrogen) are in good agreement with the relation $N(A) \sim A^{2/3}$. This conclusion was previously drawn in [44] and confirmed in [65]. This A dependence of the yield of neutral pion photoproduction can be explained with the model of elastic coherent pion formation at composite nuclei.

Model of Elastic Coherent Photoproduction of Neutral Pions at Composite Nuclei

Feinberg [66] in 1941 predicted quadratic effects ($\sim A^2$) in the interaction between mesons and nuclei. The theory of neutral pion photoproduction at nuclei was considered in detail in [67-70]. The main goal of the theory was to obtain, under certain assumptions, relations between the differential cross section of pion formation at a nucleus, the differential cross section of pion formation at hydrogen, and the characteristics of the ground state of the nucleus. The initial assumptions of the theory are as follows: (a) The direct interaction model, in which the interaction between the photon and the nucleus is expressed as interactions with the individual nucleons, can be employed to describe the photoproduction of pions at nuclei. In other words, the operator of photoproduction at the nucleus can be written as a sum

$$T = \sum_{n=1}^{A} e^{ipx_n} t_n, \tag{II.1}$$

where t_n denotes the operator of photoproduction at the nucleon; x_n denotes the spatial coordinate of the nucleon; and p denotes the momentum transferred to the nucleon. (b) Both the momentum approximation and the Born approximation can be used to treat pion photoproduction at nuclei. This assumption means that in Eq. (II.1), the operator for the bound nucleon is identical to the photoproduction operator for the free nucleon and that the wave functions of the incident photon and the pion generated are plane waves. The differential cross section of pion photoproduction at a nucleus can be represented as a sum of two terms in the lab system [69]:

$$\frac{d\sigma}{d\Omega}=\int\frac{q^2dq}{(2\pi)^2}\langle 0|T^{+}\delta(E_i-H_0)\ T|0\rangle\simeq\int\frac{q^2dq}{(2\pi)^2}\sum_{n=1}^{A}\langle 0|t_n^{+-i\bar{p}\bar{x}_n}\delta(E_i-$$

$$-H_0)e^{ipx_n}|0\rangle+\int\frac{q^2dq}{(2\pi)^2}\sum_{n\neq m}\sum\langle 0|t_m^{+}e^{-i\bar{p}\bar{x}_m}\delta(E_i-H_0)e^{i\bar{p}\bar{x}_n}t_n|0\rangle=\left(\frac{d\sigma}{d\Omega}\right)_D+\left(\frac{d\sigma}{d\Omega}\right)_{ND}, \qquad \text{(II.2)}$$

where q denotes the pion momentum; ω denotes the pion energy; k denotes the photon energy; H_A denotes the Hamiltonian of the nucleus; and W denotes the energy of the $|0\rangle$ ground state of the nucleus.

The first term, i.e., the cross section $\left(\frac{d\sigma}{d\Omega}\right)_D$ which is diagonal in the subscripts of the spatial coordinates of nucleon defines the single-particle contributions, while the second term $\left(\frac{d\sigma}{d\Omega}\right)_{ND}$, which is the nondiagonal cross section, defines the two-particle contributions to photoproduction.

According to [69], the diagonal cross section can be adequately approximated in the form

$$\left(\frac{d\sigma}{d\Omega}\right)_D=A\int\frac{d^3p_i}{(2\pi)^3}P(p_i)\int\frac{q^2dq}{(2\pi)^2}\langle 0|t^{+}t|0\rangle\,\delta\left(k+\frac{p_i^2}{2M}-\omega-\frac{p_f^2}{2M}\right), \qquad \text{(II.3)}$$

where $\bar{p}_i$ denotes the momentum of the nucleon in the initial state; $\bar{p}_f=\bar{p}_i-\bar{p}$ denotes the momentum in the final state; and $P(p_i)$ denotes the momentum distribution of the nucleons in the ground state of the nucleus.

Obviously, $\left(\frac{d\sigma}{d\Omega}\right)_D$ is equal to the cross section at the free nucleon, averaged over spin and momentum states, and multiplied by A. The nondiagonal cross section can be expressed by the transition operators for nucleons 1 and 2 [14]:

$$\left(\frac{d\sigma}{d\Omega}\right)_{ND}=A(A-1)\int\frac{q^2dq}{(2\pi)^2}\langle 0|e^{ip(x_1-x_2)}t_2^{+}t_1|0\rangle\,\delta(k-\omega). \qquad \text{(II.4)}$$

We assumed $W_0=H$ and ignored correction terms which result from the momentum dependence of the transition operator.

When we assumed that the kinematics is the same for the various cross section components and corresponds to the kinematics of photoproduction at a free nucleon, we obtain the following expressions by averaging Eqs. (II.3) and (II.4):

$$\left(\frac{d\sigma}{d\Omega}\right)_D=A\Gamma\langle 0|t_1^{+}t_1|0\rangle, \qquad \text{(II.5)}$$

$$\left(\frac{d\sigma}{d\Omega}\right)_{ND}=A(A-1)\Gamma\langle 0|e^{ip(x_1-x_2)}t_2^{+}t_1|0\rangle, \qquad \text{(II.6)}$$

where Γ denotes a quantity which is proportional to the state density. The position operator of the nucleons is denoted by $e^{ip(x_1-x_2)}$; $t_2^{+}t_1$ includes the spin operators and isotope spin operators of nucleons 1 and 2.

In the case of nuclei with filled shells (each spatial state is filled by four nuclei in accordance with the possible combinations of spin and isotope spin), the matrix elements of space and spin coordinates can be split so that the nondiagonal cross section is divided into two components, namely an uncorrelated and a correlated cross section

$$\left(\frac{d\sigma}{d\Omega}\right)_{ND} = \left(\frac{d\sigma}{d\Omega}\right)_{NC} - \left(\frac{d\sigma}{d\Omega}\right)_{C}, \tag{II.7}$$

where

$$\left(\frac{d\sigma}{d\Omega}\right)_{NC} = A^2\Gamma\,|F(p)|^2\,|K|^2, \tag{II.8}$$

$$\left(\frac{d\sigma}{d\Omega}\right)_{C} = A\Gamma G(p)\,(|K|^2 + |L|^2 + |M|^2 + |N|^2), \tag{II.9}$$

the photoproduction operator for the nucleon has the form

$$t_n = K + \mathbf{L}\sigma_n + M\tau_{3n} + \tau_{3n}\mathbf{L}\sigma_n, \tag{II.10}$$

and the form factors are

$$F(p) = \int \rho(x)\, e^{ipx}\, d^3x, \tag{II.11}$$

$$G(p) = \int \rho(\bar{x}_1)\,\rho(\bar{x}_2)\, h(\bar{x}_1, \bar{x}_2)\, e^{ip(x_1 - x_2)}\, d^3x_1\, d^3x_2. \tag{II.12}$$

The cross section $\left(\frac{d\sigma}{d\Omega}\right)_{ND}$ could be resolved by representing the density of the two particles as a sum of two terms

$$\sim \rho(\mathbf{x}_1)\,\rho(\mathbf{x}_2)\,[1 + 4h(\mathbf{x}_1, \mathbf{x}_2)], \tag{II.13}$$

where $h(\mathbf{x}_1, \mathbf{x}_2)$ denotes the correlation function which accounts for the influence of the position of one nucleon upon the space coordinate of the other nucleon (consequence of the Pauli principle). The uncorrelated part is simply the cross section of the elastic process in which the nucleus remains in the ground state, whereas the amplitudes of the transitions from the various nucleons are combined into a coherent sum.

The remaining diagonal part of the cross section can be combined with the correlated part of the nondiagonal cross section. We obtain

$$\left(\frac{d\sigma}{d\Omega}\right)_{D} - \left(\frac{d\sigma}{d\Omega}\right)_{C} = A\Gamma\,[1 - G(p)]\,(|K|^2 + |L|^2 + |M|^2 + |N|^2). \tag{II.14}$$

A reduced cross section of inelastic photoproduction of neutral pions at nuclei is the only consequence of the Pauli principle.

When the Fermi gas model is used for infinite nuclear matter, the reduction factor is

$$1 - G(p) = \begin{cases} \frac{3}{4}\left(\frac{p}{p_F}\right) - \frac{1}{16}\left(\frac{p}{p_F}\right)^3 & \text{for} \quad p < 2p_F, \\ 1 & \text{for} \quad p > 2p_F, \end{cases} \tag{II.15}$$

where $p_F^2 = 3\pi^2 A/2V$ denotes the Fermi momentum for A nucleons in the volume V. Engelbrecht [69] made calculations for He^4 and O^{16} with the wave functions of the harmonic oscillator in the independent particle model and obtained the following rough estimate of the ratio of the reduction factor of a nucleus with finite dimensions to $[1 - G(p)]_\infty$:

$$\frac{[1 - G(p)]_A}{[1 - G(p)]_\infty} = \begin{cases} \frac{1}{2}pR & \text{for} \quad pR < 2, \\ 1 & \text{for} \quad pR > 2. \end{cases} \tag{II.16}$$

We therefore obtain the following expressions for the differential cross sections from our considerations of the neutral pion photoproduction at nuclei. In the case of nuclei with filled shells, the coherent and incoherent cross sections can be written in the form

$$\left(\frac{d\sigma}{d\Omega}\right)_{\text{coh}} = \left|Zf_p + Nf_n\right|^2 |F(p)|^2 \sin^2\theta, \tag{II.17}$$

$$\left(\frac{d\sigma}{d\Omega}\right)_{\text{incoh}} = \left[Z\left(\frac{d\sigma}{d\Omega}\right)_p + N\left(\frac{d\sigma}{d\Omega}\right)_n\right][1 - G(p)], \tag{II.18}$$

where f_p and f_n denotes the amplitudes of pion photoproduction at protons and neutrons, respectively; the two quantities differ from f_p^{lab} and f_n^{lab} by the factor $\left[1 - \frac{\omega(q - k\cos\theta)}{qM}\right]$, which accounts for the form of the coherent cross section; the corresponding differential cross sections are denoted by $\left(\frac{d\sigma}{d\Omega}\right)_p$ and $\left(\frac{d\sigma}{d\Omega}\right)_n$.

The momentum approximation was relativistically generalized in [70] where the following expression was obtained for the differential cross section of the elastic coherent photoproduction of π^0 mesons at nuclei with mass number A:

$$\left(\frac{d\sigma}{d\Omega}\right)_A = \frac{A^2}{2} F_A^2(Q) \sin^2\theta \,|\mathcal{F}_2^+|^2 R, \tag{II.19}$$

where

$$R = \frac{q^3 k}{(q'k')^3}\left[1 + \frac{k}{M_N}\frac{(M_A - M_N)}{M_A}\right]^2, \tag{II.20}$$

M_N denotes the nucleon mass, M_A, the mass of the nucleus; q and k denote the momentum of the pion and the photon, respectively, in the center-of-mass of the pion–nucleus system; q' and k' denote the same quantities for the pion–nucleon system; Q denotes the momentum transferred to the nucleus; and $\mathcal{F}_2^+$ denotes the spin-independent isovector part of the amplitude of neutral pion photoproduction at nucleons. Equation (II.19) refers to a nucleus with equal number of neutrons and protons. In the general case, the spin-independent part, with the isospin quantum number included, can be written in the form

$$L_i = \frac{\mathbf{q}'[\mathbf{k}'\varepsilon]}{q'k'}[\mathcal{F}_2^+ \pm \mathcal{F}_2^0], \tag{II.21}$$

where the plus sign must be taken for protons, and the minus sign, for neutrons; $\mathcal{F}_2^+$ denotes the isovector part of the amplitude; and $\mathcal{F}_2^0$, the isoscalar part. Only the spin-dependent isovector part $\mathcal{F}_2^+$ remains for N = A − Z = Z in $L = \Sigma L_i$.

In the photon energy range close to the photoproduction threshold (~140 MeV), the momentum transferred are much smaller than $p_{\mathcal{F}}$ We may therefore assume in a first approximation that the photoproduction of neutral pions is in this energy range completely determined by elastic coherent pion formation. But since the theoretical reduction factor 1 − G(p) was defined for a nucleus of finite dimensions, we must experimentally determine the upper energy limit up to which the model of elastic coherent pion generation can be employed.

Experimental Determination of the Upper Limit to Which the Model of Elastic Coherent Photoproduction Can Be Employed in the Case of the Carbon Nucleus

The first experiments for estimating the applicability range of the presumed coherent elastic photoproduction were made with a carbon target.

Coherent photoproduction at composite nuclei was for the first time studied in detail in the work of Leiss and Schrack [71]. $\gamma-\gamma$ coincidences were used to measure the pion emission from a carbon target at three angles θ_l (0, 90, and 180°) in dependence of the maximum energy of the bremsstrahlung spectrum between the threshold and approximately 180 MeV in intervals $\Delta k_{max} = 4$ MeV. The analysis of the measurements revealed that the elastic coherent photoproduction at the C^{12} nucleus is the dominating process at least up to 180 MeV (the maximum energy reached in that experiment) and that this process can be used as a new method of investigating the nucleon density distribution in nuclei.

We came to a similar conclusion after measuring the energy dependence of the ratio $\frac{\sigma_C}{\sigma_H}(k)$, i.e., of the total cross section of neutral pion photoproduction at carbon to the total cross section of neutral pion photoproduction at hydrogen [72]. As has been mentioned above, σ_C/σ_H must decrease with increasing energy in the case of coherent elastic photoproduction and the absolute value of the ratio must be greater than A at low energies. In the case of inelastic photoproduction, the ratio must increase from 0 at low energies to A at $k \to \infty$. The energy dependence of the yield of γ quanta originating from the decay of π^0 mesons from carbon was measured at three angles in the 265-MeV synchrotron of the Lebedev Physics Institute of the Academy of Sciences of the USSR. The yields of decay photons were measured with an accuracy of 1-2%. The corresponding energy dependence of the emission cross section of the decay photons was calculated from the measured yield curves. The resulting angular dependence of the decay γ quanta was thereafter integrated to obtain the total cross section of neutral pion photoproduction at carbon (σ_C) in relative units.

The yield ratio of decay γ quanta from carbon and hydrogen (this ratio had been determined at 90° from counts of decay photons originating from hydrogen and polystyrene targets with the same geometrical form) was used to determine the ratio of the total cross sections σ_C/σ_H. The measured ratios are listed in Table 19.

The results were then compared with the conclusions of the elastic photoproduction theory of neutral pions generated at nuclei. Equation (II.17) was integrated over the solid angle to calculate the total photoproduction cross section. The electron scattering data of [73] were used for the form factor of the carbon nucleus. The spin-independent cross section $\sigma_H^{C.H}$ of neutral pion photoproduction at nucleons [this cross section is important in Eq. (II.17)] was determined from the total cross section of pion photoproduction in hydrogen targets, which was measured in [17]. The spin-independent differential cross section $(d\sigma/d\Omega)^{C.H}$ was brought into a relation to the spin-independent total cross section under the assumption that the M_1-amplitude is negligibly small:

$$\left(\frac{d\sigma}{d\Omega}\right)^{C.H} = \frac{3\pi}{8}\sigma_H^{C.H}\,|M_{1+}|^2. \tag{II.22}$$

We obtain, when we neglect the contributions of the E_{0+} and E_{1+} amplitudes to the total cross

TABLE 19. Ratio of Total Cross Section of the Photoproduction of Neutral Pions at Carbon and Hydrogen

Average photon energy, MeV	160	180	200
σ_C/σ_H (experiment)	64±9	36±5	31±4
(calculation)	71	36	25

section

$$\sigma_H^{C,H} = \frac{2}{3}\sigma_H. \quad (II.23)$$

The calculated σ_C / σ_H values are listed in the second line of Table 19.

We infer from the table strong agreement between the experimental results and theoretical calculations. This made it possible to use the coherent photoproduction process in research on nuclear structure and for obtaining information on the spin-independent amplitude of pion photoproduction at nucleons. Our ensuing experiments [47, 48, 74, 75] were undertaken for the same purpose. Similar measurements were made at the same time by Leiss and Schrack [76, 77].

Determination of the Parameters of the Nucleon Density Distribution in Nuclei

Experimental Scheme and Nuclei

The angular distributions of neutral pions produced at various nuclei by photons were measured in order to determine the parameters of the nucleon distribution inside the nucleus with the 2γ method.

Measurements were made at the maximum energies $k_{max} = 180$ and 200 MeV. The radiation pulse had a duration of 1 msec. Figure 19 shows schematically the experimental setup. Pion formation at Be, C, Al, Cu, Cd, Ta, and Pb nuclei was considered. The beryllium and carbon targets had the form of spheres with a diameter of 7 cm (density of carbon 1.66 g/cm^3, of beryllium, 1.83 g/cm^3). The other targets had the form of square plates (10 cm × 10 cm) which were placed perpendicular to the primary photon flux. The target thickness figures are listed in Table 20.

The pions produced in the target material were observed with two scintillation telescopes via the coincidence γ quanta from the decay $\pi^0 \to \gamma + \gamma$. Each telescope consisted of a 3-cm-thick getinax absorber, a lead converter (0.6 cm), and two scintillation counters with an interposed aluminum filter (1 cm). Radial sensitivity and energy dependence of the γ telescopes were determined in individual experiments. The limit values of the pion energies which could be measured in the experiments were given (E_{min}) by the angle ψ between the telescopes and

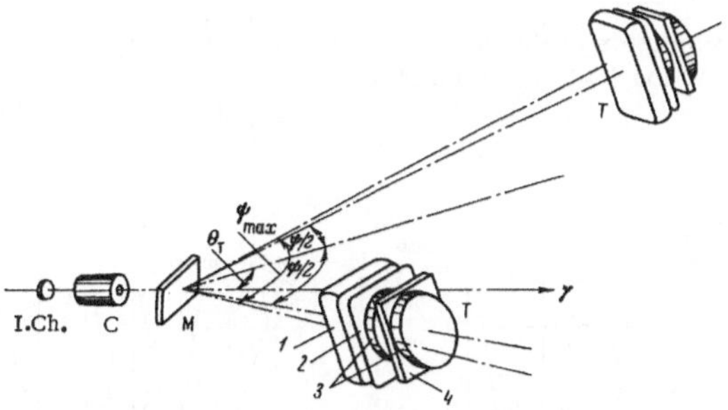

Fig. 19. Experimental setup for investigating the angular distribution of neutral pion photoproduction at nuclei. γ, bremsstrahlung beam; M, target; I.Ch., ionization chamber; C, collimator; T, γ telescope; 1, absorber; 2, converter; 3, scintillators; 4, filter.

TABLE 20. Target Thickness

Target material	Thickness		
	cm	g/cm^2	radiation units
Aluminum	0.672	1.79	0.075
Copper	0.145	1.28	0.10
Cadmium	0.106	0.886	0.10
Tantalum	0.032	0.503	0.080
Lead	0.100	1.09	0.19

the upper energy limit of the bremsstrahlung spectrum. The angle θ_T (see Fig. 19) determined the average angle of pion emission. Since the rest energy of the nuclei under consideration is much greater than the energies of the primary γ quanta, the energy spectrum of the pions recorded depends only insignificantly upon θ_T and the mass of the nucleus at a certain angle ψ between the telescopes. The angle ψ was therefore slightly adjusted in the measurements on various nuclei and amounted, on the average, to 133.7° at k_{max} = 180 MeV and 101° at k_{max} = 200 MeV. In order to establish the angular dependence of the cross section of neutral pion photoproduction, the angle θ_T was varied, i.e., the angles 0°, 30°, 45°, 60°, 75°, 90°, 110°, 130°, 150°, and 180° were used at k_{max} = 180 MeV, and 0°, 30°, 50°, 70°, 90°, 110°, 130°, 150°, and 180°, at k_{max} = 200 MeV. The curves which express the energy resolution and the angular resolution of the apparatus were established with statistical trials. Figures 20 and 21 show the results of the calculations. The energy resolution was independent of the angle at a given k_{max} value (because the correlation angle ψ depends only slightly upon θ_T). The primary photons had the following average energies: k_{max} = 153 MeV at k_{max} = 180 MeV, and k = 181.5 MeV at k_{max} = 200 MeV.

The angular resolutions which are shown in Fig. 21 for various θ_T values turned out to be practically independent of the atomic number of the target nuclei because $M \gg k_{max}$.

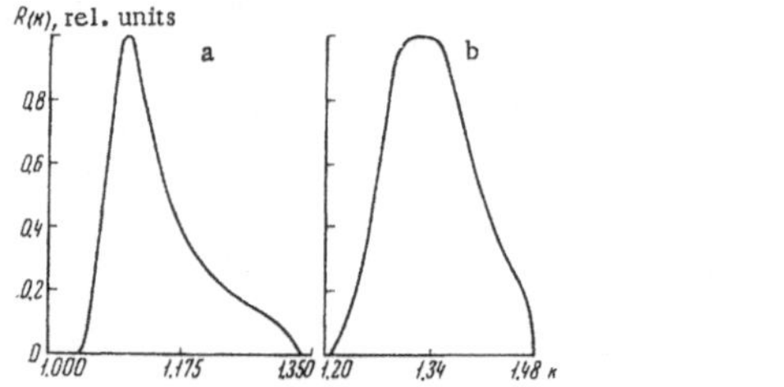

Fig. 20. Energy resolution of the apparatus at (a) k_{max} = 180 MeV and (b) k_{max} = 200 MeV.

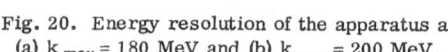

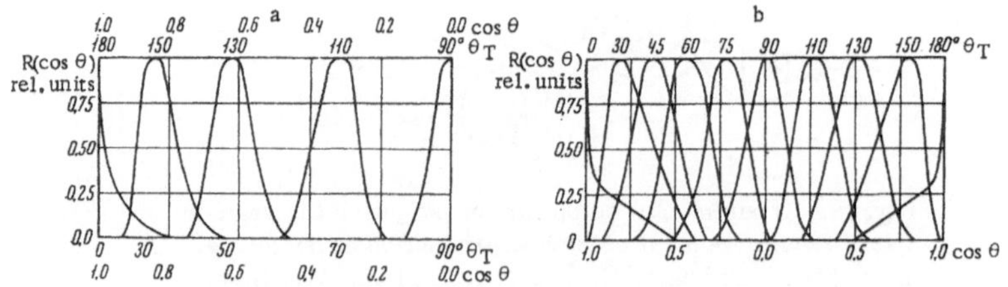

Fig. 21. Angular resolution of the apparatus at (a) k_{max} = 180 MeV and (b) k_{max} = 200 MeV.

TABLE 21. Measured Pion Yields at k_{max} = 180 MeV (per 10^{12} Effective Quanta at the Nucleus)

θ, deg	Be	C	Al	Cu	Cd	Ta
0	39.9±2.8	43.2±2.5	18.7±1.3	14.9±1.2	6.69±0,48	2.92±0,30
30	60.7±2.6	64.0±2.9	25.4±1.5	15.2±1.3	7,03±0.56	2.05±0.29
45	78.0±3.6	84,6±3.8	26.1±1.6	14.5±1.3	4.35±0.42	1.16±0,30
60	76,9±3.5	88.1±3.9	18.5±1.4	5.1±0.6	1,98±0.29	0,72±0.31
75	87.0±4.1	92.3±4.2	15,2±1.2	2.76±0.46	1.20±0.36	1.55±0,34
90	71.0±2.5	68.8±2.2	9.3±0.7	1.55±0.25	0.64±0.41	0.87±0.17
110	51,3±2.8	51.5±2.3	3,8±0.6	0.66±0.37	0,48±0.23	0.65±0.24
130	29.6±2.3	27.4±1.9	2,75±0.43	1.83±0,41	1.30±0.31	0.68±0.18
150	20.7±2.0	15.5±1.4	1.88±0.38	1.22±0.32	0,58±0.25	0,03±0,16
180	11,4±1.3	7.4±1.3	1.23±0.32	1.05±0,29	0.33±0.16	0,06±0.09

Measurements were simultaneously made with three identical "telescope" pairs for three angles. Cyclic repositioning of the telescope pairs facilitated checks of the operation. All telescope pairs gave identical results within the statistical limits of accuracy. Apart from this, stability of operation was periodically checked at certain telescope positions by inserting the carbon target.

Random coincidences accounted for less than 1% of the total effect at all angles and with all targets. The counting rate without a target was negligibly small in the case of Be, C, and Al, but significant at large angles θ_T in the case of Ca, Cd, and, particularly, Ta. This counting rate was subtracted from the counting rate obtained with the target. The background without target can be explained by pion photoproduction in the air layer onto which the γ telescopes were directed.

Parameters of the Nucleon Density Distribution in Nuclei

The measured yields of neutral pions are listed in Table 21 (k_{max} = 180 MeV) and Table 22 (k_{max}= 200 MeV) as a function of the angle θ.

The parameters which characterize the nucleon distribution in the nucleus were determined by comparing the experimental data with calculations based on the following formula, obtained from expression (II.17):

$$\left(\frac{d\sigma}{d\Omega}\right)_A = A^2 \frac{\sigma_t}{4\pi}\left|\int\limits_{\tau} \rho(\tau) \exp\left(\frac{i}{\hbar}\mathbf{qr}\right) d\tau\right|^2 \sin^2\theta, \quad (\text{II}.24)$$

TABLE 22. Measured Pion Yields at k_{max} = 200 MeV (per 10^{12} Effective Quanta at the Nucleus)

θ, degrees	Be	C	Al	Cu	Cd	Ta	Pb
0	49,5±4.3	71.7±6,8	19.0±2.0	18.2±1.1	10.0±1.1	5,58±0.67	10,8±1.0
30	124,0±6.7	141.0±8.9	34.6±3.2	29.1±2.4	10.4±1.0	6,03±0.56	5.63±0.62
50	137,3±8.4	162.5±9.0	25,2±2.3	9.2±1,0	3.74±0.91	1.34±0.45	1.06±0.34
70	119.0±8.3	139.6±5.5	12.1±1.1	1.93±0.71	2.13±0.68	1.70±0.39	1.14±0,44
90	96.7±7.2	85.5±6.0	2.18±0.71	0.94±0.40	0,87±0.35	0.43±0.28	0.41±0.19
110	46.8±6.0	38.8±3.6	2.28±0.76	1.82±0.68	0.17±0.40	0.24±0.16	0,66±0.32
130	30.2±5.4	19.9±5.4	2.33±0.53	2.05±0,63	0.56±0.32	0.41±0.20	0.39±0.30
150	13.9±4.2	9.5±2.7	1.53±0.49	0.74±0.46	0.67±0,29	0.35±0.25	0.37±0.21
180	1.2±0,7	0.9±1.0	0.22±0.30	0.12±0.19	0.19±0.23	0.04±0.17	0.07±0.15

where σ_t denotes the total cross section of pion photoproduction at the nucleon (assumed to be independent of the isotope spin). It was assumed in the derivation of Eq. (II.24) that only the M_{1+} amplitude exists in the low-energy region of the photons. The cross sections $(d\sigma/d\Omega)_A$ and σ_t were compared at identical values of transferred momentum in the center-of-mass system. The distribution of matter in the nucleus was obtained with a method which Hofstadter [78] had employed in the analysis of electron scattering experiments: several models were assumed for $\rho(\tau)$ and the least-square method was used to select among the models those which rendered optimal agreement with the yields measured in experiments. The function $\rho(\tau)$ was approximated by spherically symmetric models of the density distribution of nucleons in nuclei: homogeneous density distribution (for all nuclei), modified exponential distribution and shell configuration (in the case of Be, C, and Al), and trapezoidal distribution (in the case of Cu, Cd, Ta, and Pb). The homogeneous model is characterized by the radius R, the modified distribution and the shell configuration are characterized by a mean-square radius a, and the trapezoidal distribution, by the distance C, at which the density is reduced to half the density in the center of the nucleus, and by the thickness S of the boundary layer (in which the density decreases from 0.9 to 0.1 of the maximum value).

The unknown parameters were determined by minimizing the following sum:

$$\sum_{i=1}^{l} p_i (N_i - RN_i^T)^2 + p_k (R-1)^2, \qquad (II.25)$$

where N_i denotes the measured yields and p_i, their statistical weights; N_i^T denotes the theoretical yields which were obtained by averaging the theoretical cross sections of the angular intervals and energy ranges in which the probability of recording the photoproduction process with the apparatus is nonvanishing. The additional parameter R in the sum of (II.25) accounts for the error made in the normalization of both the theoretical calculations and experimental data. The average value is R = 1 and has the weight p_k. The weight p_k depends upon the error of the absolute value of the bremsstrahlung flux and the efficiency of the γ telescopes and of the calculation of both the recording probability and the spin-independent part of the theoretical cross section. The total error of the normalization amounted to ±10%.

The system of Eq. (II.25) was solved by successive approximation. The parameters of the charge distribution which were obtained for nuclei in electron-scattering experiments were used in the zeroth approximation. The parameters which were obtained in our analysis for the density distribution models of the nucleons in nuclei are listed in Table 23.

Two values are stated for each parameter; the upper value was obtained from the analysis of angular distributions at k = 154 MeV, the lower, from the analysis of angular distributions at k = 182 MeV (all parameter values of the table are stated in Fermi units: F = 10^{-13} cm). The F distribution with a 5% significance level was used to select models which result in a theoretical description in agreement with experimental results. The parameters of models which are inconsistent with the experimental results are indicated in parentheses in the table. Since the parameters of the distributions agree within the experimental error limits at the various energies, we state in the following the combined estimates of the nucleus dimensions (obtained at the two energies).

The angular dependence of the pion photoproduction at Be and C can be described with Eq. (II.24) if a modified exponential model is adopted for Be, and the shell model or a modified exponential model for C. The experimental results for Al are inconsistent with the theoretical description provided by any one of the models. This situation may result from the limited number of models considered in the analysis or from the fact that Eq. (II.24) is inadequate for the description of pion photoproduction at aluminum.

TABLE 23. Parameters of the Density Distribution Models
of Nucleons in Nuclei

Model	Parameters	Be	C	Al
Homogeneous model	R	(2.76$\pm$0.08) (2.64$\pm$0.11)	2.91$\pm$0.05 (2.80$\pm$0.08)	(3.73$\pm$0.09) (3.90$\pm$0.10)
Modified exponential model	a	2.48$\pm$0.09 2.44$\pm$0.09	2.70$\pm$0.08 2.65$\pm$0.08	3.75$\pm$0.20 3.84$\pm$0.13
Shell configuration	a	2.41$\pm$0.21 (2.19$\pm$0.09)	2.37$\pm$0.15 2.31$\pm$0.08	(3.07$\pm$0.23) (3.18$\pm$0.10)
Trapezoidal model	C t			

Model	Parameters	Cu	Cd	Ta	Pb
Homogeneous model	R	(4.99$\pm$0.12) 5.06$\pm$0.09	5.77$\pm$0.07 5.97$\pm$0.13	6.76$\pm$0.08 6.83$\pm$0.10	(7.23$\pm$0.16)
Modified exponential model	a				
Shell configuration	a				
Trapezoidal model	C	(5.28$\pm$0.25) 5.03$\pm$0.15	5.73$\pm$0.11 5.83$\pm$0.18	6.79$\pm$0.12 6.68$\pm$0.12	6.83$\pm$0.20
	t	(1.10$\pm$0.78) 0.35$\pm$0.71	0.46$\pm$0.52 0.85$\pm$0.42	1.48$\pm$1.42 0.82$\pm$0.22	1.89$\pm$0.29

TABLE 24. Differential Cross Section (μbarn/sr) of the
Photoproduction of Neutral Pions at Nuclei
(Average Energy 154 MeV)

cos θ	Be	C	Al	Cu	Cd	Ta	Pb
0.86	6.3$\pm$0.5	10.2$\pm$0.6	53.7$\pm$3.5	148$\pm$12	190$\pm$17	193$\pm$24	
0.75	9.4$\pm$0.4	15.2$\pm$0.7	73.1$\pm$3.7	153$\pm$13	201$\pm$18	139$\pm$18	
0.6	12.1$\pm$0.5	20.1$\pm$0.9	73.5$\pm$3.7	149$\pm$11	115$\pm$11	82$\pm$21	
0.43	11.9$\pm$0.6	20.8$\pm$0.9	58.3$\pm$3.7	71$\pm$10	59$\pm$11	62$\pm$25	
0.22	13.6$\pm$0	21.8$\pm$1.0	52.8$\pm$6.5	43.5$\pm$6.5	40$\pm$16	125$\pm$25	
0	11.0$\pm$0.4	16.4$\pm$0.6	34.0$\pm$2.5	22.5$\pm$6.0	30$\pm$17	100$\pm$21	
−0.29	8.0$\pm$0.5	12.2$\pm$0.6	12.8$\pm$2.0	8.5$\pm$3.5	28$\pm$11	61$\pm$20	
−0.52	4.6$\pm$0.4	6.5$\pm$0.5	7.8$\pm$1.4	14.5$\pm$3.0	34$\pm$8	48$\pm$14	
−0.75	3.2$\pm$0.3	3.8$\pm$0.4	5.2$\pm$2.3	11.7$\pm$3.0	22$\pm$10	2$\pm$14	
−0.86	1.9$\pm$0.3	1.7$\pm$0.3	3.3$\pm$1.2	9.5$\pm$2.3	8$\pm$4	5$\pm$11	
0.95	16.4$\pm$1.8	36$\pm$3.2	111$\pm$11	354$\pm$40	735$\pm$80	1175$\pm$115	880$\pm$70
0.82	41.2$\pm$2.4	71$\pm$4.5	204$\pm$19	587$\pm$50	750$\pm$70	855$\pm$100	450$\pm$55
0.62	45.2$\pm$2.8	81.8$\pm$4.2	154$\pm$14	188$\pm$23	196$\pm$50	190$\pm$65	87$\pm$23
0.32	39.5$\pm$3.0	70.2$\pm$2.5	79$\pm$7	44$\pm$15	125$\pm$40	280$\pm$55	125$\pm$50
0	32$\pm$2.4	43$\pm$3.3	19$\pm$6	51$\pm$23	123$\pm$50	150$\pm$93	170$\pm$75
−0.32	15.5$\pm$2.0	19.5$\pm$2.0	15$\pm$4	42$\pm$16	8$\pm$30	40$\pm$27	70$\pm$33
−0.62	10$\pm$1.8	10$\pm$2.5	14$\pm$4	40$\pm$17	32$\pm$22	57$\pm$30	33$\pm$17
−0.83	4.6$\pm$1.2	4.3$\pm$1.5	8.5$\pm$5	15$\pm$10	35$\pm$17	46$\pm$30	28$\pm$20
−0.96	0.4$\pm$1.0	0.7$\pm$1.0	1.5$\pm$3.5	3$\pm$8	7$\pm$15	6$\pm$20	5$\pm$15

The theory of coherent pion photoproduction at medium and heavy nuclei is consistent with the experimental results, if the trapezoidal model is used for $\rho(\tau)$. We note that the errors of the measurements were rather large in the case of Pb and Te.

Angular Dependences

The measured differential cross sections (lab system) are listed in Table 24 and Figs. 22-25 for the various nuclei in dependence of the pion emission angle. These figures include the theoretical differential cross section curves calculated with Eq. (II.24) and with the param-

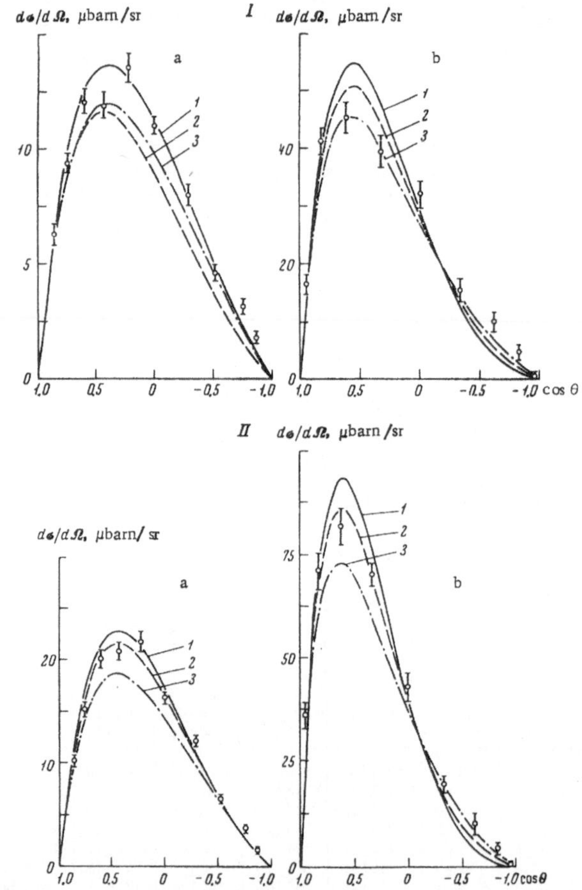

Fig. 22. Differential cross sections of the photoproduction of neutral pions at (I) beryllium and (II) carbon at (a) k = 154 MeV and (b) k = 181.5 MeV. The curves were calculated for the (1) homogeneous, (2) shell-type, and (3) modified exponential density distributions of nucleons in the nucleus. The parameter values of Table 23 were used in the calculations.

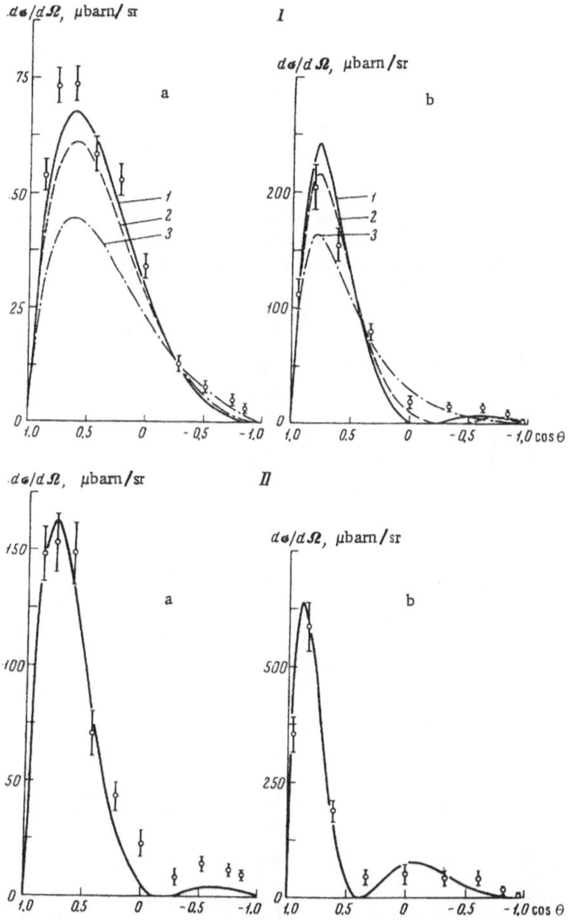

Fig. 23. Differential cross sections of the photoproduction of neutral pions at (I) aluminum and (II) copper at (a) 154 MeV and (b) 181.5 MeV. The notation for the Al curves is the same as in Fig. 22. The calculation for copper was made with the model of a homogeneous density distribution of nucleons.

eters calculated with the least-square method for the distributions of the nucleons; in addition, the statistical errors of the experimental points are indicated.

A comparison of experimental points and theoretical curves reveals that the first maximum of the angular distribution is adequately described by the model of elastic coherent photoproduction, particularly in the case of light elements (Be, C).

The experimental results are inconsistent with the model of coherent photoproduction in the angular distribution of the pions within the range of the second maximum.

Of particular interest is a comparison of the mean-square radii obtained in our experiments with the radii obtained by other methods. Table 25 and Fig. 26 show the mean-square radii $\langle a^2 \rangle^{1/2}$ of the nucleon density distributions measured in our experiments; the data are the result of averaging the two values which correspond to the two photon energies. The figure includes the electromagnetic radii of nuclei which were considered in electron-scattering experiments (black circles) [78-83]. The figure shows, in addition, the results of a joint analysis of absorption cross sections of high-energy nuclei (0.3-5 GeV) scattered at nuclei [79] and the results of other work on the photoproduction of neutral points at nuclei [76, 77]. It follows from the figure that the $\langle a^2 \rangle^{1/2}$ values obtained from the application of nuclear methods and combined methods (π^0 generation) are in good agreement. The mean-square radii of [76] are everywhere by 0.2-0.3 F below our points and the results of [79, 81].

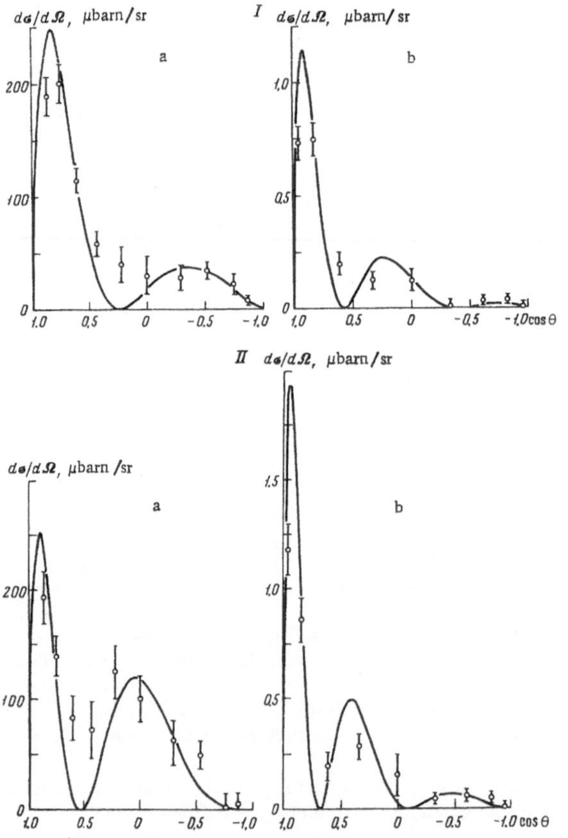

Fig. 24. Differential cross sections of the photoproduction of neutral pions at (I) cadmium and (II) tantalum at (a) k = 154 MeV and (b) 181.5 MeV. The curves were calculated with the trapezoidal density distribution of the nucleons and with the parameters listed in Table 23.

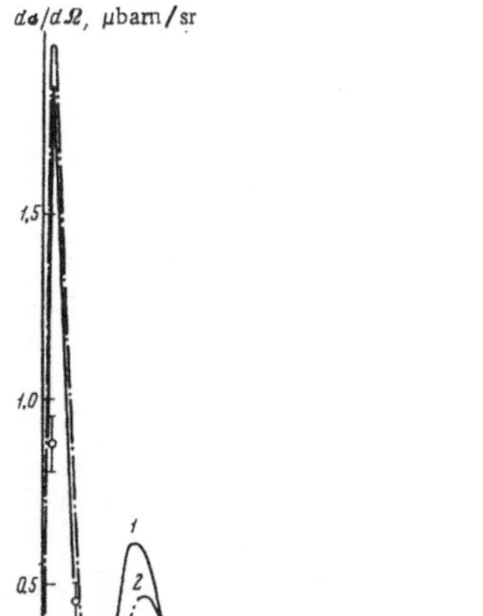

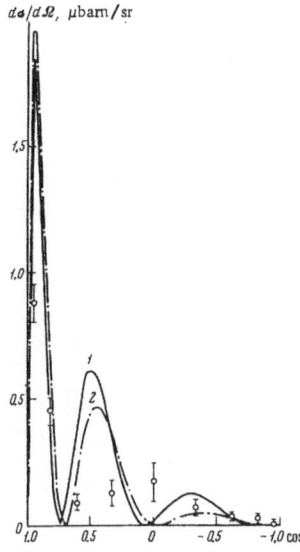

Fig. 25. Differential cross sections of the photoproduction of neutral pions at lead; k = 181.5 MeV. The curves were calculated for (1) the homogeneous and (2) the trapezoidal density distribution of the nucleons.

We used the least-square method to determine from our data the dependence of $\langle a^2 \rangle^{1/2}$ upon the mass number

$$\langle a^2 \rangle^{1/2} = (0.47 \pm 0.09) + (0.85 \pm 0.02)\, A^{1/3}. \qquad \text{(II.26)}$$

This dependence is shown in Fig. 26. The term which is independent of A in Eq. (II.26) proves that the nucleons are rather tightly packed in heavy nuclei. A comparison of the $\langle a^2 \rangle^{1/2}$ values obtained in studies of the nucleon distribution in the nucleus with the $\langle a^2 \rangle^{1/2}$ values from eA-scattering experiments proves that the results are identical to within 0.1–0.2 F in the entire A interval considered (from A = 10 to A ~ 200).

As can be inferred from Table 23 the homogeneous model, which is a very coarse approximation of the nucleon distribution in light nuclei, satisfies the experimental values obtained with Cu, Cd, and Ta. A statistical analysis proves that the radii obtained in the present work with the aid of the homogeneous model (averaging over the two energies) can be appropriately approximated by the relation

$$R = (0.37 \pm 0.07) + (1.14 \pm 0.02)\, A^{1/3}. \qquad \text{(II.27)}$$

The relation of Eq. (II.27) and the resulting R values are shown in Fig. 27. The R values were obtained from electron-scattering experiments [78, 79, 83, 84], from measurements

TABLE 25. Mean-Square Radii (in Units of 10^{-13} cm)

Nucleus	$\langle a^2 \rangle^{1/2}$	Nucleus	$\langle a^2 \rangle^{1/2}$	Nucleus	$\langle a^2 \rangle^{1/2}$
Be	2.22±0.08	Cu	4.01±0.04	Ta	5.27±0.08
C	2.32±0.07	Cd	4.50±0.08	Pb	5.51±0.16
Al	3.16±0.09				

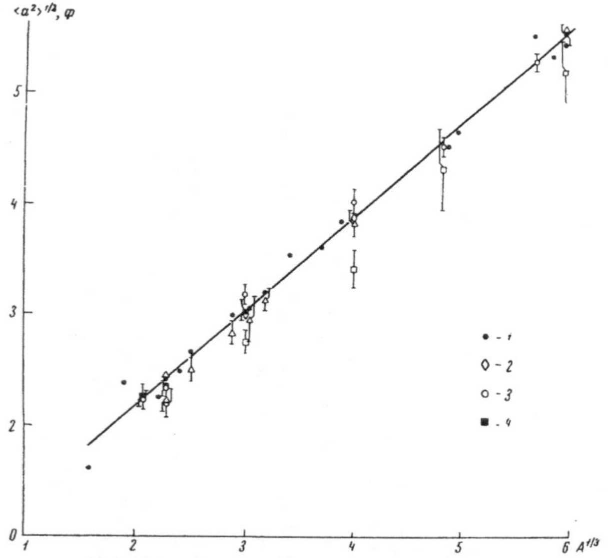

Fig. 26. Mean-square radii $\langle a\rangle^{1/2}$ of nuclei in dependence of the mass number. 1) Results of electron-scattering experiments performed on nuclei [78–83]; 2) results of nucleon-scattering experiments [79]; 3) results of our work [47]; 4) [77]. The straight line corresponds to the equation $\langle a\rangle^{1/2} = (0.47 \pm 0.09) + (0.85 \pm 0.02)\ A^{1/3}$.

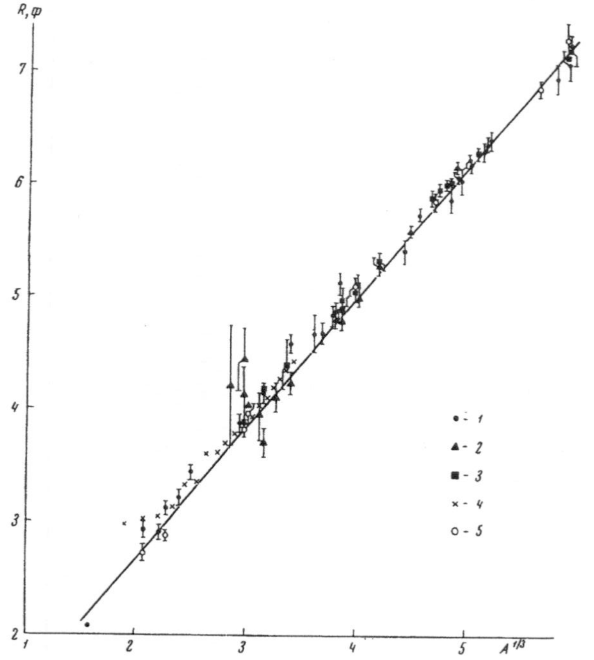

Fig. 27. Dependence of the radius of the homogeneous nucleon-density distribution in nuclei upon the mass number. 1) Results of the electron-scattering experiments [78–83]; 2, 3) results of research on μ-mesonic atoms [85, 86]; 4) results of research on the Coulomb energies of mirror nuclei [87]; 5) results of our work on π^0 photoproduction. The straight line was calculated with the least-square method from our data and is given by the equation $R = (0.37 \pm 0.07) + (1.14 \pm 0.02)\ A^{1/3}$.

of the energy of the $2\mathcal{P}_{3/2} - 1S_{1/2}$ transition of the muon in μ-mesonic atoms [85, 86], and from measurements and calculations of the difference in the Coulomb energies of mirror nuclei [87]. The figure indicates good agreement between the radius R of our experiment and the results obtained with various electromagnetic methods. Our values in the range $A^{1/3} \sim 2$-3.5 are situated slightly below (by 0.1-0.15 F) the electromagnetic radii.

It follows from calculations that the photoproduction of neutral pions is extremely sensitive to the radius value of the nucleon density distribution in nuclei so that the contribution of incoherent processes (which may amount to as much as 5-10%) to the measured yield values has almost no influence upon the determination of $\langle a^2\rangle^{1/2}$ and R. Since the greatest contribution of incoherent photoproduction occurs at large angles, it appeared interesting to check whether differences are observed between the mean-square radii obtained from the analysis of all points of the angular distributions and the experimental data only in the region of the first maximum of the angular distribution (the first five points). The corresponding analysis revealed agreement (within the statistical error limits) of the mean-square radii obtained with the two methods. Our estimates of $\langle a^2\rangle^{1/2}$ and R are therefore rather reliable. The mean-square electromagnetic radius $\langle a^2\rangle^{1/2}_{\mathrm{e.m.}}$ is determined in the electron-scattering experiments, whereas the mean-square radius $\langle a^2\rangle^{1/2}_{\mathrm{c.n}}$ of the distribution of nucleon centers is obtained in the π^0-photoproduction experiments. But we have

$$\langle a^2\rangle_{\mathrm{e.m.}} = \langle a^2\rangle_{\mathrm{c.p.}} + \langle a^2\rangle_{\mathrm{e.m.p.}} \tag{II.28}$$

where $\langle a^2\rangle_{\mathrm{c.p.}}$ denotes the square of the mean-square radius of the distribution of proton centers; and $\langle a^2\rangle_{\mathrm{e.m.p.}}$ denotes the square of the electromagnetic radius of the proton, with $\langle a^2\rangle^{1/2}_{\mathrm{e.m.p.}} \simeq 0.8$ F. The difference between $\langle a^2\rangle^{1/2}_{\mathrm{e.m.}}$ and $\langle a^2\rangle^{1/2}_{\mathrm{c.p.}}$ amounts to 0.13 F in the case of the carbon nucleus, to about 0.1 F in the case of Al, and to 0.1 F in the case of heavier nuclei. According to a comparison between the $\langle a^2\rangle^{1/2}_{\mathrm{c.n.}}$ values obtained from experiments on the photoproduction of neutral pions and the $\langle a^2\rangle^{1/2}_{\mathrm{c.n.}}$ values from electron-scattering experiments, the difference between the two values is less than 0.1-0.2 F. This means that the mean-square radius of the neutron distribution in nuclei is very close to the radius of the proton-center distribution. Electron-scattering experiments and the elastic coherent photoproduction experiments resulted in different values of the parameter s. When we set $C = r_1A^{1/3}$ and assume the parameter t = s to be independent of A, a joint analysis can be made by using the trapezoidal model for all nuclei considered. The following figures are obtained for the parameters of the trapezoidal model: $r_1 = 1.20 \pm 0.01$ and $s = t = 1.07 \pm 0.16$. These figures disagree with the values r = 1.07 and s = 2.4 from the scattering experiments with high-energy electrons [78, 79]. The reason is that the cross section of elastic coherent photoproduction is less sensitive to the higher moments of the distribution of nuclear matter. The estimates of t can therefore not be accepted with the same confidence as the estimates of R and $\langle a^2\rangle^{1/2}$. We relate the difference between the t values of our work and those of the electron experiments to the fact that the model of a purely elastic coherent photoproduction of neutral pions can only inadequately describe the region of the second maximum in the angular distribution.

The interaction between the pions in the final state was tentatively used in [76] to explain the excessively large experimental cross sections in the region of the second maximum of the coherent photoproduction at k = 165 MeV.

We are of the opinion that the interaction of pions in the final state does not suffice for explaining the discrepancies described above. These effects must depend strongly upon the pion energy and must increase with increasing energy. As can be inferred from Figs. 22-25, a similar increase in the discrepancies between experimental points and theoretical curves is not observed at increasing energies in our experiments.

In order to analyze the possible importance of inelastic photoproduction for the observed discrepancies, we calculated the differential cross sections of inelastic pion generation. The

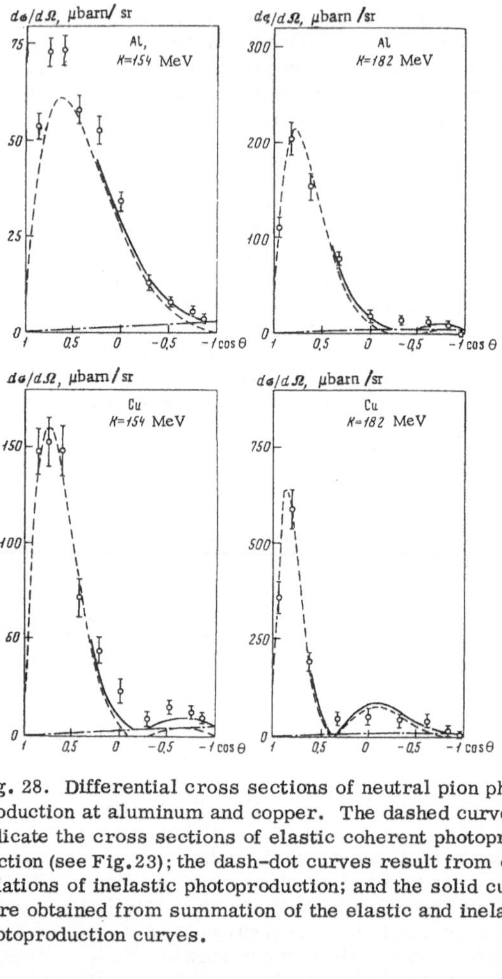

Fig. 28. Differential cross sections of neutral pion photoproduction at aluminum and copper. The dashed curves indicate the cross sections of elastic coherent photoproduction (see Fig. 23); the dash-dot curves result from calculations of inelastic photoproduction; and the solid curves were obtained from summation of the elastic and inelastic photoproduction curves.

reduction factor was calculated with Eqs. (II.15) and (II.16). The differential cross sections of the neutral pion photoproduction at protons and neutrons were determined with the multipole amplitudes listed in [8]. The interaction of the pions in the final state was taken into account in the case of the inelastic process. When this effect was taken into account, the cross section of the inelastic photoproduction had to be multiplied by some function $\varphi(R/\lambda)$, where

$$\varphi(R/\lambda) = 1.5\,\frac{\lambda}{R}\left\{1 - 2\left(\frac{\lambda}{R}\right)^2\left[1 + \left(1 + \frac{R}{\lambda}\right)e^{-R/\lambda}\right]\right\}. \qquad \text{(II.29)}$$

The nuclear radius is denoted by R, and the mean free path of the pion in nuclear matter, by λ. We used the λ values of [88] in our calculations. Figure 28 displays the calculated inelastic cross sections along with the measured points and the theoretical curves obtained with the model of purely elastic coherent photoproduction. The solid curves represent the summation of the elastic and inelastic photoproduction curves. It follows from the figure that the agreement between experimentally measured and theoretically calculated cross sections is considerably improved when the contribution of inelastic photoproduction is taken into account. Moreover, the relative importance of inelastic photoproduction does not greatly increase with increasing energies, because the mean free path length of the pion in nuclear matter decreases rapidly (from 10-12 F at 15-20 MeV) with increasing energy (to 2-3 F at 50-70 MeV). This means that the inelastically produced pions must essentially be emitted from surface layers of the nuclei at high energies. Furthermore, the contribution of inelastic photoproduction is comparable to the contribution of the elastic coherent photoproduction in the region of large pion emission angles. Disregarding inelastic photoproduction may therefore distort the parameters of a nucleon-density distribution, which produce the second maximum of the angular distribution.

Since the contribution of the inelastic process and pion interaction effects in the final state are currently calculated with poor accuracy, it is not possible to take these processes fully and quantitatively into account and to improve the values obtained for the parameters C vs s. Moreover, it is then desirable to increase both the accuracy and the number of experimental data, i.e., one should have 9-10 points at each maximum in an angular distribution with an accuracy of 3-5%.

Dependence of the Total Cross Section of Coherent Photoproduction upon the Mass Number

The dependence of the total cross section of elastic coherent photoproduction upon the mass number was discussed in detail in [89]. In order to avoid ambiguities which are the consequence of the conversion to absolute units, the ratio of the total cross section of neutral pion photoproduction at a nucleus with mass number A to the total cross section of photoproduction at carbon (σ_C) was considered. It is easy to derive from Eq. (II.24)

$$\frac{\sigma_A}{\sigma_C} = \frac{A^2}{144} \frac{\int_0^\pi F_A^2 (QR) \sin^3\theta \, d\theta}{\int_0^\pi F_C^2 (QR) \sin^3\theta \, d\theta}. \tag{II.30}$$

At the threshold energy of the primary photon, the momentum transferred to the nucleus can assume a unique value which is equal to the photon momentum Q = k so that we obtain in this case:

$$\frac{\sigma_A}{\sigma_C} = \frac{A^2}{144} \frac{F_A^2 (QR)}{F_C^2 (QR)}. \tag{II.31}$$

R denotes in this case the mean-square radius of nucleus A. Since $M_A \gg M_\pi$, the threshold value depends only insignificantly upon the mass number A so that the ratio σ_A/σ_C as a function of A has the same form as the function $A^2F^2(QR)$. We consider for the sake of simplicity the homogeneous density model. The form factor is in this case proportional to

$$F(QR) - \frac{3}{QR} j_1(QR), \tag{II.32}$$

where $j_1(QR)$ denotes a spherical Bessel function of the first order, which is regular at the point QR = 0 [90]. The zeros of the function $j_1(QR)$ are defined by the equation

$$\tan QR = QR. \tag{II.33}$$

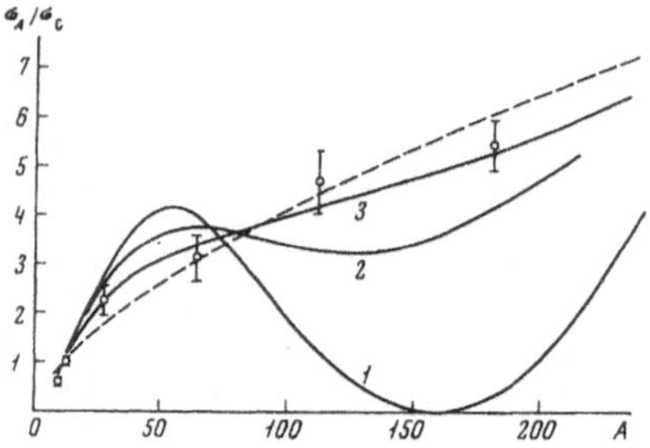

Fig. 29. Dependence of the total cross section of elastic coherent photoproduction of neutral pions upon the mass number. 1) For the threshold value of the photon energy k; 2) for k = 145 MeV; 3) for k = 154 MeV. All curves are normalized to unity at A = 12. The dashed curve represents the function $A^{2/3}$. The experimental points were obtained at k = 153 MeV.

The first zero is situated at QR = 4.49. Curve 1 of Fig. 29 shows the calculated σ_A/σ_C dependence for the threshold values of the primary photons. Obviously, Eq. (II.32) is satisfied for the mass number 158. The particular model which is adopted for the nucleon distribution in the nucleus has no strong influence upon the result. Thus, when the form factor for the trapezoidal density model with parameters determined in electron-scattering experiments is used, we obtain for σ_A/σ_C values which coincide with curve 1 up to A = 60, whereupon the values decrease somewhat more slowly to zero which is reached at A ~ 180.

When the energy of the primary photons is increased, the form factor depends not only upon k but also upon the emission angle θ. In order to determine the total cross sections in this case, we must measure the differential cross sections in the angular interval between 0° and 180°. The measurements comprise a range of momenta imparted to the nucleus in place of the precise QR value of the threshold. Figure 29 shows in addition the calculated σ_A/σ_C values for the primary photon energies 145 MeV (curve 2) and 154 MeV (curve 3). The $A^{2/3}$ dependence is denoted by the dashed curve. We recognize from the figure that the σ_A/σ_C curve becomes smoother and smoother at increasing photon energies. The σ_A/σ_C curves get closer to the $A^{2/3}$ curve. The relation $\sigma_A/\sigma_C \rightarrow A^{2/3}$ holds in the limit $k \rightarrow \infty$. The figure includes the σ_A/σ_C ratio measured at the primary photon energy 154 MeV. The experimental points are in excellent agreement with the theoretical curves, particularly in the region of small A values (beryllium and aluminum). The contribution of inelastic processes and the pion interaction in the final state are unimportant for the photoproduction at k = 154 MeV in the case of the light nuclei. As has been mentioned above, the theoretical calculations are practically independent of the form of the model employed for the nucleon-density distribution in the nucleus. The agreement between experimental and theoretical values is therefore an experimental confirmation of the diffraction-like behavior of the total cross section of the elastic coherent photoproduction of neutral pions as a function of the mass number at low energies of the primary photons. An even more pronounced diffraction-like A dependence of the total cross section must be expected in the case of η photoproduction. Since the mass of the η meson is relatively large (548.3 MeV), we obtain a value for the form factor in the range of the third peak of the carbon nucleus even at Q values as low as the threshold value. σ_A/σ_C measurements in dependence of A at the threshold (in the interval A = 12 to A ~ 200) result in four diffraction maxima in the model of a homogeneous nucleon-density distribution. Of course, we must bear in mind that we speak of η meson generation in the p state at a nucleus which has spin zero. This cross section can be very small near the threshold.

Spin-Independent Isovector Part of the Amplitude of Neutral Pion Photoproduction at Nucleons

When we calculated the above-mentioned parameters of the nucleon-density distributions in nuclei, we related the $|\mathcal{F}_2^+|$ amplitude to the total cross section σ_t of neutral pion photoproduction at protons and used, in essence, the experimental σ_t values. However, in the analysis of experimental results with the least-square method, $|\mathcal{F}_2^+|$ can be considered an additional parameter, which can be determined along with the parameters characterizing the ground state of the particular nucleus under consideration. The analysis of the carbon-nucleus cross sections measured at the energies 154 and 181.5 MeV leads to the $\mathcal{F}_2^+$ amplitudes denoted by the circles in Fig. 30, when the generalized shell model with the parameters $a = 2.3f$ and $\alpha = 4.3$ is used. The figure includes the curve which was obtained in [8] with the aid of dispersion theory. According to the figure, the measured and calculated $\mathcal{F}_2^+$ amplitudes agree to within 5-10%. Dispersion theory predicts that the $\mathcal{F}_2^+$ amplitude is almost completely determined by the dispersion integral or, more precisely, by the contribution of the (3/2, 3/2) isobar to the dispersion integral. The experimental determination of this amplitude is therefore an accurate check of the validity of dispersion-integral estimates in the case of pion photoproduction.

The results of the first successful $\mathcal{F}_2^+$ amplitude determination from measurements of the pion photoproduction at carbon nuclei made us continue the measurements so that we would obtain more reliable information on the energy dependence of the amplitude.

The energy dependence of the differential cross section of neutral pion photoproduction was measured at 70° in the energy interval between the threshold and 190 MeV [75]. The pion formation in a 2-cm-thick carbon target was studied in the bremsstrahlung beam of the 265-MeV synchrotron of the Physics Institute of the Academy of Sciences of the USSR. The reaction $\gamma + C^{12} \rightarrow C^{12} + \pi^0$ was studied by recording two-photon coincidences produced by the decay of the π^0 meson. The photons were detected with telescopes consisting of four scintillation counters each. The γ telescopes were placed in a plane which included a 70° angle with the primary photon beam. The measurements were made at various ψ_k and k_{max} values; measurements were made at three different ψ_k values for each particular k_{max}.

The measurements rendered the yields N of the $\gamma-\gamma$ coincidence counts per unit energy interval of the primary beam at certain k_{max} and ψ_k values. The calculated probabilities ε (k cos θ) of detecting a pion which is formed by a photon with the energy k and emitted under an angle θ, along with the measured yields, were used to determine the differential cross sections of Fig. 31. In order to check the influence of high-frequency "tails" of R(k), we also calculated cross sections with the method proposed in [91] for narrowing R(k). The differential cross sections calculated in this fashion are indicated by the black dots in Fig. 31. There is obviously good agreement between the energy dependence of the cross section calculated with two different methods from the measured yields.

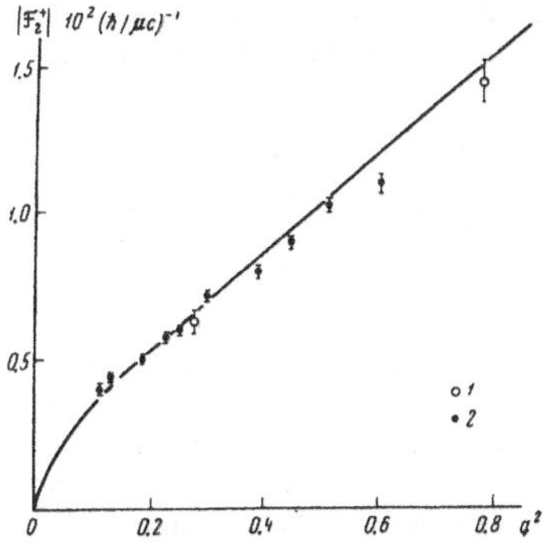

Fig. 30. Spin-independent isovector part of the amplitude of neutral pion photoproduction at nucleons. The curve represents the results of calculations according to dispersion theory [8]; 1) results of measurements of the angular distributions of pions [47]; 2) results of measurements of the energy dependence of the differential cross section at 70° (lab system). The points were obtained for the generalized shell model of the nucleon-density distribution in the carbon nucleus and with $a = 2.3 \cdot 10^{-13}$ cm.

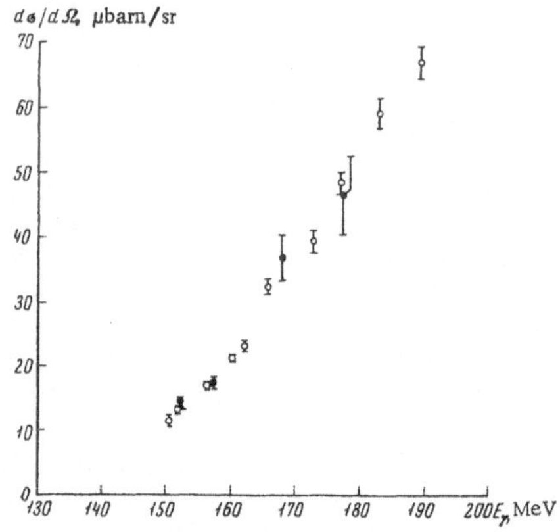

Fig. 31. Differential cross sections of the neutral pion photoproduction at carbon; 70° angle in the lab system. The black dots were obtained after improving the energy resolution [see (II.35)].

If we use the $F_C^2(Q)$ values which were calculated with some model consistent with the charge-distribution obtained in electron-scattering experiments, we can determine $\mathscr{F}_2^+$. Figure 30 shows the calculated $\mathscr{F}_2^+$ amplitudes. We have used as $F_C(Q)$ a form factor calculated with the generalized shell model for the mean-square radius $a = 2.2$ F. We see from the figure that the experimentally measured energy dependence of $\mathscr{F}_2^+$ is in excellent agreement with the theoretical curve. The absolute values begin to coincide when rather realistic values of the mean-square radius of the nucleon distributions in the carbon nucleus are used, i.e., with $a = 2.2$–2.25 F.

Details of the photoproduction of neutral pions at nuclei were established in our work for the near-threshold energy range.

1. In the near-threshold energy range, the principles of neutral pion photoproduction at nuclei can be rather well described with the theory of elastic coherent photoproduction, whereby the momentum approximation is used.

2. Measurements of the angular distribution of pion photoproduction at nuclei are a new accurate technique of determining the mean-square radius of the nucleon-density distribution in nuclei. The measurements which were made at two energies on Be, C, Al, Cu, Cd, Ta, and Pb nuclei helped to determine the mean-square radii with an accuracy of ±3% and to approximate the radius function $\langle a^2 \rangle^{1/2} = (0.47 \pm 0.09) + (0.85 \pm 0.02)\,A^{1/3}$.

3. The mean-square radii $\langle a^2 \rangle^{1/2}$ of the density distribution of nuclear matter differ by less than 0.1–0.15 F from the radii of the charge-density distributions obtained in high-energy electron scattering experiments.

4. The nucleon-density distribution in light nuclei (Be, C) is almost identical with the charge-density distribution. In order to establish the details of the nucleon-density distribution (size of the straight region, etc.) in heavy nuclei, more detailed future measurements and more accurate theoretical calculations of the reaction (consideration of the pion interaction in the final state and inelastic photoproduction) are necessary.

5. In the near-threshold region, the total cross section of elastic coherent photoproduction (cross section referred to carbon) has a diffraction-like dependence upon the mass number. The $A^{2/3}$ dependence is rapidly approached with increasing energy. The calculated cross section values are in good agreement with the results of measurements. We suggest that exact measurements be made in order to determine the cross section of photoproduction and of η

meson production as a function of the mass number in the near-threshold energy range, since these measurements help to determine the structure of atomic nuclei.

6. Measurements of the elastic coherent photoproduction at a nucleus with spin zero (C^{12}) provide additional information on the photoproduction of pions at nucleons and hence can replace, to some extent, polarization experiments. For example, our measurements of the energy dependence of the differential cross section of neutral pion photoproduction at the carbon nucleus (emission angle 70°, energy interval between the threshold and 190 MeV) made it possible to determine the energy dependence of the spin-independent isovector part of pion photoproduction at nuclei for the $\mathcal{F}_2^+$ amplitudes.

Moreover, our research indicates that future more detailed measurements of the angular dependencies (8–10 points at each maximum) and more accurate theoretical calculations of inelastic photoproduction and of the pion interaction in the final state are necessary. The contributions of inelastic and elastic processes should be experimentally established. Experiments of this type will provide new interesting information on the structure of atomic nuclei and on the pion photoproduction mechanism.

References

1. G. F. Chew, M. L. Goldberger, F. Low, and Y. Nambu, Phys. Rev., 106:1345 (1957).
2. I.S.Ball, Phys. Rev., 124:2014 (1961).
3. G. T. Hoff, Phys. Rev., 122:665 (1961).
4. M. I. Moravcsik, Phys. Rev., 125:1088 (1962).
5. B. B. Govorkov and A. I. Lebedev, FIAN Preprint A-48 (1963).
6. A. Müllensiefen, Z. Phys., 188:199 (1965).
7. A. Müllensiefen, Z. Phys., 188:238 (1965).
8. L. D. Solov'ev (Solovyev), Nucl. Phys., 5:256 (1958).
9. K. M. Watson, Phys. Rev., 95:228 (1954).
10. M. Gell-Mann and K. M. Watson, Uspekhi Fiz. Nauk, 59:399 (1956).
11. A. M. Baldin and B. B. Govorkov, Nucl. Phys., 13:193 (1959).
12. G. F. Chew and F. Low, Phys. Rev., 101:1579 (1956).
13. G. F. Chew, M. L. Goldberger, F. Low, and Y. Nambu, Phys. Rev., 106:1337 (1957).
14. E. A. Knopff, R. Kenney, and V. Perez-Mendez, Phys. Rev., 114:605 (1959).
15. R. G. Vasil'kov, B. B. Govorkov, and V. I. Gol'danskii, Zh. Eksp. Teor. Fiz., 37:11 (1959).
16. L. I. Koester and F. E. Mills, Phys. Rev., 105:1900 (1957).
17. R. G. Vasil'kov and B. B. Govorkov, Zh. Eksp. Teor. Fiz., 37:317 (1959).
18. R. G. Vasil'kov, B. B. Govorkov, and A. V. Kutsenko, Pribory Tekhn. Eksp., No. 2, 23 (1960).
19. B. B. Govorkov, V. I. Gol'danskii, O. A. Karpukhin, A. V. Kutsenko, and V. V. Pavlovskaya, Dokl. Akad. Nauk, 111:988 (1956).
20. B. T. Feld, Ann. Phys., 4:189 (1959).
21. A. M. Baldin, Zh. Eksp. Teor. Fiz., 38:579 (1960).
22. A. M. Baldin and A. I. Lebedev, Nucl. Phys., 40:44 (1963).
23. A. M. Baldin, Trudy Fiz. In-ta Akad. Nauk, 19:3 (1963).
24. K. Dietz, G. Höhler, and A. Müllensiefen, Z. Phys., 159:77 (1959).
25. A. M. Baldin, Proc. 1960 Rochester Conf., 325.
26. N. V. Demina, V. A. Evteev, V. A. Kovalenko, L. D. Solov'ev, R. A. Khrenova, Ch'en Jung-mo, Zh. Eksp. Teor. Fiz., 44:272 (1963).
27. L. D. Solov'ev and G. N. Tentyukova, OIYaI Preprint D-728 (1961).
28. A. Donnoshie and G. Shaw, Nucl. Phys., 87:556 (1967).
29. I. M. McKinley, Univ. Illinois Report, No. 38 (1962).
30. A. I. Lebedev and S. P. Kharlamov, 12th Intern. Conference on High-Energy Physics (Dubna), 1:988 (1966).
31. W. Schmidt, Z. Phys., 182:76 (1964).
32. G. Höhler and W. Schmidt, Ann. Phys., 28:34 (1964).
33. M. Gourdin and P. Salin, Nuovo Cim., 27:193 (1963).
34. P. Salin, Nuovo Cim., 28:1294 (1963).

35. W. Hitzeroth, Proc. Internat. Sympos. on Electron and Photon Interactions at High Energies, Hamburg (1965), 12th Intern. Conf. on High-Energy Physics (Dubna), 1:836 (1966); Heidelberg Intern. Conf. on Elementary Particles (1967).
36. A. S. Belousov, S. V. Rusakov, E. I. Tamm, and L. S. Tatarinskaya, Zh. Eksp. Teor. Fiz., 43:1550 (1962).
37. D. B. Miller and E. H. Bellamy, Proc. Phys. Soc., 81:343 (1963).
38. G. Bernardini, A. O. Hanson, A. C. Odian, I. Yamagata, L. B. Auerbach, and I. Filosofo, Nuovo Cim., 18:1203 (1960).
39. A. Odian, G. Stoppini, and I. Yamagata, Phys. Rev., 120:1468 (1960).
40. K. H. Althoff, K. Kramp, H. Matthay, and H. Piel, Preprint, Bonn, 1-001 (1965); 1-007, 1-008 (1966).
41. D. J. Drickey and R. F. Mozley, Phys. Rev. Lett., 8:291 (1962); Phys. Rev., 136:B543 (1964).
42. G. Barbiellini, G. Bologna, J. De Wice, G. Diambrini, G. P. Murtas, and G. Setle, Proc. Sienna Intern. Conf. on Elementary Particles, 516 (1964); 12th Intern. Conf. on High-Energy Physics (Dubna), 1:838 (1966).
43. P. D. Luckey, L. S. Osborne, and I. I. Russell, Phys. Rev. Lett., 3:240 (1959).
44. W. K. H. Panofsky, J. N. Steinberger, and J. S. Steller, Phys. Rev., 76:180 (1952).
45. A. S. Penfold and J. E. Leiss, Phys. Rev., 114:1332 (1959).
46. V. P. Agafonov, B. B. Govorkov, S. P. Denisov, and E. V. Minarik, Pribory Tekhn. Eksp., No. 5, 47 (1962).
47. B. B. Govorkov, S. P. Denisov, and E. V. Minarik, Zh. Eksp. Teor. Fiz., 44:878 (1963); 44:1780 (1963).
48. B. B. Govorkov, S. P. Denisov, A. I. Lebedev, and E. V. Minarik, Zh. Eksp. Teor. Fiz., 44:1463 (1963).
49. B. B. Govorkov, S. P. Denisov, and E. V. Minarik, Yadernaya Fizika, 6:597 (1967).
50. B. B. Govorkov, S. P. Denisov, and E. V. Minarik, Yadernaya Fizika, 4:371 (1966).
51. L. I. Lapidus and Chou Huan-chao, Zh. Eksp. Teor. Fiz., 39:112 (1960).
52. A. M. Baldin, B. B. Govorkov, S. P. Denisov, and A. I. Lebedev, Yadernaya Fizika, 1:92 (1965).
53. B. B. Govorkov, S. P. Denisov, A. I. Lebedev, E. V. Minarik, and S. P. Kharlamov, Zh. Eksp. Teor. Fiz., 47:1199 (1964).
54. J. Hamilton and W. S. Woolcock, Rev. Mod. Phys., 35:737 (1963).
55. A. Donnachie and J. Hamilton, Phys. Rev., 138:B678 (1965); A. Donnachie, J. Hamilton, and A. T. Lea, Phys. Rev., 135:B515 (1964); P. Anvil, A. Donnachie, A. T. Lea, and C. Lovelace, Phys. Lett., 12:76 (1964); L. D. Roper, R. W. Wright, and B. T. Feld, Phys. Rev., 138:B190 (1965).
56. M. I. Adamovich, V. G. Larionova, A. I. Lebedev, S. P. Kharlamov, and F. R. Yagudina, Trudy Fiz. Inst. Akad. Nauk, 34:57 (1966) [English translation: Photomesic and Photonuclear Processes, D. V. Skobel'tsyn, ed., p. 49, Consultants Bureau, New York (1967)].
57. L. D. Solov'ev and Ch'en Jung-mo, OIYaI Preprint D-728 (1961).
58. J. H. Christenson, J. W. Cronin, V. L. Fitch, and R. Turloy, Phys. Rev. Lett., 13:138 (1964).
59. L. B. Okun', Uspekhi Fiz. Nauk, 89:603 (1966).
60. K. Dietz and G. V. Gehlen, Proc. Intern. Symposium on Electron and Photon Interactions at High Energies, 2:297 (1965).
61. T. Ebata and K. Hiida, Progr. Theoret. Phys., 35:432 (1966).
62. B. B. Govorkov, FIAN Preprint (1967).
63. B. B. Govorkov, V. I. Gol'danskii, O. A. Karpukhin, A. V. Kutsenko, and V. V. Pavlovskaya, Dokl. Akad. Nauk, 112:37 (1957).
64. D. J. Anderson, R. W. Kenney, and C. A. McDonald, Phys. Rev., 100:1798 (1955).
65. A. S. Belousov, E. I. Tamm, and E. V. Shitov, Abstracts of the Reports of the All-Union Conference on High-Energy Particle Physics [in Russian], 93 (1956).
66. E. L. Feinberg, Phys. USSR, 5:177 (1941).
67. S. M. Berman, Nuovo Cim., 21:1020 (1961).
68. R. H. Traxler, Preprint, Berkeley, Calif. (1962).
69. C. A. Engelbrecht, Phys. Rev., 133:B988 (1964).
70. V. A. Tsarev, Yadernaya Fizika, 4:11 (1966).
71. J. E. Leiss and R. A. Schrack, Rev. Mod. Phys., 30:456 (1958).
72. R. G. Vasil'kov, B. B. Govorkov, and V. I. Gol'danskii, Zh. Eksp. Teor. Fiz., 38:1149 (1949).
73. J. H. Fregean and R. Hofstadter, Phys. Rev., 99:1503 (1955).
74. B. B. Govorkov, S. P. Denisov, and E. V. Minarik, Zh. Eksp. Teor. Fiz., 42:1010 (1961).
75. B. B. Govorkov, S. P. Denisov, and E. V. Minarik, Yadernaya Fizika, 5:190 (1967).
76. R. A. Schrack, J. E. Leiss, and S. Penner, Phys. Rev., 127:1772 (1962).
77. R. A. Schrack, Phys. Rev., 140:B897 (1965).
78. R. Hofstadter, Electromagnetic Structure of Nuclei and Nucleons [Russian translation], Foreign Literature Press (1958).
79. L. R. Elton, Nuclear Sizes, Oxford University Press, New York (1961).
80. V. Meyer-Berkhout, K. W. Ford, and A. E. S. Green, Ann. Phys., 8:119 (1959).
81. H. F. Ehrenberg, R. Hofstadter, V. Meyer-Berkhout, D. C. Ravenhall, and S. E. Sobottka, Phys. Rev., 113:656 (1959).

82. R. H. Helm, Phys. Rev., 104:1461 (1956).
83. B. Hahn, D. G. Ravenhall, and R. Hofstadter, Phys. Rev., B7:865 (1965).
84. V. I. Startsev, Moscow, Dissertation [in Russian] (1966).
85. H. L. Anderson, C. S. Johnson, and E. P. Hinks, Phys. Rev., 130:2468 (1963).
86. D. Whitmann, R. Engler, V. Hevel, P. Brix, G. Backenstoss, K. Goebel, and B. Stadler, Nucl. Phys., 51:609 (1964).
87. P.C. Sood and A. E. S. Green, Nucl. Phys., 4:274 (1957).
88. R. M. Frank, J. L. Gammel, and K. W. Watson, Phys. Rev., 101:891 (1956).
89. B. B. Govorkov, Yadernaya Fizika, 6:116 (1967).
90. L. I. Schiff, Quantum Mechanics, 2nd ed., McGraw-Hill, New York (1956).
91. J. D. Prentice, E. H. Bellamy, and W. S. C. Williams, Proc. Phys. Soc., 74:124 (1959).

PHOTOPRODUCTION OF π^0 MESONS AT HELIUM AND AT PHOTON ENERGIES OF 160–240 MeV

A.S. Belousov, S.V. Rusakov, E.I. Tamm, L.S. Tatarinskaya, V.A. Tsarev, and P.N. Shareiko.

Experimental values of the differential and total cross sections of elastic π^0 meson photoproduction at helium were obtained in the photon-energy interval 160-240 MeV. The measured differential cross sections agree with the calculations based on the already developed theory of the relativistic momentum approximation in the bremsstrahlung energy range $E_\gamma \leq 200$ MeV. When corrections for the π^0 meson interaction in the final state are introduced, one can obtain agreement with the available experimental data up to energies $E_\gamma \leq 300$ MeV. The experimental value of the amplitude $\mathcal{F}_{20}^{+}$ of π meson photoproduction at nucleons was found for $\langle r \rangle_{He} = 1.4$ F.

Introduction

The development of a theory of strong interactions has been one of the most interesting and important problems in the nuclear physics of the last few years. Naturally, experiments on the N$-\pi$ interaction have been an important incentive for theoretical work in this field. Experiments on the photoproduction of pions at nucleons are, from the theoretical viewpoint, the simplest experiments in which strong interactions can be observed. These experiments can be termed simple, because the well-investigated electromagnetic interactions play an important role in photoproduction processes. Interactions of this type can be very accurately described by quantum electrodynamics. The treatment of photomesonic processes and their theoretical interpretation are therefore greatly simplified.

The possible theoretical predictions are particularly reliable in the energy range in which the energy of the primary photons is close to the threshold of single π^0 meson generation ($E_\gamma \leq$ 220 MeV). Both a phenomenological analysis [1] of experimental data with the aid of general quantum-mechanical conservation laws and an interpretation with dispersion relations [2-4] based on fundamentals of field theory are feasible in this energy interval.

Of greatest value in this respect are experiments on pion photoproduction at protons [5], because this reaction is strongly dependent upon the characteristics of the N$-\pi$ interaction. Apart from this, the relative simplicity of these experiments is an advantageous methodological aspect. But in order to determine the total set of characteristics which describe the pion photoproduction process at nucleons, investigations with polarized beams of primary γ quanta and with polarized targets are required. Experiments of this kind are at the present time difficult from the technical viewpoint, particularly in the energy range near the threshold of π meson photoproduction, since the cross sections are small at those energies.

Determination of individual features of the nucleon–pion interaction from investigations of the π meson photoproduction at nuclei [6-8] is one of the possible ways of circumventing the difficulties. The experimental data are most easily interpreted in the case of pions because whatever the energy of neutral pion photoproduction, corrections for the Coulomb interaction between mesons and nuclei in the final state need not be introduced. For example, investigations of the reaction $\gamma + d \rightarrow \pi^0 + d$, which was considered in several experiments, provide additional information on multipole and isotope amplitudes of π meson photoproduction at nucleons [9]. In particular, elastic photoproduction of π^0 mesons at helium facilitates direct estimates of the $\mathcal{F}_2^+$ photoproduction amplitude of π mesons at nucleons, this amplitude being one of the most important parameters of the N–π interaction. Studies of this process are of independent interest, because the data on π^0 meson photoproduction at helium can provide insight into the structure of the He^4 nucleus, the influence of the nucleon bond upon the photoproduction process, etc. In addition, of interest are quantities such as the total (σ_{tot}^{He}) and differential cross sections of π^0 meson photoproduction at He^4 at various energies E_γ and meson emission angles θ_π, because these quantities and the particular reaction can be used to measure the linear polarization of the γ quanta [10].

The present article relates to the experimental investigation of π^0 meson photoproduction at He^4 nuclei and the information obtained on the above-mentioned quantities.

CHAPTER 1

π^0 Meson Photoproduction at Composite Nuclei

Less experimental work has been done on π^0 meson photoproduction at nuclei [6-8, 11-14] that on pion photoproduction at nucleons, since, on the one hand, the theoretical treatment is difficult, and, on the other, the process is hard to identify in experiments. The problem is somewhat simpler when the π^0 photoproduction is considered in the energy range of primary γ quanta between the threshold and 200 MeV. As has been shown in [7, 12-14], the experimental results in this energy interval can be satisfactorily described by a model which is based only on elastic coherent photoproduction, with the nuclecus remaining in the ground state, i.e., the wave functions in the initial and final states of the nucleus are the same. The following simplifying assumptions were used in calculations of the elastic coherent photoproduction: 1) photon and meson, which participate in the reaction, can be represented by plane waves; 2) the total amplitude of π^0 meson photoproduction at a nucleus can be stated as sum over the amplitude of photoproduction at individual nuclei; 3) the nuclear binding forces are negligible during the photoproduction process; and 4) the photoproduction operator is independent of the isotope spin of the nucleon.

The shortcomings of this model are obvious: the influence of inelastic processes is ignored (the accuracy of this approximation has not been estimated); assumptions 1-4) are not always justified since the errors related to these assumptions may sometimes be substantial [15].

The available experimental results [12-14], however, do indicate that up to E_γ values close to 200 MeV, the model is correct and describes adequately the experimental results.

The importance of inelastic π^0 meson photoproduction, in which the target nuclei are split or excited, increases with increasing energy and must be determined either by experiments or by theories. Inelastic photoproduction at, say, He^4 is characterized by a threshold value which is 20 MeV greater than the threshold of elastic photoproduction [16]. This difference in the threshold values is determined by the individual properties of nuclei, e.g., the position of the excited levels, the binding energy, and the momentum distribution of the nucleons in the nucleus.

The photoproduction of π^0 mesons at nuclei has been used in some instances to investigate the structure of those nuclei and to check various related models. As a matter of fact, the differential cross section of the reaction $\gamma + A \rightarrow A + \pi^0$ can be written in the form of [7]:

$$\left[\frac{d\sigma_\pi}{d\Omega}\right] = \frac{q_A}{k_A} | H |^2, \tag{I.1}$$

where q_A and k_A denote the momenta of the meson and the photon, respectively; and H denotes the matrix element of photoproduction. We obtain in the momentum approximation

$$H = \left\langle \Psi_f \middle| \sum_{j=1}^{A} T_j \middle| \Psi_i \right\rangle = A \langle \Psi_j | T_j | \Psi_i \rangle, \tag{I.2}$$

where A denotes the number of nucleons in the nucleus, T_j denotes the photoproduction operator which acts upon the nucleon variables; and Ψ_i and Ψ_f denote the wave functions of the initial and final states of the nucleus, respectively.

The operator T_j can be written in the form

$$T_j = \mathbf{K}_j \sigma_j + L_j, \tag{I.3}$$

where $\mathbf{K}_j \sigma_j$ denotes the spin-dependent part, and L_j, the spin-independent part.

In the case of nuclei with nonvanishing spin, operator $\mathbf{K}_j$ renders an A times smaller contribution than operator L_j [7]. In the case of nuclei with vanishing spin, this contribution must be completely missing.

Photon and meson are described by plane waves in the Born approximation, and the operator of spin-independent photoproduction can be written in the form $L_j e^{i(Qr)}$, where $Q = \hbar^{-1} (k^2 + q^2 - 2qk \cos \theta_\pi^*)$ denotes the momentum transferred to the nucleus. Since L_j is independent of the nucleon variables, we obtain

$$H = AL \langle \Psi_f | e^{-i(Qr)} | \Psi_i \rangle. \tag{I.4}$$

In the case of elastic coherent photoproduction, $\Psi_f = \Psi_i$ and $\rho = \langle \Psi_i | \Psi_f \rangle$. The quantity $\langle \rho (r) e^{-i(Qr)} \rangle = F (Qr)$ is termed form factor of the nucleus and refers to the distribution of nuclear matter. Therefore,

$$H = ALF (Qr), \tag{I.5}$$

$$\left[\frac{d\sigma_\pi}{d\Omega}\right]_A = A^2 L^2 F^2 (Qr) \frac{q_A}{k_A}. \tag{I.6}$$

When we assume the form factor for the proton equal to unity [14], we obtain the spin-independent part of the cross section in the form

$$\left[\frac{d\sigma}{d\Omega}\right]_H = L^2 \frac{q_H}{k_H}. \tag{I.7}$$

When only the M_{1+} amplitude, which results from the $(^3/_2, {}^3/_2)$ resonance for the $N - \pi$ system is taken into account in the matrix element of photoproduction at the nucleon, we can set [5]

$$\left[\frac{d\sigma}{d\Omega}\right]_H = \sigma_{tot}^H \frac{\sin^2\theta^*}{4\pi}, \tag{I.8}$$

i.e., we have

$$\left[\frac{d\sigma}{d\Omega}\right]_A = A^2 \frac{\sigma_{tot}^H}{4\pi} F_A(Qr)\sin^2\theta_\pi^*, \tag{I.9}$$

where σ_{tot}^H denotes the total cross section of π^0 meson photoproduction at hydrogen. In order to establish a relation between π^0 meson photoproduction at a nucleon and a nucleus, we assumed that θ_π^* denotes the meson emission angle in the center of mass of the photon–nucleon system.

A slightly different expression was obtained in [17] for the cross section of coherent π meson production at nuclei. In particular, in the case under consideration, the cross section assumes the form

$$\left[\frac{d\sigma}{d\Omega}\right]_{He} = 8\frac{q_{He}}{k_{He}}|\mathcal{F}_2^+|^2 RF^2(Qr)\sin^2\theta_\pi^*, \tag{I.10}$$

where

$$R = \left[\frac{w^2}{E_\gamma E_\pi E_i E_f}\right]_N \left[\frac{E_\gamma E_\pi E_i E_f}{w^2}\right]_{He}. \tag{I.11}$$

The subscripts N and He denote the center-of-mass system of the meson (photon)–nucleon and meson (photon)–nucleus configurations, respectively; q_A, E_π, and k_A, E_γ denote the momenta and energies of the meson and photon, respectively, in the center of mass of the meson (photon)–nucleus system; E_i and E_f denote the total energy of the system; and $\mathcal{F}_2^+$ denotes one of the amplitudes of π meson photoproduction at nucleus (see Chapter 2).

Since the quantity σ_{tot}^H is known from experiments with an accuracy of about 5% in the entire range of the first pion–nucleon resonance [18, 19], Eq. (I.9) can be used to study the structure of a nucleus A. Investigations of this kind were made in [6-8] and [11-14]. On the other hand, when a particular form of $F_A(Qr)$ has been adopted on the basis of electron-scattering experiments [20, 21], certain conclusions concerning the amplitude L can be drawn [8]. All results concerning this quantity were obtained from experiments on composite nuclei. Interpretation of the results is then complicated by the need for taking into consideration the typical structural features of each nucleus, e.g., spin, excited states, shell structure, and interactions between the meson and the nucleus in the final state.

In order to simplify the discussion of these factors, it is important to study π^0 meson photoproduction at light nuclei such as deuterium, tritium, and helium. The He^4 nucleus is evidently particularly suitable for the following reasons:

1) The He^4 nucleus consists of a small number of nucleons (A = 4) and has a more compact structure than, say, the deuterium nucleus, i.e., we have $\langle r \rangle = 1.4$ F;

2) the total spin of the He^4 nucleus is zero;

3) no excited He^4 states exist near the threshold of π^0 meson photoproduction (137 MeV);

4) the wave function of the ground state of the He^4 nucleus is relatively well known [22, 23];

5) the relatively small number of nucleons in this nucleus allows more rigorous conclusions concerning the applicability of the model with the momentum approximation. These de-

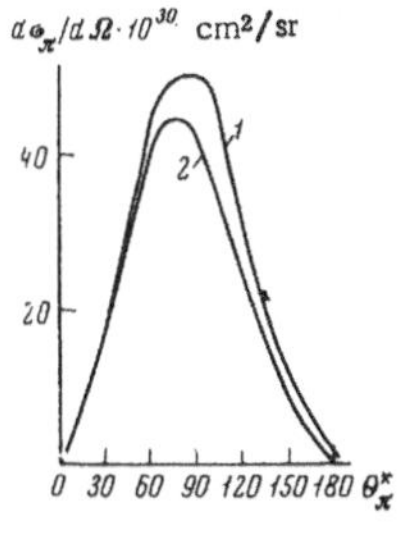

Fig. 1. Differential cross section calculated by Yamaguchi [24] (1) by the "completeness" approximation and (2) according to the elastic π^0 meson photoproduction at He^4.

tails (except for the last one) of the reaction

$$\gamma + He^4 \rightarrow He^4 + \pi^0 \tag{I.12}$$

were emphasized by Goldwasser, Koester, and Mills [16] and by Yamaguchi [24]. Yamaguchi used the phenomenological approach, calculated the cross section of reaction (I.12) by the momentum approximation, and compared the result with the total cross section (elastic and inelastic) calculated by the "completeness" approximation.

It follows from Fig. 1, which shows the results of this work, that even at E_γ = 200 MeV, elastic π^0 meson photoproduction must provide the main contribution to the total cross section. But the absolute value of $d\sigma_\pi/d\Omega$, which is indicated, is much too large because the mean-square radius $\langle r \rangle$ = 0.99 F was assumed in the calculations.

In an ensuing theoretical paper, Cook [25] studied the dependence of the differential cross section upon the form factor F(Qr). He could show that up to energies $E_\gamma \leq$ 700 MeV, the form factor at a fixed radius has little influence upon the calculated cross section. Cook also analyzed the experimental results of [26, 27]. The mean-square radius obtained in this analysis for the He^4 nucleus was (1.5 ± 0.1) F. This result agrees with the value $\langle r \rangle$ = 1.61 F obtained from electron-scattering experiments and the photon fission of the He^4 nucleus. Cook came to the conclusion that multiple scattering of π^0 mesons in the final state must be taken into account in the experimental cross section of reaction (I.12).

Goldwasser, Koester, and Mills [16] investigated for the first time the $He^4(\gamma\pi^0)He^4$ reaction in experiments. They measured the π^0 meson yield in dependence of the maximum energy of the primary photons in the energy interval 150-190 MeV. The π^0 mesons were detected via the two γ quanta from the decay. The angle between the plane of the counter and the beam axis was 80°. The correlation angle between the axes of the γ telescopes was varied from 180° to 90° in 30° steps. The counting rates measured at E_γ = 150, 155, and 160 MeV led the researchers to the conclusion that π^0 mesons are produced at He^4 in an elastic process. A qualitative estimate of the importance of elastic and inelastic processes at the energy E_γ = 190 MeV was the second result of that work. The thresholds of the correlation curves for π^0 mesons produced in the two processes differ in this case by 15°. The experimentally determined dependence of the yield upon the correlation angle agreed fully with the characteristics of an elastic process. The authors therefore came to the conclusion that the elastic process still plays the essential role at 190 MeV. This conclusion agrees with the theoretical estimates of [24].

De Smissure and Osborn [28] worked on these processes, in which they recorded the He^3 and He^4 recoil nuclei with nuclear emulsions. Table 1 lists the energy distribution which they recorded in the center-of-mass system at the angle θ^*_π = 90°. The maximum of the distribution corresponds to primary photon energies close to 270 MeV. The peak appears at θ_π = 65° in the angular distribution. But attempts to determine the radius of the He^4 nucleus from these results gave the value $\langle r \rangle$ = (0.93 ± 0.12) F, which is much smaller than any of the previously

TABLE 1. Differential Cross Section of the Reaction $\gamma + He^4 \rightarrow \pi^0 + He^4$ at the Angle $\theta_\pi^* = 90°$ in the Center-of-Mass System

E_γ, MeV	$\frac{d\sigma_\pi}{d\Omega} 10^{30}$, cm²/sr	$\frac{d\sigma_\pi}{d\Omega} 10^{30}$, cm²/sr
210	81.1±12.4	
230	89.0±13.6	
250	89.5±14.6	
270	130.2±18.5	15.8±3,52
280		13.65±2.35
290	120.4±16.5	
310	77.2±29.4	
320		7.52±3.53
330	87.0±33.3	

known values. The difference was probably caused by the relatively large absolute value of the experimentally obtained cross section of the $He^4(\gamma, \pi^0) He^4$ reaction. The differential cross sections of the elastic process, which were measured in [26] by Palit and Bellamy, confirm this conclusion. According to Table 1, the $[d\sigma/d\Omega]_{He}$ values stated in this article for $\theta_\pi^* = 90°$ are approximately ten times smaller than the values of [28]. Palit and Bellamy obtained in their experiments cross sections at 70°, 115°, and 135° in the energy range beyond the first resonance of the N−π interaction. In order to interpret the results, the momentum approximation is used in the present work, along with corrections for multiple meson scattering in the final state, and an analysis without these corrections.

The influence of multiple scattering upon the differential cross section of elastic π^0 meson photoproduction at deuterium was for the first time calculated by Chappelear [15]. He could show in his work that multiple scattering almost halves the cross section of π^0 meson photoproduction at all angles in the energy range near $E_\gamma = 300$ MeV. This conclusion was confirmed in several experiments [29-31]. No such calculations were made for the energy range of γ quanta energies close to the threshold of π^0 meson photoproduction at nucleons. But the experimental results of [9] prove that multiple scattering in this energy range has no substantial influence upon the π^0 meson photoproduction at deuterium. Multiple scattering was considered in Stoodley's dissertation in the case of π^0 meson photoproduction at He^4 under 90° [32]. Estimates of this effect were made in the present work for an increased range of energies (from the threshold to 300 MeV) and angles (0-180°).

The Physics Institute of the Academy of Sciences of the USSR made recently new measurements of the cross section of the elastic photoproduction of π^0 mesons at He^4 nuclei. The results of the measurements will be discussed below.

CHAPTER II

Cross Section of π^0 Meson Photoproduction at He^4

We consider first the $\mathcal{F}_2^+$ amplitude of π meson photoproduction at free nucleons, since this amplitude determines the absolute value of the cross section of reaction (I.12).

In the center-of-mass system, the amplitude of π meson photoproduction at nucleus has the following form [2]:

$$\mathcal{F} = (K\sigma + L) = i(\sigma\epsilon)\mathcal{F}_1 + \frac{(\sigma q)(\sigma[k\epsilon])}{kq}\mathcal{F}_2 + \frac{(\sigma k)(q\epsilon)}{kq}\mathcal{F}_3 + \frac{(\sigma q)(q\epsilon)}{q^2}\mathcal{F}_4, \qquad (II.1)$$

where K denotes the spin-dependent part of the amplitude; L denotes the spin-independent part of the amplitude; σ is the Pauli matrix; $\in$ denotes the polarization vector of the incident photon; q denotes the meson momentum in the center-of-mass system; and $\mathcal{F}_i$ denote functions of both the energy and the emission angle of the meson in the center-of-mass system.

In our ensuing discussion, we will be concerned only with the $\mathcal{F}_2$ amplitude because it characterizes that portion of the amplitude $\mathcal{F}$ which is unrelated to spin changes.

It follows from an expansion of the function $\mathcal{F}_i$ in terms of the amplitudes of electric $E_{l\pm}$ and magnetic $M_{l\pm}$ transitions that $\mathcal{F}_2$ implies only the magnetic transitions M_{l+} and M_{l-} into a state of the N$-\pi$ system with the orbital moment l and the total moment $\mathcal{J} = l \pm 1/2$, i.e., we have

$$\mathcal{F}_2(q, \cos\theta^*) = \sum_{l=1}^{\infty} [(l+1) M_{l+} + l M_{l-}] P_l'(\cos\theta^*), \tag{II.2}$$

where $P_l'(\cos\theta^*)$ denotes the first derivative of Legendre polynomials.

When the concepts of isotope invariance are employed to describe photoproduction processes the amplitude of any of the four principal pion-photoproduction processes at nucleons

$$\gamma + p \to p + \pi^0; \tag{II.3}$$

$$\gamma + p \to n + \pi^+; \tag{II.3a}$$

$$\gamma + n \to p + \pi^-; \tag{II.3b}$$

$$\gamma + n \to n + \pi^0 \tag{II.3c}$$

can be expressed in the isospin space of the system by the three amplitudes

$$\begin{gathered} \left\langle T_f = \frac{1}{2} \left| S \right| T_i = \frac{1}{2} \right\rangle = \mathcal{F}^0, \\ \frac{1}{3}\left\langle \frac{1}{2} \left| V \right| \frac{1}{2} \right\rangle + \frac{2}{3}\left\langle \frac{3}{2} \left| V \right| \frac{1}{2} \right\rangle = \mathcal{F}^+, \\ \frac{1}{3}\left\langle \frac{1}{2} \left| V \right| \frac{1}{2} \right\rangle - \frac{1}{3}\left\langle \frac{3}{2} \left| V \right| \frac{1}{2} \right\rangle = \mathcal{F}^-. \end{gathered} \tag{II.4}$$

where T denotes the isospin and S and V denote the isoscalar and isovector parts of the operator of pion photoproduction. We therefore obtain for processes (II.3) and (II.3a)-(II.3c):

$$\begin{gathered} \mathcal{F} = \mathcal{F}^+ + \mathcal{F}^0, \\ \mathcal{F} = \mathcal{F}^+ - \mathcal{F}^0, \end{gathered} \tag{II.5}$$

where $\mathcal{F}^0$ denotes the isoscalar and $\mathcal{F}^+$ and $\mathcal{F}^-$, the isovector amplitudes, respectively.

In the case of reactions (II.3a) and (II.3b), the quantity $\mathcal{F}$ contains $\mathcal{F}^-$ in space of $\mathcal{F}^+$ in Eq. (II.5). Thus, in order to fully characterize processes (II.3) and (II.3a)-(II.3c), we must know twelve amplitudes $\mathcal{F}_i^\alpha$ where i = 1, 2, 3, 4, and α = +, −, 0.

We realize that in the case of elastic π^0 meson photoproduction at nuclei with equal numbers of protons and neutrons, only the isovector part of the photoproduction amplitudes remains. The $\mathcal{F}_2^+$ values obtained by Kharlamov and Lebedev [4, 33] are used in the discussion of the experimental data obtained in our work.

Figure 2 shows the results of various modifications of the $\mathcal{F}_{20}^+$ amplitude calculations. The real part of that amplitude was in all cases calculated with the dispersion relations of [2]

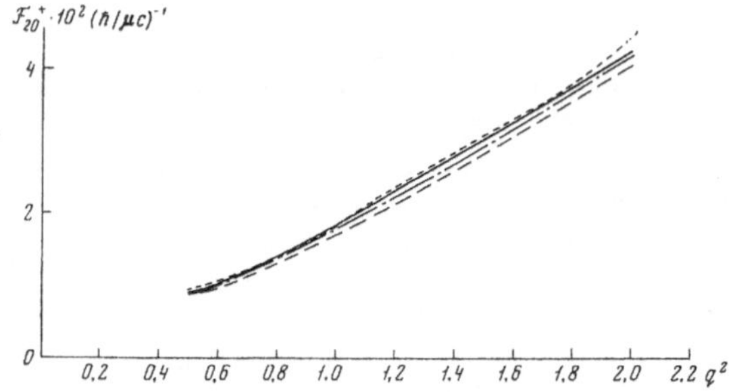

Fig. 2. The quantity $\mathcal{F}_{20}^{+}$ in dependence of the square of the meson momentum in the center-of-mass system. The curves correspond to various calculations of this quantity in the theory of dispersion relation [4, 11].

[see (II.15)]

$$\mathrm{Re}\,\mathcal{F}_2^{+} = \mathcal{F}_2^{+B} + \frac{1}{\pi}\int\limits_{M+1}^{\infty} dW \sum_{j=1}^{4} \Phi_{2j}(s, s', t)\,\mathrm{Im}\,\mathcal{F}_j^{+}(st), \tag{II.6}$$

where $\mathcal{F}_2^{+B}$ denotes the Born part of the $\mathcal{F}_2^{+}$ amplitude (the Born part includes the charge of the nucleon, the interaction constant, and the magnetic moment of the nucleon [2]); Φ_{2f} (s, s', t) denotes a kinematic function with the singularities; W' denotes the total energy in the center-of-mass system; s, s', k, q, p_1, and p_2 denote the momenta of the photon, the meson, and the nucleon in the initial and final states, respectively.

The Born term in Eq. (II.6) was explicitly calculated without multipole expansion; the full form of this term was used when the function Φ_{2j} was integrated. The integration interval was restricted to 1 BeV, i.e., the integration interval comprised the first $(P_{3/2},\, _{3/2})$ and second $(D_{3/2},\, _{3/2})$ resonance of photoproduction. The constant of the N$-\pi$ interaction was assumed as $f^2 = 0.08$. Only the imaginary part of the magnetic dipole amplitude of the resonance [2] was considered in the dispersion integral of the first, second, and fourth version of the calculations:

$$\begin{aligned}
\mathrm{Im}\,\mathcal{F}_1^{+} &= 3\frac{2}{3}\,\mathrm{Im}\,M_{1+}^{3/2}\cos\theta_\pi^*, \\
\mathrm{Im}\,\mathcal{F}_2^{+} &= 2\frac{2}{3}\,\mathrm{Im}\,M_{1+}^{3/2}, \\
\mathrm{Im}\,\mathcal{F}_3^{+} &= -\,3\frac{2}{2}\,\mathrm{Im}\,M_{1+}^{3/2}, \\
\mathrm{Im}\,\mathcal{F}_4^{+} &= 0,
\end{aligned} \tag{II.7}$$

where

$$\mathrm{Im}\,M_{1+}^{3/2} = |M_{1+}^{3/2}|\sin\alpha_{33}. \tag{II.8}$$

The experimental value sin α_{33} of [34] was used up to energies E_π = 525 MeV. At energies E_π > 525 MeV, the phases were described with the method proposed in [35], where

$$\frac{\sin^2\alpha_{33}}{q^3} = A\exp\left[-\frac{\omega}{\sigma}\right], \tag{II.9}$$

and A = 47.5; σ = 0.397; $\omega = W - M$; W denotes the total energy in the center-of-mass system, and $W = \sqrt{1+q^2} + \sqrt{M^2+q^2}$; ($\hbar = \mu = c = 1$).

In the first version (Fig. 2, long-dash curve) the $|M_{1+}^{3/2}|$ value appearing in Eq. (II.8) was determined from the experimentally obtained total cross section of the reaction $\gamma + p \to \pi^0 + p$ [19, 36]:

$$|M_{1+}^{3/2}| = \sqrt{\frac{9}{32}\frac{k}{q}\sigma_{tot}^{\pi^0}} \tag{II.10}$$

The small contribution from nonresonance multipoles has been ignored in this formulation.

In the second version (see Fig. 2, dash-dot curve), the small contributions from the M_{1+}^{s} and $M_{1+}^{1/2}$ amplitudes have been included. Static dispersion relations with corrections of the order 1/M were used to describe these amplitudes [37]. We obtain in this case

$$\sigma_{tot}^{\pi^0} = 8\pi\frac{q}{k}\left|\frac{2}{3}M_{1+}^{3/2}e^{i\alpha_{33}} + e^{i\alpha_{31}}\left[\frac{1}{3}M_{1+}^{1/2} + M_{1+}^{s}\right]\right|^2. \tag{II.11}$$

The value of the nonresonance α_{31} phase of pion–nucleon scattering corresponded to the value obtained in [34].

In the third version of the calculations (Fig. 2, solid curve) the imaginary part of the amplitude in the dispersion integral of Eq. (II.6) was selected in a fashion similar to the preceding version, but the second resonance of the N–π interaction and the M_{1-} multipole amplitude were taken into account. It was assumed that the resonance is caused by the $E_{2-}^{1/2}$ amplitude which has isovector structure, and that the other multipole amplitudes are immaterial. We can therefore represent $|E_{2-}^{1/2}|$ in the form

$$\frac{1}{3}|E_{2-}^{1/2}| = \sqrt{\frac{k}{8\pi}[\sigma_{tot}^{\pi^0} - \sigma_{P_{3/2},3/2}]}, \tag{II.12}$$

where $\sigma_{P_{3/2},3/2}$ denotes the cross section resulting from the $P_{3/2},{}_{3/2}$ resonance.

Finally, it was assumed that the usual relation exists between photoproduction and scattering and that the α_{15} phase passes through 90° in the resonance region ($E_\pi \simeq 600$ MeV). We have

$$\frac{1}{3}\operatorname{Im}E_{2-}^{1/2} = \frac{\frac{k}{8\pi q}[\sigma_{tot}^{\pi^0} - \sigma_{P_{3/2},3/2}]}{\sqrt{\frac{k}{8\pi q}[\sigma_{tot}^{\pi^0} - \sigma_{P_{3/2},3/2}]}}. \tag{II.13}$$

The fourth and final version of calculating the amplitude $M_{1+}^{3/2}$ (Fig. 2, short-dash curve) is similar to the first version, but the imaginary part of the resonance amplitude $M_{1+}^{3/2}$ is described by the "effective length" formula of [3]:

$$\operatorname{Im}M_{1+}^{3/2} = \frac{k}{q}\frac{\mu_p - \mu_n}{2f}\frac{q^6}{q^6 + \Gamma(s-s_r)(s-M^2)^2}, \tag{II.14}$$

where $\Gamma = 3.5 \cdot 10^{-4}$ and s = 76.6; $s_r = W^2$ denotes the square of the total energy; $\mu_p' = 1.78$ and $\mu_n = -1.91$ denote the anomalous magnetic moments of proton and neutron, respectively; and $\mu_p = \mu_p' + 1$.

The quantity Im F_2^+ which is part of the total amplitude $\mathcal{F}_2^+$ corresponded in each version to an expression for the absorbed part of the amplitude $\mathcal{F}_2^+$ which was used to calculate the dispersion integrals in Eq. (II.6). The relative importance of the real and imaginary parts of the quantity $\mathcal{F}_2^+$ can be inferred from Fig. 3 which displays the energy dependence of the

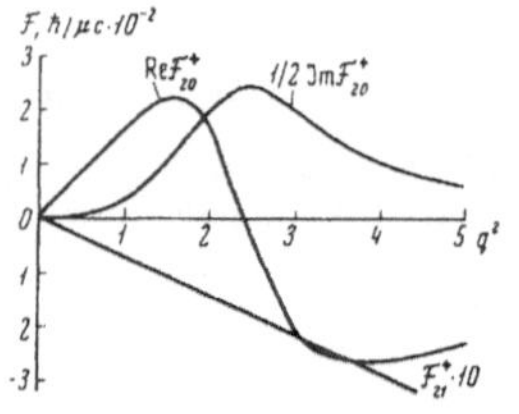

Fig. 3. The first terms of the expansion of the $\mathcal{F}_2^+$ amplitude in Legendre polynomials $\mathcal{F}_{20}^+$, and $\mathcal{F}_{21}^+$ and imaginary part $\operatorname{Im}\mathcal{F}_{20}^+$.

coefficients of the expansion of the amplitude $\mathcal{F}_2{}^+$ in a power series of $\cos\theta_\pi^*$ to the fourth power, i.e., we have [3]:

$$\mathcal{F}_2^+ = \mathcal{F}_{20}^+ + \mathcal{F}_{21}^+ \cos\theta_\pi^*. \tag{II.15}$$

We consider now the differential cross section of π^0 meson photoproduction at a nucleus A. It is well known that the momentum approximation [38, 39] can be used to express the amplitude of the interaction between an incident particle and a nucleus by the amplitude of the interaction between that particle and a single nucleon of the nucleus, with the amplitude averaged over the wave function Ψ of the nucleus

$$U = \int d\varkappa \Psi'(\varkappa) K(\varkappa) T(\varkappa) \Psi(\varkappa) \tag{II.16}$$

(where K denotes a kinematic factor and $\varkappa$, the momentum produced by the relative motion of nucleons in the nucleus). Application of the momentum approximation to the calculation of the cross section of π^0 meson photoproduction at nuclei was considered in detail in [24, 25, 38, 40, 41].

A method based on dispersion relations has been recently used in several papers [42, 43] to calculate relativistic corrections to the momentum approximation. Unfortunately, attempts to develop thereafter a relativistic theory encounter considerable difficulties because the contributions of a large number of complicated diagrams must be taken into consideration.* Noteworthy in this connection is a relativistic approach in which a way of making the momentum approximation relativistic is obtained only from the simplest diagrams [44-46]. For example, an approximate expression for the amplitude of π^0 meson scattering at deuterons was obtained in the work of Cutcovsky [44]; this expression includes the first irregular singularity related to a triangular diagram in the dispersion relations for the transferred momentum. As has been shown, this expression is then the relativistic generalization of the momentum approximation in the case of the asymptotic wave function of the deuteron.

We have employed Cutcovsky's method to examine the elastic coherent photoproduction of π^0 mesons at nuclei [45]. For the sake of simplicity, we restricted ourselves to nuclei with spin s = 0 (and the isospin I = 0) and generalized the method. To do this, we considered spin particles in an intermediate state and introduced the influence of remote singularities by replacing the asymptotic form factor with a more accurate form factor.

The amplitude of π^0 meson photoproduction at a nucleus A with spin zero can be written in the form

$$\langle p_2 q \mid S \mid p_1, k\rangle = \frac{1}{(2\pi)^2} \frac{\delta(p_1 + k - p_2 - q)}{4\sqrt{p_{10}p_{20}q_0k}} U. \tag{II.17}$$

* A systematic introduction to nonrelativistic Feynman diagrams and their application to direct nuclear reactions was presented in I. S. Shapiro's work [47].

It follows from relativistic and gradient-type invariance that

$$U = \varepsilon_{\mu\nu\lambda\sigma}\, \epsilon_\mu q_\nu p_{1\lambda} p_{2\sigma} \Phi(s, t), \qquad (II.18)$$

where k, q, p_1, and p_2 denote the four-dimensional momenta of the photon, meson, and initial and final nuclei, respectively; **ε** denotes the polarization vector of the photon; $\varepsilon_{\mu\nu\lambda\sigma}$ denotes an antisymmetric tensor, and $s = (p_1 + k)^2$ and $t = (q - k)^2$. In the center-of-mass system of $A + \pi$, Eq. (II.18) can be rewritten in the form

$$U = (\tilde{q}\,[\tilde{k}\tilde{\epsilon}])\,\mathcal{F}, \quad \mathcal{F} = qkW\Phi, \qquad (II.19)$$

where

$$q \equiv |\mathbf{q}|, \quad k \equiv |\mathbf{k}|, \quad \tilde{q} = \frac{\mathbf{q}}{q}, \quad \tilde{k} = \frac{\mathbf{k}}{k}$$

denote the total energies in the center-of-mass system of $A + \pi$.

The differential cross section of meson photoproduction by unpolarized photons can be expressed by $\mathcal{F}$

$$\frac{d\sigma}{d\Omega} = \frac{q}{k}\,\frac{|\mathcal{F}|^2}{(8\pi W)^2}\,\frac{\sin\theta_\pi^*}{2}, \quad \theta_\pi^* = (\widehat{q, k}). \qquad (II.20)$$

As has been shown by Cutcovsky [44], taking into account the irregular singularity corresponding to the diagram of Fig. 4 is equivalent to using the momentum approximation. The amplitude can in this case be written in the form

$$U = AF_0(t) \oint \frac{d\sigma}{2\pi i}\,\frac{R}{\sqrt{(\sigma_+ - \sigma)(\sigma_- - \sigma)}}, \qquad (II.21)$$

where A denotes the number of nucleons in the nucleus; $F_0(t)$ denotes the form factor of the nucleus (for details see below); $\sigma = (n_1 + k)^2$ is an invariant quantity which in the center-of-mass system of $N + \pi$ has the meaning of the square of the total energy $\omega = n_{10} + k$; $\sigma_\pm(s, t)$ denotes the largest and smallest σ values which, at given s and t, can be transferred in agreement with the conservation laws (upper block of the diagram); and R is a function which depends upon the amplitude T of π^0 meson photoproduction at the nucleon and will be defined below. The integration path of Eq. (II.21) comprises the point $\sigma_\pm$. We obtain from the geometry of the dual graph

$$\sigma_\pm = \sigma_0 \pm \Delta, \qquad (II.22)$$

where

$$\sigma_0 = s + m^2 - \frac{2}{4M^2 - t}(M^2 + m^2 - N^2)\left(s + M^2 - \frac{\mu^2}{2}\right), \qquad (II.23)$$

$$\Delta = \frac{2m}{4m^2 - t}\sqrt{t^{-1}(t - t_0)\Lambda(s, t)}. \qquad (II.24)$$

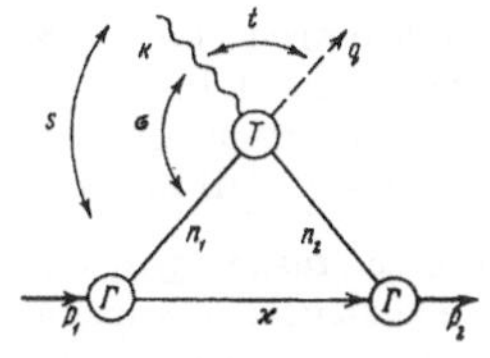

Fig. 4

In this formula,

$$t_0 = m^{-2}[(m+N) - M^2][M^2 - (m-N)^2] \quad (II.25)$$

denotes the irregular singularity corresponding to the diagram; M, N, and m denote the masses of nucleus A, nucleon N, and core fragment A', respectively, so that $m + N - M = \varepsilon$ denotes the binding energy;

$$\Lambda(s, t) = st(2M^2 + \mu^2 - s - t) - tM^2(M^2 - \mu^2) - \mu^4 M^2, \quad (II.26)$$

where the equation $\Lambda(s, t) = 0$ defines the limit of the range of the reaction $A(\gamma, \pi^0)A'$. We have in the center-of-mass system of $A + \pi$:

$$\Delta = (4M^2 - t)^{-1} 4mkqW \sin\theta_\pi^* \sqrt{t^{-1}(t - t_0)}. \quad (II.27)$$

The following remark must be made. In real calculations, the expression for the amplitude U in the momentum approximation of Eq. (II.16) is normally simplified to the so-called form factor approximation by putting K and T before the integral at some average value $\varkappa = \varkappa_{av}$:

$$U_{av} = K(\varkappa_{av})\, T(\varkappa_{av})\, F. \quad (II.28)$$

An ambiguity arises from the selection of average values for K and T. The relative motion of the nucleons is ignored in the usual approximation. As has been shown, when π^0 mesons are scattered at deuterons (this effect has not been studied in photoproduction), this approximation means satisfactory agreement between theoretical differential cross sections and experimental cross sections at low meson energies, i.e., far from the 3–3 resonance. But, as outlined in [48–50], the theoretical cross section becomes with increasing energies much greater than the experimental cross section at small scattering angles. This situation seems to result from ignoring the relative nucleon motion [49].

An attempt to introduce the relative motion of the nucleons in the case of π–d scattering has been made in [51]. K and T were expanded for this amplitude in a power series of $(\varkappa_{av} - \varkappa)$; The optimal $\varkappa_{av}$ value was obtained from the condition that the second expansion term must vanish. When $\varkappa_{av}$ is used in Eq. (II.6), there appears an additional angular dependence which does not improve the agreement with the experimental results, because this additional angular dependence does not modify the cross sections at small angles but increases the cross sections at large angles.

The approximation which was made in deriving Eq. (II.21) is an analog to the transition from Eq. (II.16) to Eq. (II.28) and means that R and $\sigma_\pm$ are drawn before the dispersion integral (for $t' \geq t_0 > 0$ which are meaningless in terms of physics) when the transferred momentum $t' = t < 0$, which is meaningful in terms of physics. The form factor approximation is obtained. When the relative motion is taken into account in this approximation of π meson scattering (see Appendix), the cross section of small-angle scattering in the resonance region is reduced (and, consequently, the resonance in the differential cross section of the reaction $\pi + A \rightarrow \pi + A$ is broadened at small angles).

In other words, Cutcovsky's method produces an additional angular dependence of the cross section. It is of definite interest to check this dependence against experimental results in order to test Cutcovsky's model. But we must bear in mind that the additional dependence is distorted by multipole scattering which we will discuss in Chapter 4.

We consider apices Γ formed by the three rays, which correspond to the disintegration of a nucleus A with spin zero into a nucleon N and a nuclear fragment A' with spin 1/2. It is easy to show that the spin part of the apex Γ can be written in the form

$$\Gamma_s = \frac{\hat{p}_1 + M}{2M} \frac{\gamma_5 C}{\sqrt{2}}. \quad (II.29)$$

where $\hat{p} = \gamma_\mu p_\mu$ and C denotes the charge-conjugation operator. In the rest system of the nucleus, Γ_s is reduced to the matrix

$$\frac{1}{2}\begin{pmatrix}\sigma_y & 0\\ 0 & 0\end{pmatrix},$$

which forms the singlet state of A from the two spinors corresponding to N and A'. When we ignore ε relative to C, we can introduce the approximation

$$\Gamma_s \simeq \frac{1}{\sqrt{2}}\gamma_5 C. \tag{II.30}$$

To take into account the first irregular singularity (when all internal particles are situated on the mass surface in the diagram) means to replace an apex Γ by a constant Γ_0, or, to take only the asymptotic form of the wave function of the nucleus

$$\Psi \sim \frac{e^{-\delta r}}{r}, \quad \delta = \sqrt{\frac{2mN}{m+N}}. \tag{II.31}$$

It is generally accepted that the wave function of (II.31) describes only poorly the structure of the nucleus. But when we assume that replacement of the constant apex Γ_0 with the accurate Γ in the diagram (i.e., consideration of the singularities but t_0) affects only the form factor but does not substantially change the T dependence of U (i.e., the form of the function R), we can modify Eq. (II.31) by replacing the "asymptotic" form factor F_0 by some more accurate form factor F which was obtained either from experiments or nuclear models. This replacement can be regarded as an approximate introduction of remote singularities.

The amplitude of π^0 meson photoproduction at the nucleon of Eq. (II.21) is determined as follows:

$$\langle n_2 q \,|\, S \,|\, n_1 k\rangle = \frac{N}{(2\pi)^2}\frac{\delta(n_1 + k - n_2 - q)}{2\sqrt{n_{10} n_{20} q_0 k}}\,\bar{u}(n_2)\,Tu(n_1), \tag{II.32}$$

$$T = \sum_{i=1}^{4} r_i(\gamma, k, q, Q)\,T_i(\sigma, t), \tag{II.33}$$

$$\begin{aligned} r_1 &= \gamma_5 \hat{k}\hat{\epsilon},\\ r_2 &= 2\gamma_5\,[(Q\epsilon)(qk) - (Qk)(q\epsilon)],\\ r_3 &= \gamma_5\,[\hat{k}(q\epsilon) - \hat{\epsilon}(qk)];\\ r_4 &= 2\gamma_5\,[\hat{k}(Q\epsilon) - \hat{\epsilon}(Qk)];\\ Q &= (n_1 + n_1)/2. \end{aligned} \tag{II.34}$$

In the $N-\pi^0$ center-of-mass system, the amplitude is conveniently written in the form

$$\bar{u}(n_2)\,Tu\,(n_1) = \frac{4\pi\omega}{N}\chi_2\{i(\sigma\epsilon)\mathcal{F}_1 + (\sigma\tilde{q}')(\sigma[\tilde{k}'\epsilon])\mathcal{F}_2 + i(\sigma\tilde{k}')(\tilde{q}'\epsilon]\mathcal{F}_3 + i(\sigma\tilde{q}')(\tilde{q}'\epsilon)\mathcal{F}_4\}\chi_1, \tag{II.35}$$

where k' and q' denote the momenta of photon and meson in the center of mass of the N + π system; the amplitudes $\mathcal{F}_i$ are related to T_i by simple relations. The fact that the isospin of the nucleus is zero means that only the isovector paper $T^{(+)}$ of the amplitude will contribute to U.

With Eqs. (II.30) and (II.33), we rewrite R as follows:

$$R = \frac{M}{N}\frac{1}{4mN}\frac{1}{2}\sum_{i=1}^{4} R_i T_i, \tag{II.36}$$

$$R_i = Sp\{r_i(\hat{n}_1 + N)(\hat{\varkappa} + m)(\hat{n}_2 + N)\}. \tag{II.37}$$

M/N of Eq. (II.36) denotes a factor normalizing $\Phi_0(0)$ to unity; the coefficient $(4mN)^{-1}$ accounts for the usual relation between the binding constants of particles A, A', and N when A' and N have spins 0 and 1/2. By calculating the spur and ignoring terms of the order of $(\mu/N)^2$, $(q/N)^2$, and $(k/N)^2$, we obtain (see [45]):

$$R = \varepsilon_{\mu\nu\lambda\sigma}\epsilon_{\mu}q_{\nu}p_{1\lambda}p_{2\sigma}\frac{8\pi\mathcal{F}_2^+(\sigma, t)}{q'(\sigma)\,k'(\sigma)}\left[1+\frac{k(M-N)}{MN}\right],$$
$$\Phi = 8\pi AF_0(t)\left[1+\frac{k(M-N)}{M+N}\right]\oint\frac{\mathcal{F}_2^+(\sigma, t)}{q'(\sigma)\,k'(\sigma)}\frac{d\sigma}{\sqrt{(\sigma_+-\sigma)(\sigma_--\sigma)}}. \tag{II.38}$$

As can be shown by calculations, in the range of low photon energies we can set with good accuracy $\sigma_+ \simeq \sigma_- = \sigma_0$ (see Appendix), whereupon the integral of Eq. (II.38) can be easily calculated

$$\Phi = 8\pi AF_0(t)\frac{\mathcal{F}_2^+(\sigma_0 t)}{q'(\sigma_0)\,k'(\sigma_0)}\left[1+\frac{k(M-N)}{MN}\right]. \tag{II.39}$$

By substituting Eq. (II.39) into Eq. (II.20), we obtain the following expression for the differential cross section of meson photoproduction:

$$\frac{d\sigma}{d\Omega} = \frac{A^2}{2}\frac{q}{k}F^2(t)\sin^2\theta_\pi^*\,|\mathcal{F}_2^{(+)}|^2 L, \tag{II.40}$$

where

$$L = \left(\frac{qk}{q'k'}\right)^2\left[1+\frac{k(M-N)}{MN}\right]^2. \tag{II.41}$$

σ_0 is in these formulas related to s and t by Eq. (II.23) which can be approximately stated (with $\varepsilon \ll N$ and $t \ll 4N^2$) as follows:

$$\sigma_0 \simeq \sigma_0' = 2NE_\gamma + N_2 + \frac{M-N}{2M}\mu^2. \tag{II.42}$$

where E_γ denotes the energy of the γ quantum in the lab system of the nucleus. We note that

$$\sigma_0 = \frac{1}{2}(\sigma' + \sigma''). \tag{II.43}$$

where $\sigma' = 2NE_\gamma + N^2$ and $\sigma'' = \sigma' + \frac{M-N}{M}M^2$ denotes the σ values which are obtained for a particular s (or E_γ) under the assumption that the nucleon is not bound in the nucleus and moves, before and after the reaction, with the same velocity as the nucleus.

We can use the wave functions proposed in [22, 23]

$$\Psi_s = \text{const}\cdot\exp\left[-\alpha\left(\sum_{i>j}^{4} r_{ij}^2\right)^{\frac{1}{2}}\right]. \tag{II.44}$$

as radial wave functions of He^4 in the initial and final states. The form factor can therefore be written in the form

$$F_{He}(Qr) = \frac{1}{\left[1+\frac{3Q^2}{64\alpha^2}\right]^5}. \tag{II.45}$$

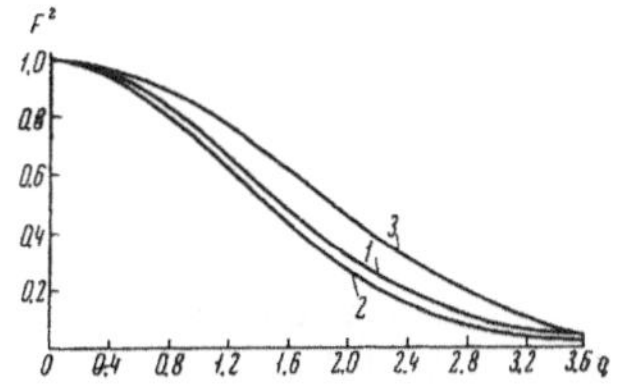

Fig. 5. Dependence of the form factor F_{He} upon the transferred momentum. 1) Eq. (II.45); 2) Eq. (II.46); 3) Eq. (II.47) ($\langle r\rangle = 1.4$ F was used in calculating F_{He}).

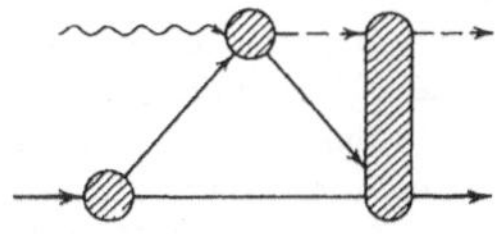

Fig. 6

where $\alpha = \frac{32}{45}\langle r_{He}\rangle^2$. The dependence of F_{He} upon the transferred momentum is shown in Fig. 5 (curve 1).

A form factor like a Gauss function (Fig. 5, curve 2) was used in [52]:

$$F_{He}(Qr) = e^{-\frac{Q^2a^2}{6}}. \qquad (II.46)$$

where a denotes the mean-square radius in $\hbar/\mu c$ units.

When we derived Eq. (II.40), we disregarded the interaction between the meson and the target nucleus. This interaction is described by the diagram of Fig. 6, which refers to multiple scattering of the meson generated inside the nucleus. As far as our approximation is concerned, this contribution is conveniently introduced by an appropriate modification of the form factor of the nucleus by means of the distorted wave method. A "distorted" wave, which is the solution of the Schrödinger equation for π meson scattering at the nucleus, is used in place of a plane wave to describe the emitted meson. The Schrödinger equation is formulated with the optical potential.* As has been shown in [40], in the simple nuclear model of a homogeneous sphere with radius R the optical potential results in a replacement of the form factor

$$F_{hom} = \frac{3}{p^2R^2}\left(\frac{\sin pR}{pR} - \cos pR\right) \qquad (II.47)$$

by the distorted form factor

$$\tilde{F}_{hom} = \int_0^1 \frac{3r}{2}\mathcal{J}_0(\beta\gamma)\frac{K(r) + iL(r)}{R(\gamma + i\alpha - i\delta)}dr, \qquad (II.48)$$

where

$$\begin{gathered}\alpha = k\cos\theta - q; \quad \beta = k\sin\theta; \quad \gamma = {}^1/_2\lambda;\\ \delta = q(n-1);\\ K[(1-r^2)^{1/2}] = \cos\alpha Rr - e^{2\gamma Rr}\cos[(\alpha - 2\delta)Rr];\\ L[(1-r^2)^{1/2}] = \sin\alpha Rr + e^{-2\gamma Rr}\sin[(\alpha - 2\delta)Rr];\end{gathered} \qquad (II.49)$$

* It was shown in [53] that even in the case of nuclei with few nucleons, such as He^4, the optical potential can be used to describe π meson scattering.

n denotes the diffraction coefficient of the nucleus, and λ denotes the mean free path of the mesons in nuclear matter.

In order to take into consideration multiple scattering, we introduce the factor

$$\xi(E_\gamma \theta_\pi^*) = |\tilde{F}|^2/F^2. \quad \text{(II.49')}$$

Naturally, the description of real nuclei (e.g., He^4) with the homogeneous sphere model is not always adequate. Figure 5 includes for comparison the form factors of Eq. (II.47) (curve 3) and of Eqs. (II.45) and (II.46) (curves 1 and 2). Curve 3 differs evidently to a large extent from curves 1 and 2 (which are similar and describe the He^4 nucleus rather well). But since the ratio of form factors $\tilde{F}$ and F calculated for the same model appears in ξ, there is reason to hope that the difference in the absolute value of the form factors only insignificantly influence ξ. Of greater importance in this respect is the difference in the q^2 dependence of the form factors, because this difference may increase the inaccuracies in calculations of the effect. For example, assuming that at $\theta_\pi^* = 90°$ and $E_\gamma = 180$ MeV multiple scattering at He^4 was introduced with 20% accuracy by ξ calculated with Eqs. (II.47)-(II.49), the error increases to about 40% at the same angle and at the energy $E_\gamma = 300$ MeV.

CHAPTER III

Experiments

Apparatus

Elastic photoproduction of π^0 mesons at He^4 nuclei can be identified, at known energies of the primary γ quanta, by recording either the recoil nuclei or the π^0 mesons. Both the energy of these reaction products and the angle under which they are emitted relative to the incident photon beam must be recorded.

The first recording technique was used in the experiments of [28] and [26]. The recoil nuclei were recorded with thick nuclear emulsions in [28]. The energy range between 240 MeV and 320 MeV of the primary photons was studied. Inelastic π^0 meson production under emission of He^3 nuclei and neutrons occurred in addition to reaction (I.12). The complicated splitting of the He^3 and He^4 nuclei greatly complicated the identification of the reaction and did not permit the authors to obtain reliable results. The problem of He^4 and He^3 splitting was satisfactorily solved with the aid of a double ionization chamber [27] which allowed determinations of both the energy and emission angle of He^4 and He^3. In the first case, a section inside the container filled with gaseous He^4 under normal pressure was used as the target; the container held also the nuclear emulsions. A gas target was used in the second case under a pressure of 2-5 atm. Due to the low energy of the recoil nuclei, this technique cannot be employed at lower primary photon energies E_γ.

Goldwasser and Koester [16, 54] used π^0 meson recordings to identify reaction (I.12). They did their research work in the energy interval of primary photons between the threshold (137 MeV) and 200 MeV and recorded first π^0 mesons via two γ quanta from the decay and then via one γ quantum. A target with liquid He^4 can be employed and reaction (I.12) can be studied near the threshold though the cross section of the reaction is very small.

We studied in our work π^0 meson photoproduction at He^4 placed in the bremsstrahlung beam of the Physics Institute of the Academy of Sciences of the USSR, the beam having a maximum energy of 264 MeV [55]. The π^0 mesons were recorded via the decay γ quanta. A VGM-1 target [56] with liquid helium was used in the experiment (see Fig. 7). The design of the cham-

Fig. 7. Liquid helium target.

ber (nitrogen shielding, radiation shielding, heat insulation by vacuum) considerably reduced the heat transfer from the external walls to the helium tank and made it possible to perform an experiment during 7-12 h without interrupting the helium influx due to vacuum pressure variations in the chamber.

In contrast to work with hydrogen, several difficulties arise from the use of liquid helium. As can be inferred from Table 2, helium has a lower boiling point than all known liquefied gases; the extremely small heat of evaporation and the low density of helium imply extensive evaporation losses of the liquid even at a small heat influx.

TABLE 2. Parameters of Liquefied Gases [57]

Gas	Boiling point, °C	Density of the liquid, kg/liter	Heat of evaporation of the liquid, kcal/kg	Ratio of the volume of 1 g of gas under standard conditions to the volume of 1 g of liquid under normal pressure
Nitrogen	—195.8	0.808	38.5	645
Hydrogen	—252.8	0.071	7.54	786
Helium	—268.9	0.125	0.71	700

Accordingly, the preparation of the target for filling with the liquid helium is greatly complicated. The target is cooled first by filling with liquid nitrogen and then with liquid hydrogen. After complete evaporation of these liquids, helium is poured in.

The γ quanta produced in the decay of the π^0 mesons were recorded with scintillation telescopes (Fig. 8) which comprised four counters when the π^0 mesons were registered via single γ quanta, and of three counters, when two decay photons were recorded. As in the usual arrangement, each counter comprised a plastic scintillator (diameter 120 mm, thickness 20 mm, p-terphenyl + POPOP in polystyrene) and an FEU-33 photomultiplier with a voltage divider and an anode limiter stage. The scintillator was filled into an aluminum container with approximately 15-mm-thick side walls which served simultaneously as shield against the background of scattered electrons and γ quanta. The face wall had a thickness of about 1 mm. The internal surface of the container was polished and acted as a reflector. The photomultiplier was placed in a cylindrical steel screen which provided simultaneously a magnetic shield from stray fields in the accelerator building. Despite its shortcomings (e.g., small curvature) [58], the FEU-33 photomultiplier proved to be little sensitive to the voltage divider used and could therefore be easily replaced by other photomultipliers of the same type. The relatively high stability of this multiplier type (compared to the FEU-36 multiplier) at medium currents facilitated its use in long (about 120 h) measurements.

The characteristics of γ telescope operation, i.e., the absolute efficiency of photon recording, were measured in our work with the experimental method proposed in [59].

The energy resolution of the apparatus, which has been described in [59], was about 2%. The efficiency of the telescope is depicted in Fig. 9 and can be expressed by the empirical

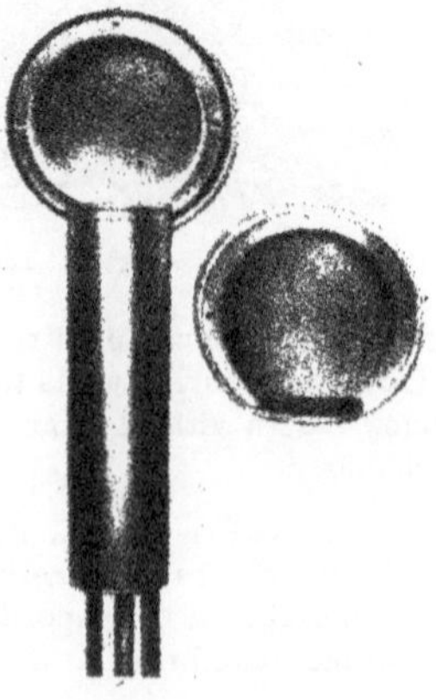

Fig. 8. Scintillation γ telescope.

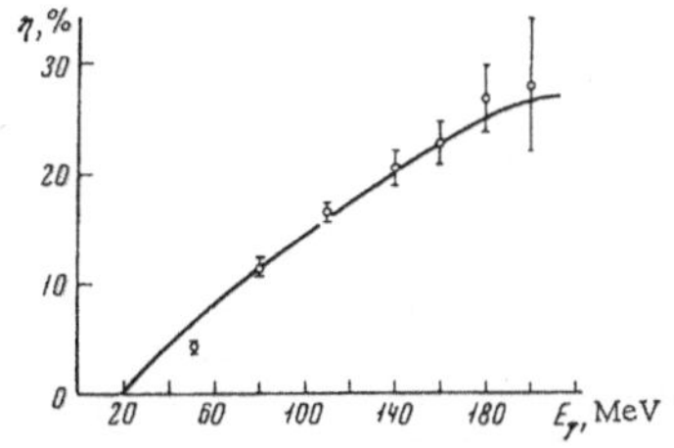

Fig. 9. Energy dependence of the efficiency of the γ telescope. The curve corresponds to Eq. (III.1).

formula

$$\eta(E_\gamma) = \eta_\infty\left[1 - \exp\left(-\frac{E_\gamma - E_n}{E_1}\right)\right], \tag{III.1}$$

where η_∞ denotes the efficiency limit for $E_\gamma \to \infty$; E_1 denotes a parameter which determines the form of the curve; and E_n denotes the threshold of the energy of γ quanta recorded by the telescope. In our case, $E_n = 20$ MeV, $\eta_\infty = 0.45$, and $E_1 = 200$ MeV.

Differential Cross Section of the Photoproduction of γ Quanta Produced in the Decay of π^0 Mesons at Incident Photon Energies between 160 and 260 MeV

The first qualitative experiments [16, 54] and theoretical estimates [24] lead to the conclusion that at γ quantum energies of the interval 150-200 MeV; the main γ quanta flux from the target results from π^0 mesons of the elastic process (I.12). As a matter of fact, the elastic photoproduction processes of π^0 mesons

$$\begin{array}{ll} & E_\pi,\ \text{MeV} \\ \gamma + He^4 \to \pi^0 + H^3 + p & 158, \\ \gamma + He^4 \to \pi^0 + H^3 + n & 159, \\ \gamma + He^4 \to \pi^0 + d^2 + d^2 & 162 \end{array} \tag{III.2}$$

have thresholds which are about 20 MeV higher than the threshold of reaction (I.12). In some energy range of primary photons near the threshold of elastic π^0 meson photoproduction at He^4, there will be a total effect not exceeding the experimental accuracy limits (5-6%). The limits of the range in which this effect occurs are not exactly known.

The contribution of γ quanta due to Compton scattering at He^4 nuclei is in this energy range very small, because the cross section of Compton scattering is 1.5-2 times smaller than that of reaction (I.12). But we must bear in mind that for photon-emission angles smaller than 45° (angle included with the beam), there will appear processes which are related to inelastic photon scattering by the Coulomb field of the nucleus [60].

Thus, our first problem is to establish the energy range in which reaction (I.12) dominates at the particular photon energy.

Let us consider the details of π^0 meson recording via the detection of a single decay photon.

The authors of [61] mentioned for the first time that information on the angular dependence of the cross section of π^0 meson production in the center-of-mass system can be obtained by detection of single decay photons, provided that the cross section can be represented in the form

$$\frac{d\sigma_\pi}{d\Omega} = \sum_{i=0}^{n} A_n \cos^n \theta_\pi^*. \qquad (\text{III}.3)$$

In research on the π^0 meson photoproduction at nucleons, Cocconi and Silverman [62], Koester and Mills [18], and Vasil'kov, Govorkov, and Gol'denstein [63] used this technique. Their work resulted in relations between $d\sigma_\pi/d\Omega$ in the center-of-mass system and the γ quantum yield in the lab system at the angle under consideration and for n = 2 in Eq. (III.3). Kaplon and Yamaguchi [64] considered the general form of the angular distribution of γ quanta produced in the decay of π^0 mesons. But the authors of that work disregarded the real recording efficiency which substantially complicates the calculations (see Appendix).

Angular Resolution and Energy Resolution in the π^0 Meson Recording Technique via Single γ Quanta from the Decay

The energy of a decay γ quantum which is emitted at an angle $\varkappa$ with respect to the direction of π^0 meson emission in the lab system can be calculated with energy and momentum conservation laws and is given by the formula

$$\mathscr{E}_\gamma = \frac{m_\pi c^2}{2\gamma} \frac{1}{(1 - \beta_\pi \cos\chi)}, \qquad (\text{III}.4)$$

in which m_π denotes the mass of the π^0 meson; $\beta_\pi c$ denotes the velocity of the π^0 meson in lab system, and $\gamma = (1 - \beta_\pi^2)^{-1/2}$.

The probability that a γ quantum, which is emitted in the π^0 meson decay with the velocity $\beta_\pi c$ at an angle θ_{π^0} relative to the bremsstrahlung beam, is incident on a telescope situated at the angle θ_γ relative to the beam is

$$P(\chi)\,d\Omega_\gamma = \frac{1}{2\pi} \frac{1 - \beta_\pi^2}{(1 - \beta_\pi \cos\chi)^2}\,d\Omega, \qquad (\text{III}.5)$$

where χ denotes the angle between the direction of π^0 meson motion and the direction of the decay photon incident on the detector. The relation between the angle of π^0 meson emission ($\theta^2_{\pi^0}$), the angle of the decay γ quantum (θ_γ) relative to the bremsstrahlung beam, and the angle χ has the form $\chi = \cos\theta_\gamma \cos\theta_{\pi^0} + \sin\theta_\gamma \sin\theta_{\pi^0} \cos\varphi$ as shown in [65], where φ denotes the angle between the plane of π^0 meson emission and the horizontal ($\varphi = 0$).

This means that at fixed θ_γ and θ_{π^0}, cos χ depends upon φ, i.e., upon the π^0 meson detection efficiency which is related to $\mathscr{E}_\gamma$ in the case of π^0 mesons emitted at an angle θ_π relative to the beam; ε depends also upon φ. After averaging over the angle φ, the probability of recording π^0 mesons emitted at an angle θ_π relative to the direction of the bremsstrahlung beam has the following form in the lab system

$$\Phi(\theta_\gamma, \theta_\pi, E_\pi) = \frac{1}{2\pi} \int_0^{2\pi} \frac{\in(\beta_\pi \varphi)(1 - \beta_\pi)^2\,d\varphi}{[1 - \beta_\pi(\cos\theta_\gamma \cos\theta_\pi + \sin\theta_\gamma \sin\theta_\pi \cos\varphi)]^2}. \qquad (\text{III}.6)$$

Equation (III.6) practically determines the angular distributions of this technique of π^0 meson detection. But we used the quantity

$$W(\theta_\gamma, \theta_\pi) = 2\pi\Phi(\theta_\gamma, \theta_\pi, E_\pi)\sin\theta_\pi \qquad (\text{III}.7)$$

as angular resolution function, because the cross section of γ quantum production by π^0 mesons

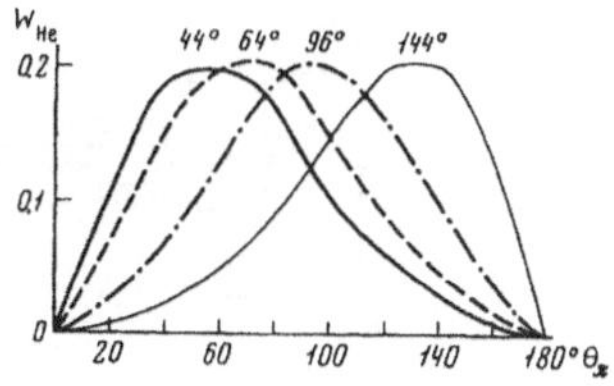

Fig. 10. Angular resolution function of π^0 meson recordings via detection of single decay photons emitted under the angles θ_γ = 44°, 64°, 96°, and 144° relative to the bremsstrahlung beam in the lab system.

has the form

$$\frac{d\sigma_\gamma}{d\Omega}(\theta_\gamma) = \int_0^\pi \frac{d\sigma_\pi}{d\Omega} W(\theta_\gamma, \theta_\pi)\, d\theta_\pi. \qquad \text{(III.8)}$$

Figure 10 shows the functions $W(\theta_\gamma, \theta_\pi)$ for the angles θ_γ = 44 °, 64°, 96°, and 144° and the energy $\overline{E}_\gamma$ = 159 MeV of the bremsstrahlung beam.

As can be inferred from Fig. 10, the angular resolution of this technique is characterized by a rather broad distribution. Nevertheless, the coefficients of the expansion of γ in terms of powers of cos θ_π^* can be determined from the measured angular distribution of photons produced in the decay of π^0 mesons [see Eq. (III.3)]. The Appendix outlines in detail the expressions for solving $d\sigma_\pi / d\Omega$ from $[d\sigma_\gamma/d\Omega]_{He}$.

No energy resolution related to the kinematics of the process under consideration exists in the proposed method of π^0 meson recordings, because the telescope detects at an angle θ_γ decay photons from π^0 mesons having some energy above the detection threshold E_n.

The maximum energy of the π^0 mesons which can be detected by the apparatus depends in this case upon the maximum energy of the photons in the bremsstrahlung beam. In order to separate mesons which were produced by primary γ quanta in a certain energy interval, the yields must be measured at least at the two maximum beam energies which correspond to the beginning and the end of the energy interval of interest.

When a photon beam with time "stretching" is used, the number of energy intervals can be increased and the yield curve of the γ quanta emitted from the target can be directly measured in a single experiment. The energy resolution is then determined by the width of the energy interval in which a constant yield can be assumed [66].

Block Scheme of the Apparatus

The electrons were deflected to the internal target of the synchrotron by switching off the accelerating electric field while the magnetic field strength still increases. When the electric field is switched off gradually, i.e., when the accelerating voltage in the resonator is slowly reduced, particles are gradually ejected onto the target and the bremsstrahlung beam is "stretched." This method allows smooth changes in the maximum energy of the bremsstrahlung spectrum during 3000 μsec in place of the 25 μsec of usual accelerator operation. While

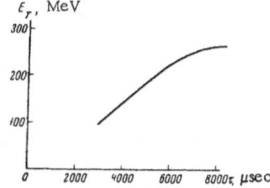

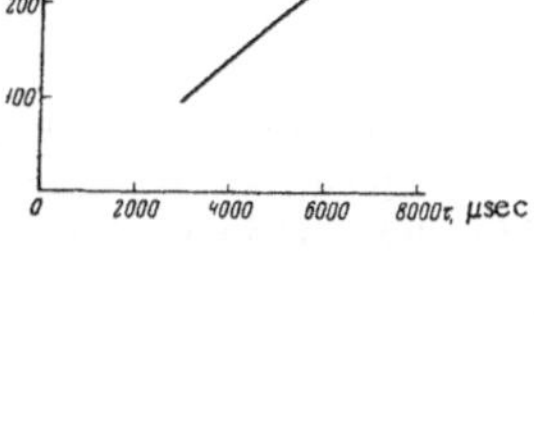

Fig. 11. Dependence of the energy of the accelerated electrons upon the acceleration time.

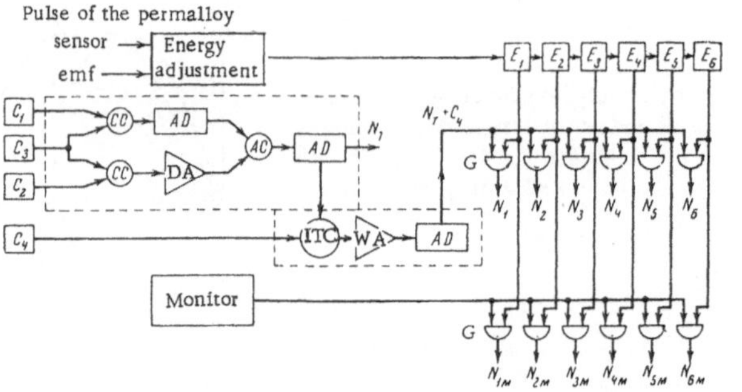

Fig. 12. Block diagram of the apparatus for recording π^0 mesons via single decay photons. C_1, C_2, C_3, C_4) scintillation counters; CC) coincidence circuit; AC) anticoincidence circuit; AD) amplifier discriminator; DA) distributed amplifier; ITC) intertelescope coincidence circuit; WA) wide-band amplifier; E_1–E_6) pulse shaping circuits for the energies E_1–E_6; G) gates.

the magnetic field strength increases, there exists a unique relation between the energy of the accelerated electrons and the time during which the electrons are on their orbits. Figure 11 shows this relation for the energy $E_\gamma = 260$ MeV.

Thus, by analyzing the effect as a function of time within the stretched interval, we can determine the energy dependence. Since the maximum energy of the photons in the bremsstrahlung beam is given by the magnetic field strength on the orbits in the target area, the moment at which the time analyzer is triggered must be related to the actual field strength.

Let us mention another important detail of this technique. The photon density as a function of time is inhomogeneous in the "stretched" bremsstrahlung beam. This effect is related to the form of the rear front of the high-frequency pulse applied to the resonator and to the acceleration effect. The intensity distribution in the beam fluctuates all the time during a series of measurements. In order to measure during a pulse the relative average distribution of the intensity of the primary photon beam, a differential intensity monitor must be employed.

The block diagram of this apparatus (Fig. 12) shows 1) a γ telescope and a sampling circuit for the γ quanta which are related to the effect under consideration; 2) a time-selection circuit for energy analysis of the effect; 3) a differential monitor counter with an amplifier and discriminator channel, which responds only to γ quanta or relativistic electrons from the bremsstrahlung beam of the synchrotron; 4) a time-selection circuit for the analysis of the relative intensity flux in the stretched energy interval; and 5) a pulse generator for controlling the time selection circuit; this generator can be synchronized by a sensor with the magnetic field strength in the target region.

The Telescope for Detecting γ Quanta and the Related Electronic Equipment

The γ telescope consisted of four counters (C_1, C_2, C_3, C_4) of the type described above. A lead converter was inserted between C_1 and C_2. The electron–positron pairs generated in the converter were recorded with counters C_2, C_3, and C_4. Counter C_1 was used to trigger the anticoincidence circuit. The pulses derived from the outputs of counters C_2 and C_3 were limited and then applied to a fast diode–capacitor coincidence circuit (CC) [67] with a resolving time $\tau = 4$ nsec and a sampling coefficient of 6-7. The pulses corresponding to the coincidence signals from C_2 and C_3 were thereafter amplified with a distributed amplifier (amplification coefficient 20-40 at a recovery time of 3 nsec) and fed into an anticoincidence circuit (AC). The blocking pulse of the anticoincidence circuit was generated with an analog fast coincidence circuit CC triggered by the pulses from C_1 and C_3 and conventional amplifier stages with a recovery time of 10-15 nsec. This recovery time is dictated by the need for increasing the efficiency of the anticoincidence circuit. The anticoincidence circuit has the resolution time $\tau = 0.2$ μsec and an efficiency of almost 100%. The efficiency of the anticoincidence circuit was experimentally determined with monochromatic electrons.

Fast-response low-threshold discriminators in a balanced bridge circuit [68] were used in channels $C_1 + C_3$ and $C_2 + C_3$. A temperature compensation, which greatly improved the response-point stability of the discriminators [69] was introduced in the operating discriminator models with semiconductor diodes.

In order to facilitate the adjustment of channels $C_1 + C_3$ and $C_2 + C_3$, each of the channels was provided with an output trigger connected to a scaling device. In the coincidence and discriminator circuits we used semiconductor diodes in which the times required for establishing and restoring the forward and backward resistances were investigated and which were selected according to the method of [70]. These diodes had therefore improved pulse characteristics. Wide-band amplifiers were used in the channels. It became therefore possible to obtain pulses with minimum distortion at the output of the telescope circuit so that these pulses could be used in the ensuing fast coincidence circuits. A detailed description of the circuit (Fig. 13) has been given in [71, 72].

The conditions of the experiment called for an additional counter C_4 (indicated by the dashed line in Fig. 17). The pulse from the output of this counter was applied to an intertelescope double coincidence circuit (ITC).* A pulse from the "fast" output of the telescope circuit was applied to the other input of this circuit. A diode–capacitor coincidence circuit was used as sampling device in the intertelescope circuit, but a pulse shaping to about 10-nsec-long pulses was introduced at the output. The resolving time was in this case determined by the duration of the pulse from the "fast" output and amounted to 0.01-0.02 μsec during actual operation. The amplified pulse, which after the sampling element corresponds to $C_2 + C_3 + C_4 - C_1$ was applied to a fast response integral discriminator [69] with a secondary emission tube. Stability of the response threshold was given particular attention in this circuit element. A pulse with a duration of 1-3 μsec was obtained at the output of the trigger circuit following the discriminator and applied to the input of the time analyzer.

The coincidence circuit together with the intertelescope circuit eliminated almost completely random coincidences which may originate from large overloads of the counters and insufficient resolution of the coincidence circuits.

Figure 14 shows the time analyzer circuit. The principal circuit element is the selection-pulse generator which consists of monovibrators triggered in succession one after the other. The number of these monovibrators corresponded to the number of energy channels, in the particular case six. The width of each channel implied simultaneous coarse adjustment of the pulse duration. Independent correction of the width of each channel was additionally provided. This correction is necessary since the characteristics of the various tubes vary and no strictly linear relationship exists between the photon energy E_γ and the time of electron ac-

* This term stems from the two-photon π^0 meson detection technique in which pulses of two different telescopes are applied to the circuit.

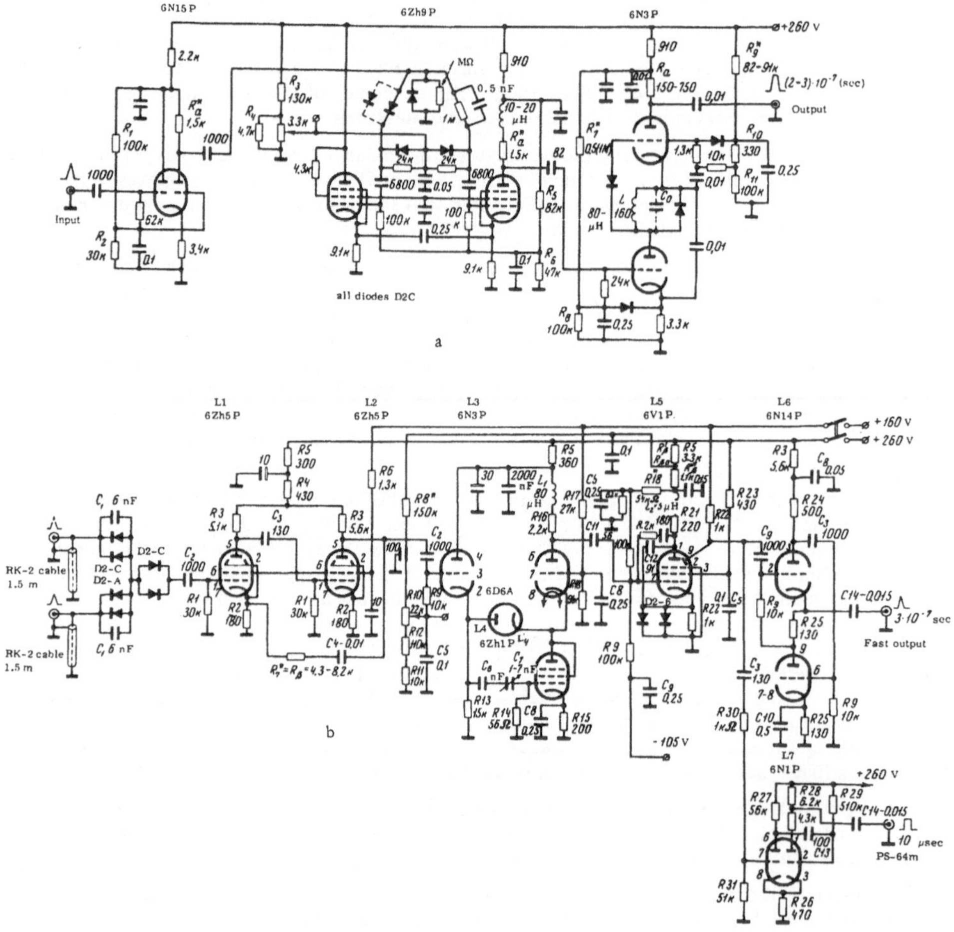

Fig. 13. a) Fast coincidence and anticoincidence circuits with discriminators and low-threshold trigger circuits; b) coincidence circuit and discriminator with sliding threshold.

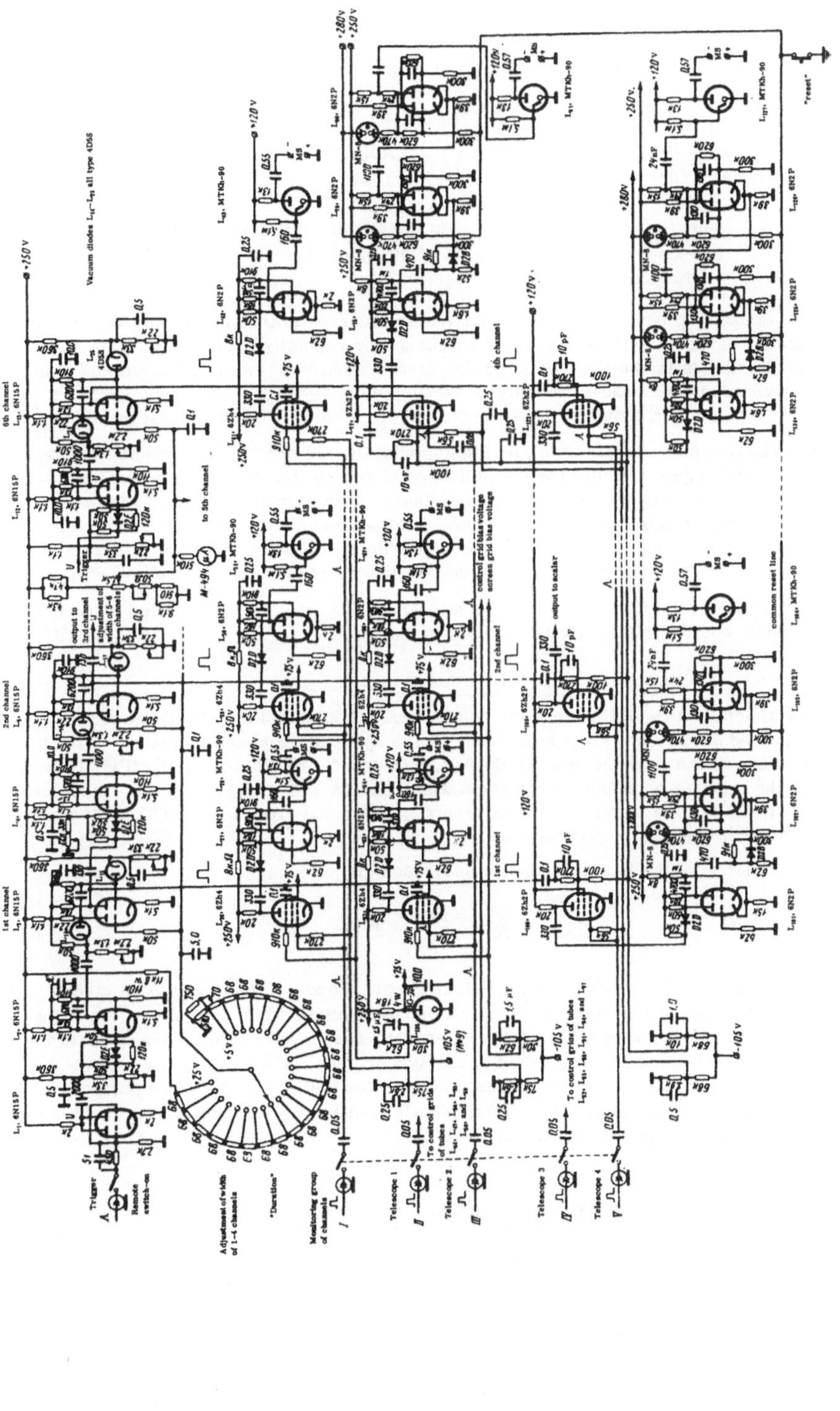

Fig. 14. Time analyzer.

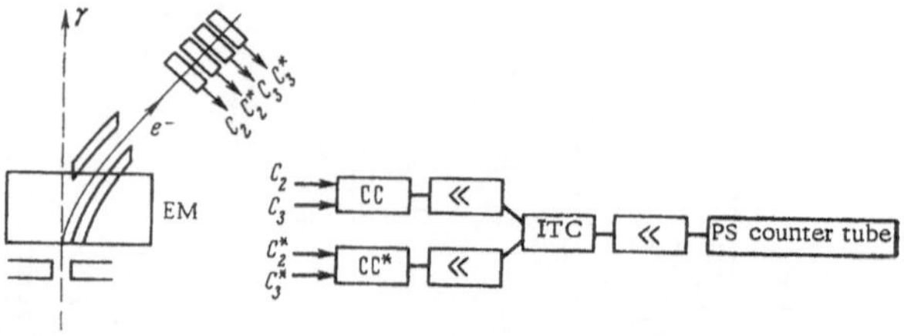

Fig. 15. Setup for determining the absolute efficiency of the γ telescopes and block diagram of the electronic equipment. EM electromagnet.

celeration. Stability received once more particular attention in the case of this circuit element. Stability of the selection pulses was increased by replacing the semiconductor diodes [71] by vacuum diodes in the time-determining circuit section of the monovibrators. Compared to previous circuits, the reliability of successive triggering of the monovibrators was improved by decoupling the monovibrators with the aid of additional pulse shaping amplifiers.

The selection pulse obtained in this fashion with the aid of time-determining monovibrators controlled gate circuits for the pulses produced by the effect under consideration and for the pulses characterizing the intensity distribution in the stretched section. The pulses produced by the effect under consideration were "redistributed" over the channels corresponding to the various energy intervals of the bremsstrahlung spectra. The pulses characterizing the intensity flux in the stretched section were similarly distributed.

In the actual equipment of the experiments, three identical sets were used since measurements were made simultaneously at three angles θ_γ.

Determination of the Efficiency of the Electronic Circuits

The efficiency of the intertelescope coincidence circuits and the efficiency of the time analyzer must be established after adjustment of the apparatus and determination of the absolute efficiency of the γ telescope. To do this, four counters C_2C_3 and $C_2^*C_3^*$ (Fig. 15) were inserted in the monochromatic electron beam which had an $\pm 2\%$ energy spread. The pulses which came from the fast outputs of the telescope coincidence circuit for each counter pair were applied to the intertelescope circuit.

The efficiency of the intertelescope coincidence circuit was

$$E_{ITC} = \frac{N_{ITC}}{N_{C_2^*C_3^*}} = 98\%. \tag{III.9}$$

The efficiency of the analyzer was measured with the aid of an analog main circuit. The pulses from the output of the intertelescope circuit were applied to one of the inputs of the time analyzer. The triggering of the time analyzer corresponded to the beginning of an intensive pulse in the bremsstrahlung beam of the synchrotron. The same pulse triggered an oscilloscope which displayed a mark corresponding to the end of the stretched period in order to check the duration of the intensive beam pulse. The efficiency of the analyzer and the efficiency of the intertelescope circuit are given by

$$E_{analyzer} = \frac{\sum_{i=1}^{6} (N_i)_{analyzer}}{N_{C_2^*C_3^*}} = 96\%. \tag{III.10}$$

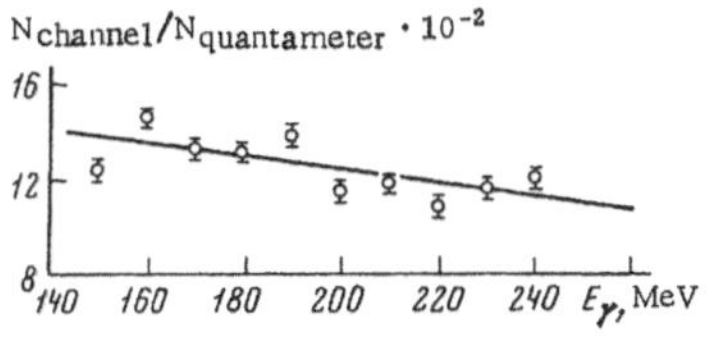

Fig. 16. Relative energy dependence of the sensitivity of the monitoring counter.

The efficiency at the analyzer output is slightly smaller than $E_{\Pi C}$ because the finite time intervals during which the monovibrator pulses rise imply a dead time of the analyzer. A correction was introduced in the final result in order to reach 100% analyzer efficiency.

In addition to the channels used for the time analysis of the pulses supplied by the three γ telescopes, the analyzer included still another set of channels for monitoring the relative intensity distribution in the stretched bremsstrahlung pulse of the synchrotron.

A Cherenkov counter with a radiation cylinder of Plexiglas (diameter 50 mm and height 70 mm) and an FEU-33 photomultiplier was used as the monitor. The pulses from this counter were amplified and applied to the input of the monitor channel of the analyzer. The output of each monitoring channel was provided with a PC-4 counter tube, because the response of the usual mechanical counters was too slow.

In order to measure the energy dependence of the monitor sensitivity, the stretching of the beam pulse was made equal to the width of one channel (10 MeV) and shifted from channel to channel. The number of counts in each channel was measured at a certain number of counts of the quantum meter [73], i.e., for a certain flux of the primary photon energy. The beam intensity was in this case intentionally reduced in order to avoid saturation of the photomultiplier. The energy dependence of the sensitivity of the monitor counter (relative sensitivity of the channels) is represented by the curve of Fig. 16.

The energy limits of the analyzer channels were calibrated before each series of measurements. These limits of the energy intervals could be maintained with an accuracy of 3-5%. The variations had practically no influence upon the average energy in each interval or upon the number of photons entering into cross section calculations.

Execution of the Experiment

The dependence of the yield of γ quanta produced in liquid helium or liquid hydrogen and in the empty target (background) upon the primary photon energy was studied in the experiment. The schematic arrangement of the apparatus is shown in Fig. 17. The photons recorded in the i-th analyzer channel corresponded to the rate of counting the effect produced by primary photons with energy $\bar{E}_\gamma$ at a fixed emission angle θ_γ in the lab system. The yield was referred to a unit reading of the quantum meter ($5.25 \cdot 10^{12}$ MeV/reading), i.e., to a relative energy unit of the bremsstrahlung incident on the i-th channel. The energy depends upon the intensity dis-

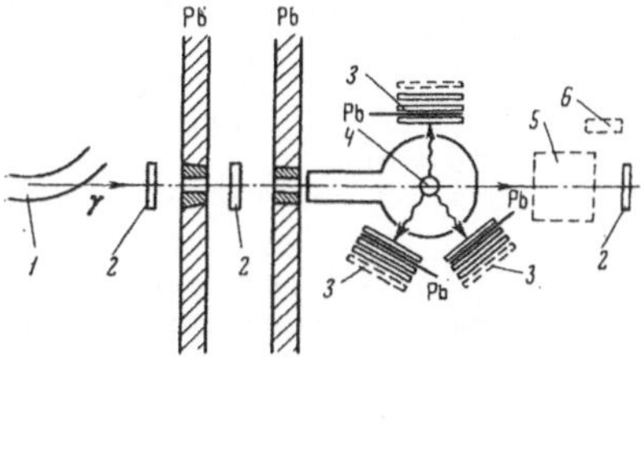

Fig. 17. Position of the apparatus in the accelerator building. 1) Accelerator target; 2) thin-wall monitor chamber; 3) telescopes for detecting γ quanta; 4) active volume of the target; 5) quantum meter; 6) monitoring Cherenkov counter.

tribution over the channels, $\frac{(N_{ik})_{mon}}{\sum_{i=1}^{6}(N_{ik})_{mon}}$ corrected by the inhomogeneity of the energy sensitivity ξ of the monitor as a function of the total bremsstrahlung energy N_q, i.e.,

$$(N_{ik})_q = N_q \frac{(N_{ik})_{mon}\xi_{ik}(E_{\gamma i})}{\sum_{i=1}^{6}(N_{ik})_{mon}\xi_{ik}(E_{\gamma i})}. \tag{III.11}$$

The measured yield is in this case

$$y_{ik} = \frac{N_{ik}}{(N_{ik})_q}, \tag{III.12}$$

where i denotes the number of the measurement in a group of measurements, and k, the number of the group.

Yield measurements at the angles $\theta_\gamma = 64°$, 96°, and 144° were made in two stages at energies between 150 MeV and 260 MeV. The energy ranged from 150 MeV to 210 MeV in the first case. Three series of measurements for He^4 and for H were made. The measurements with hydrogen were designed for the absolute calibration of the apparatus. This method made it possible to analyze the data with regression analysis [74, 75]. We determined at first the average for a group from several measurements and multiplied the average with a weight which corresponds to the bremsstrahlung intensity in the particular channel:

$$\bar{y}_k = \frac{\sum_{i=1}^{6} y_{ik}(N_{ik})_q}{\sum_{i=1}^{6}(N_{ik})_q}, \tag{III.13}$$

where $(N_{ik})_q$ denotes the dose in the particular channel. The error $\bar{y}_k$, which is equal to S_k, was determined from the deviation from the average of the elementary y_{ik}, as well as from the relation between the average and the dispersion; the latter relation is obtained from the transition from a Poisson distribution to a normal distribution. The maximum of the two calculations was then used. Checks whether the averages of the various groups are consistent were made with the t distribution (significance level $\alpha = 5\%$). The consistency of the S_k with the combined estimate of σ was checked for each group of measurements at the 5% significance level.

One series of measurements was considered in the second stage when measurements were made at an energy $E_{\gamma m}$ of the interval 210-260 MeV.

The inverse matrix technique [76] was used to calculate from the experimental y_d values the cross sections $d\sigma_\gamma/d\Omega$ of γ quanta production at He^4 or H nuclei:

$$\frac{d\sigma_\gamma}{d\Omega}(k_m^\Delta) = \frac{k_m}{\eta\int_{(E_\gamma)\lim}^{E_{\gamma m}}\frac{\Phi(E_\gamma,k)}{k}dk}\left[Y_1(E_{\gamma m}) + \frac{B_2(E_{\gamma m},\Delta,E_{\gamma m}-\Delta)}{B_1(E_{\gamma m},\Delta,E_{\gamma m})}Y_2(F_{\gamma m-1}) + \right.$$

$$\left. + \frac{B_3(E_{\gamma m},\Delta,E_{\gamma m}-2\Delta)}{B_1(E_{\gamma m},\Delta,E_{\gamma m})}Y_3(E_{\gamma,m-2}) + \ldots + \frac{B_n(E_{\gamma m},\Delta,E_{\gamma m}-\Delta(n-1))}{B_1(E_{\gamma m},\Delta,E_{\gamma m})}Y^n\right]\frac{\text{cm}^2}{\text{photon}\cdot\text{nucleus}\cdot\text{sr}}, \tag{III.14}$$

where $\eta_{He} = 2.15 \cdot 10^{21}$ nuclei $\cdot$ sr/cm^2; $\eta_H = 4.82 \cdot 10^{21}$ nuclei $\cdot$ sr/cm^2; $E_{\gamma i}$ denotes the upper limit of the photon spectrum in the i-th energy interval; Δ denotes the width of the energy interval (Δ = 10 MeV); $\Phi(F_{\gamma k}) = \frac{137k}{32z^2r_e^2}\sigma_{\text{bremsst}}(E_{\gamma k})$ denotes a function defining the intensity of the spectrum [77]; $\Phi_{\text{str}}(E_{\gamma k})$ denotes the same function with a correction for beam stretching [78]; $Y_i(E_\gamma) = \bar{y}(E_{\gamma i})\int_0^{E_{\gamma i}}\Phi(E_{\gamma ik})\,dk\,\frac{1}{E_m(E_{\gamma i})}$ denotes the reduced (per electron of the accelerator) yield of the reaction in the i-th channel; k denotes the photon energy; $E_m(E_{\gamma i})$ denotes the energy in the photon spectrum having the limit energy $E_{\gamma i}$ (corresponding to a single indication of the quantum meter); and k_m^Δ denotes the average energy of the photons which is defined as follows in the interval under consideration:

$$k_m^\Delta = \frac{\int_{(E_\gamma)\lim}^{E_{\gamma m}}\Phi_{\text{str}}(E_\gamma, k)\,dk}{\int_{(E_\gamma)\lim}^{E_{\gamma m}}\Phi_{\text{str}}(E_\gamma\ k)\frac{dk}{k}} \quad \text{for} \quad E_{\gamma i} = E_{\gamma m}, \tag{III.15}$$

$$k_{m-p}^\Delta = k_m^\Delta - (m-p)\Delta; \quad (p = m-1, m-2, \ldots, 2, 1) \quad \text{for} \quad E_{\gamma i} < E_{\gamma m}.$$

$$B_1 = \frac{1}{\Phi_{\text{str}}(E_{\gamma m}, k_m^\Delta)}; \quad B_2 = -B_1\frac{\Phi_{\text{str}}(E_{\gamma m}, k_{m-1}^\Delta)}{\Phi_{\text{str}}(E_{\gamma m}-\Delta, k_{m-1}^\Delta)};$$

$$B_3 = -B_1\frac{\Phi_{\text{str}}(E_{\gamma m}, k_{m-2}^\Delta)}{\Phi_{\text{str}}(E_{\gamma m}-2\Delta, k_{m-2}^\Delta)} - B_2\frac{\Phi_{\text{str}}(E_{\gamma m}-\Delta, k_{m-2}^\Delta)}{\Phi_{\text{str}}(E_{\gamma m}-2\Delta, k_{m-2}^\Delta)}; \tag{III.16}$$

$$\cdots\cdots\cdots\cdots\cdots\cdots\cdots\cdots$$

$$B_n = -B_1\frac{\Phi_{\text{str}}(E_{\gamma m}, k_{m-(n-1)}^\Delta)}{\Phi_{\text{str}}(E_{\gamma m}-(n-1)\Delta, k_{m-(n-1)}^\Delta)} - B_2\frac{\Phi_{\text{str}}(E_{\gamma m}-\Delta, k_{m-(n-1)}^\Delta)}{\Phi_{\text{str}}(E_{\gamma m}-(n-1)\Delta, k_{m-(n-1)}^\Delta)} - \cdots$$
$$\cdots - B_{n-1}\frac{\Phi_{\text{str}}(E_{\gamma m}-(n-2)\Delta, k_{m-(n-1)}^\Delta)}{\Phi_{\text{str}}(E_{\gamma m}-(n-1)\Delta, k_{m-(n-1)}^\Delta)}.$$

When the cross section was determined for the next interval, the set of coefficients B_i was calculated once more, because $E_{\gamma m}$ in this case. The value $(E_\gamma)_{\lim}$ for i = m is determined by the half-width of the difference in the spectra, $\frac{\Phi_{\text{str}}(E_{\gamma m}, k)}{k}$ and $\frac{\Phi_{\text{str}}(E_{\gamma m}-\Delta, k)}{k}$. The following value $(E_\gamma)_{\lim}$ of the limit energy is obtained as $(E_{\gamma i})_{\lim} = (E_{\gamma m})_{\lim} - (m-p)\Delta$. Figure 18 depicts the relation between $E_{\gamma m}$, $E_{\gamma i}$, $(E_{\gamma i})_{\lim}$ and k_m^Δ. Figures 19 and 20 depict the energy dependence of the differential cross sections $\frac{d\sigma_\gamma}{d\Omega}(\theta_\gamma)$ of photon production by the decay of π^0 mesons; the values refer to helium and hydrogen and were obtained with the above method at the three angles θ_γ = 64°, 96°, and 144°. These $\frac{d\sigma}{d\Omega}(\theta_\gamma)$ values were obtained under

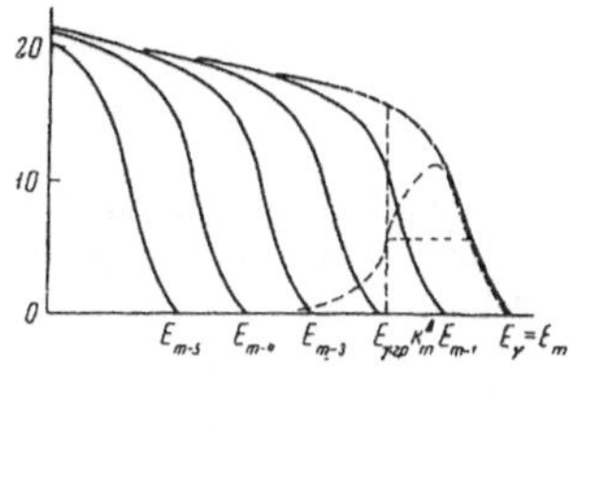

Fig. 18. Relation between the quantities $E_{\gamma m}$, $E_{\gamma i}$, $(E_{\gamma i})_{\lim}$, k_m^Δ.

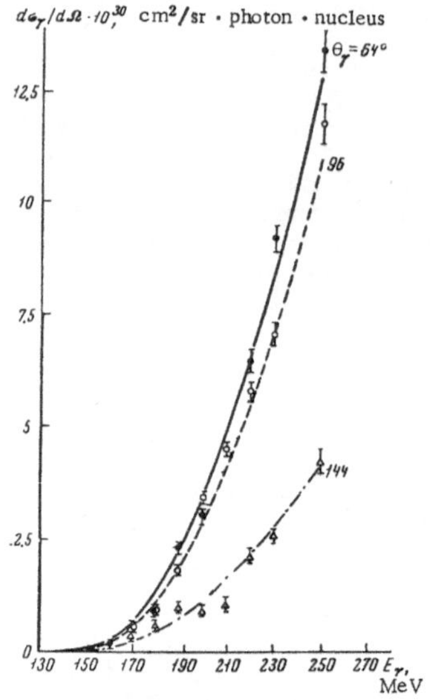

Fig. 19. Cross section of the formation of γ quanta in the decay of π^0 mesons from the reaction $\gamma + He^4 \rightarrow \pi^0 + He^4$. The curves were plotted with the least-square method.

the assumption that the γ quanta of the decay of π^0 mesons were recorded with 100% efficiency. The real efficiency was introduced in the ensuing data processing when the cross sections of π^0 meson photoproduction at H and He were calculated in the center-of-mass system. The solid curves of Figs. 19 and 20 were plotted with the least-square method. The curves were calculated on a computer using a standard program compiled according to [74].

Only statistical errors of the quantity $d\sigma_\gamma/d\Omega$ were taken into consideration in this calculation.

Measurements made on a hydrogen-filled target were used for absolute calibration of the setup as a whole, as in the preceding work [78]. As is known from [79], the differential

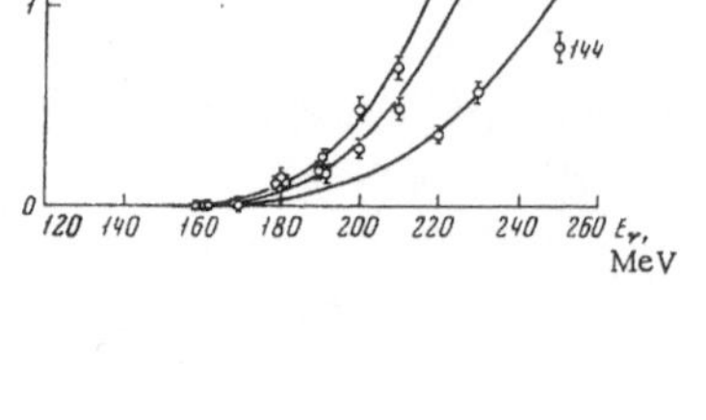

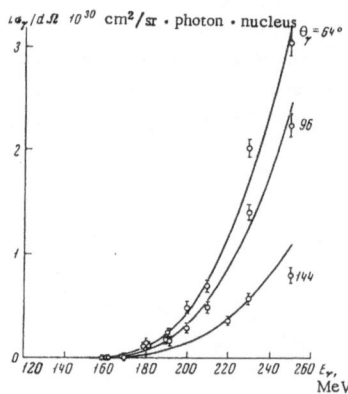

Fig. 20. Cross section of the formation of γ quanta in the decay of π^0 mesons from the reaction $\gamma + p \rightarrow \pi^0 + p$.

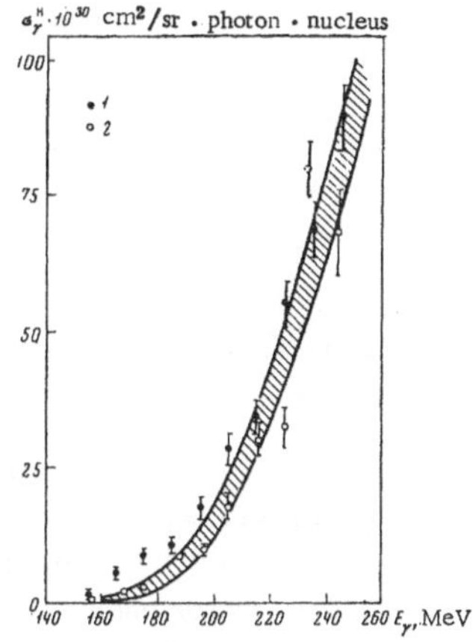

Fig. 21. Energy dependence of the total cross section of π^0 meson production at hydrogen. The shaded area corresponds to the error interval; 1) data of [15]; 2) data of [18].

cross section of π^0 meson photoproduction at hydrogen can be represented in the form

$$\left[\frac{d\sigma_\pi}{d\Omega}\right]_H = A + B\cos\theta_\pi^* + C\cos^2\theta_\pi^*. \qquad (III.17)$$

in the center-of-mass system. The total cross section has the form

$$\sigma_{tot}^H = 4\pi\left(A + \frac{1}{3}C\right). \qquad (III.18)$$

The quantities A, B, and C were determined from the experimental $\left[\frac{d\sigma_\gamma}{d\Omega}\right]_H$ values for the above-mentioned three angles of γ-quanta emission from meson decay; the method proposed in [61, 63] was used.

The errors which were made in the determination of these coefficients make it impossible in our case to directly compare our coefficients with the available experimental results. But the energy dependence of $\sigma_{tot}^H = f(E_\gamma)$ which we obtained agrees, within the limits of experimental errors, with the results of [18] and [19] (Fig. 21). We can therefore assume that the experimental setup does not introduce substantial systematic errors in the results obtained with He^4 and, moreover, that our $\frac{d\sigma_\gamma}{d\Omega}(\theta_\gamma)$ values can be used to derive the differential cross section of π^0 meson photoproduction at helium.

Differential Cross Section of the Elastic Photoproduction of π^0 Mesons at He^4

Angular Resolution and Energy Resolution of the Technique of π^0 Meson Recording via Two Decay Photons

The technique of recording π^0 mesons via the two decay photons makes it possible to directly obtain differential cross sections of process (I.12) from experiments, the values being valid for some average angles of meson emission and average energies. The kinematics of the reaction, the parameters of both the counters and the target, and the conditions of accelerator operation fully determine the reaction characteristics. Let us consider in detail the kinetics on which the recording technique is based.

When a π^0 meson decays into two photons, a unique relation exists between the angle (ψ) of photon emission in the lab system and the emission angle (χ) of one of the photons in the meson rest system, the angles being referred to the direction of π^0 meson motion in the lab system:

$$\sin\frac{\psi}{2} = (\cos^2\chi^* + \gamma^2\sin^2\chi^*)^{-1/2}. \tag{III.19}$$

This means that there exists some ψ_{min} for $\chi^* = \pi/2$, which is defined by the relation

$$\sin\frac{\psi_{min}}{2} = \gamma^{-1} = \sqrt{1-\beta_\pi^2}, \tag{III.20}$$

where β_π denotes the velocity of the meson in the lab system. The decay is under this condition "symmetric" in the lab system, i.e., $\psi_1 = \psi_2 = \psi/2$.

Thus, at a fixed meson energy, the photons are emitted under an angle which is not smaller than ψ_{min}.

The probability of decay photons being emitted under an angle ψ amounts to

$$P(\psi)\,d\psi = \frac{\sin\psi\, d\psi}{[(1-\cos\psi)\gamma^2 - 2]^{1/2}\beta_\pi\gamma(1-\cos\psi)^{3/2}} \tag{III.21}$$

in the case of monoenergetic π^0 mesons. The maximum of $P(\psi)$ is situated at $\psi = \psi_{min}$, i.e., the maximum probability of a decay in which the photons are emitted at a certain angle ψ is obtained with π^0 mesons having the energy defined by Eq. (III.20). Mesons of lower energies do not decay into γ quanta which are emitted under the particular angle ψ.

All these properties of the two-photon decay of the π^0 meson were used in the program to calculate both energy resolution and angular resolution of the apparatus under the real experimental conditions.

The rate of counting π^0 mesons which are formed in the target by photons of the bremsstrahlung beam (maximum photon energy $E_{\gamma m}$) and are recorded via the two decay photons emitted under the angle ψ can be written in the form

$$y(\psi, \psi_1, E_{\gamma m}) = \eta \int\limits_{E_\gamma}\int\limits_{\Omega_\pi} \frac{d\sigma_\pi}{d\Omega}(E_\gamma, \theta_\pi)\, \mathrm{E}(E_\gamma \Omega_\pi) \frac{\Phi(E_\gamma, E_{\gamma m})}{E_\gamma}\, dE_\gamma\, d\Omega_\pi, \tag{III.22}$$

where ψ_1 denotes the emission angle of one of the γ quanta from the decay of a π^0 meson, the angle being measured from the beam in lab system coordinates; $d\sigma_\pi/d\Omega$ denotes the differential cross section of reaction (I.12); $\Phi(E, E_{\gamma m})/E_\gamma$ denotes the intensity of the bremsstrahlung spectrum per unit energy (spectrum with maximum energy $E_{\gamma m}$ denotes the number of nuclei per cm² active area of the target; η denotes the probability of detecting a π^0 meson which was formed by a γ quantum having its energy in the interval E_γ, E_γ + EdE_γ and emitted in the angular interval Ω_π, $\Omega_\pi + d\Omega_\pi$.

We assume that the differential cross section is essentially constant in the particular interval of primary photon energies $\bar{E}_\gamma$, i.e., the differential cross section can be considered constant for some meson-emission angle $\bar{\theta}_\pi$ and average energy $\bar{E}_\gamma$ in the lab system. We have in this case

$$y(\psi, \psi_1, E_{\gamma m}) = \eta \frac{d\bar{\sigma}}{d\Omega}(\bar{\theta}_\pi, \bar{E}_\gamma) \int\limits_{E_\gamma}\int\limits_{\Omega_\pi} \mathrm{E}(E_\gamma, \Omega_\pi) \frac{\Phi(E_{\gamma m} E_\gamma)}{E_\gamma}\, dE_\gamma\, d\Omega_\pi, \tag{III.23}$$

where

$$\bar{E}_\gamma = \frac{\int\limits_{\Omega_\pi} E_\gamma \, \mathrm{E}(E_\gamma, \Omega_\pi)\, d\Omega_\pi}{\int\limits_{\Omega_\pi} \mathrm{E}(E_\gamma, \Omega_\pi)\, d\Omega_\pi}, \tag{III.24}$$

$$\cos\bar{\theta}_\pi = \frac{\int\limits_{E_\gamma} \cos\theta_\pi \frac{\Phi(E_{\gamma m}, E_\gamma)}{E_\gamma} \mathrm{E}(E_\gamma, \Omega_\pi)\, dk}{\int\limits_{E_\gamma} \frac{\Phi(E_{\gamma m}, E_\gamma)}{E_\gamma} \mathrm{E}(E_\gamma, \Omega_\pi)\, dk}. \tag{III.25}$$

In order to record the π^0 meson, the two decay photons must be emitted within the limits defined by the geometry of the γ telescopes and must be registered. This means that the electron probability must depend (1) on the efficiency of the γ telescope as a function of the energy and the angle; (2) on the transformation of the solid angles in the transition of the rest system of the π^0 meson to the lab system; (3) on the point at which the π^0 mesons are formed in the target; and (4) on the absorption of decay photons in the target before the photons can be detected.

Points 3 and 4 have practically no influence upon the probability of detection within the accuracy limits of the experiment (about 10%), when our target volumes and solid angles of the counters are used.

Since the bremsstrahlung beam causes a chain of stochastic events which determine the quantity $\mathrm{E}(E_\gamma, \Omega_\pi)$, the Monte Carlo method can be used for modeling [7, 80].

We indicate below a computer flowchart for calculating the probability of detecting a π^0 meson via two decay photons.

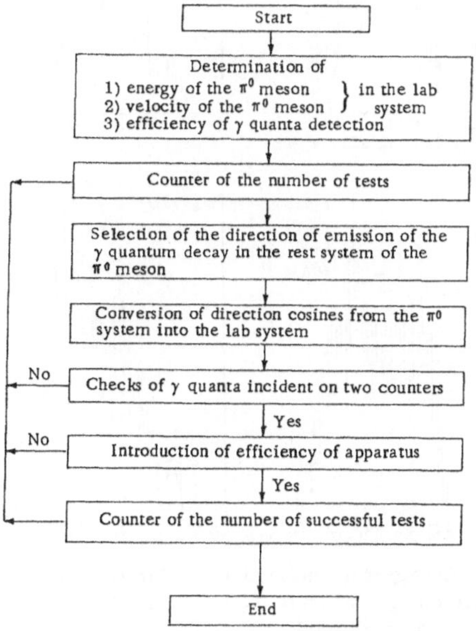

All calculations were made on a computer. The final result is stated in the form of the energy resolution and angular resolution for the geometry of the experimental setup. The resolutions are given by the relations

$$W(E_\gamma) = \frac{\Phi(E_\gamma, E_{\gamma m})}{E_\gamma} \int_{\Omega_\pi} \mathrm{E}(E_\gamma, \Omega_\pi)\, d\Omega_\pi, \qquad \text{(III.26)}$$

$$W(\Omega_\pi) = \int_{E_\gamma} \frac{\Phi(E_\gamma, F_{\gamma m})}{E_\gamma} \mathrm{E}(E_\gamma \Omega_\pi)\, dE_\gamma. \qquad \text{(III.27)}$$

Figures 22a and 22b show the resolutions.

An interesting detail is obtained in an analysis of the energy resolution for the relatively close average energies (we considered 233.9 and 214.19 MeV) but greatly different maximum energies of the bremsstrahlung spectrum ($E_{\gamma m}$ = 250 and 240 MeV, respectively).

When the upper limit of the spectrum is lowered, the energy resolution of the apparatus can be substantially narrowed (from ΔE_γ = 38 to ΔE_γ = 16 MeV), though the average energy

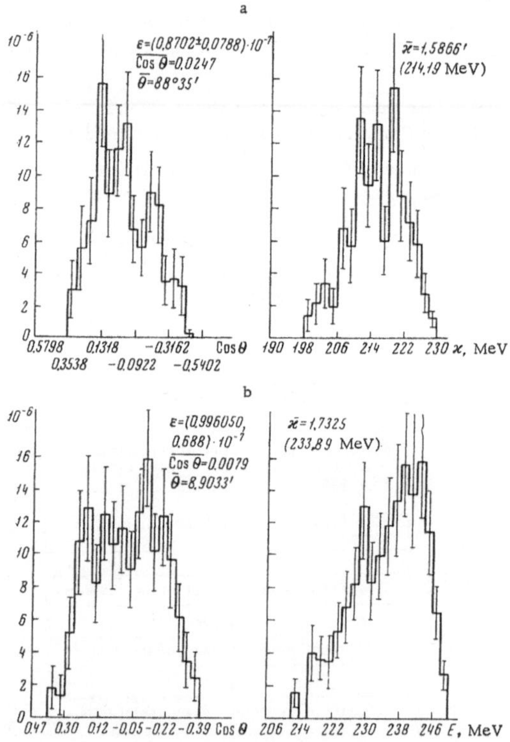

Fig. 22. Energy resolution and angular resolution for the geometric configuration of the experiment; π^0 detection via the two γ quanta of the decay.

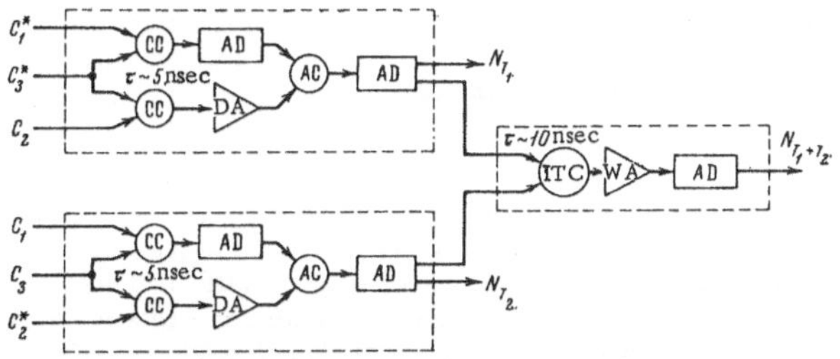

Fig. 23. Block diagram of the electronic equipment for detecting π^0 mesons via two decay photons.

remains almost unchanged in our case. In other words, by appropriately selecting the upper limit of the bremsstrahlung spectrum and the correlation angle, it is possible to detect only π^0 mesons which stem from elastic photoproduction reactions at the He^4 nucleus (the threshold of inelastic photoproduction of π^0 mesons is approximately 20 MeV lower than the threshold of the elastic process). But when we make use of this possibility, we reduce the detection probability, i.e., the yield of the process is greatly affected. The proposed method can therefore be used only in an energy range in which the cross section of the reaction is rather large. The method was used in our case to estimate the influence of inelastic photoproduction at He^4 ($E_\gamma \simeq$ 230-260 MeV) upon the final results of the experiments.

Block Diagram of the Apparatus

Figure 17 depicts the position of the equipment in the accelerator building. All components indicated by dashed lines are not used in this part of the experiment. The principal corrections and basic electronic circuits are the same as in the case of π^0 meson detection via a single decay photon. The following modifications were made: the active volume of the target was increased to 270 cm^3; the photons produced in the decay of π^0 mesons were detected with telescopes consisting of three counters of the previously described type; the block diagram of the electronic sets was partially modified (Fig. 23); the stretching of the bremsstrahlung beam to a length of 1000 μsec or 2000 μsec introduced an inaccuracy of about 4-6 MeV into the determination of the upper limit of the spectrum; and the beam of bremsstrahlung γ quanta was monitored with a thin-walled ionization chamber, which was calibrated with the quantum meter.

Electronic Sets

In addition to the separate adjustment of the two fast telescope circuits (described above) we determined in this part of our work the compensating time delay between the fast outputs of the telescope circuits and selected the discrimination point in the intertelescope coincidence circuits (ITC). One pair of telescopes was mounted for final adjustment on a special stand, with the counter telescopes arranged one behind the other and including an angle of about 90° with the bremsstrahlung beam. Carbon was used as the target. Two telescope pairs were used at the same time in our work. We found, after selection of optimum operation conditions of the circuitry, that the counting rates resulting at the ITC outputs from true coincidences differed less than 5-6% for the two telescopes.

An increase in the active volume of the target considerably increased the load of the counting channels of the telescopes. When the resolution of the CC and ITC circuits remained unchanged, the number of random coincidences increased considerably. It became necessary to add a counter C_2 and C_2^* to each pair of telescopes recording decay photons, i.e., the telescopes included the counters C_1, C_2^*, C_3, and C_1^*, C_2, C_3^*. This rearrangement resulted in an almost complete suppression of random coincidences.

Execution of the Experiment

The differential cross section of π^0 meson photoproduction was measured at meson-emission angle $\overline{\theta}_\pi$ close to 90° and at the average energies $\overline{E}_\gamma$ = 214 and 233 MeV of the primary photons. The photons emitted in the decay of the π^0 mesons were detected with the two telescope pairs in a geometrical configuration which was symmetric relative to the beam ($\overline{\theta}_\pi = 90°$) in the plane $\varphi = 0$, i.e., the counting rates of two independent series were measured at $\overline{\theta}_\pi = 90°$.

In addition to the yield measurements of correlated γ quanta on helium, similar measurements were made on hydrogen. The results were used to establish the absolute values of the data obtained on helium, because the differential cross section of the reaction $\gamma(p\pi^0)p$ is rather accurately known at the angle $\theta_\pi = 90°$ [81].

In each series, measurements on the helium-filled target were followed by measurements of the background produced by the empty target and the yield produced by hydrogen. The counting rate obtained from the background of the empty target amounted to 3-5% of the counting rate obtained with helium.

Evaluation of the Measurements

The counting rates produced by the two coordinated photons from the decay of π^0 mesons emitted at the angle $\overline{\theta}_\pi$ with respect to the bremsstrahlung beam of average energy $\overline{E}_\gamma$ were reduced to the unit indication of the thin-walled chamber and used to derive the differential cross sections of γ photoproduction at He^4 and H. Measurements were made with different pairs of γ telescopes in the various series. This data sampling allowed estimates of the average on the 5% significance level.

The differential cross section can be written in the form

$$\frac{d\sigma_\pi}{d\Omega}(\bar{\theta}_\pi, \overline{E}_\gamma) = \frac{y(\theta_\pi, \overline{E}_\gamma)\,\mathscr{E}(E_{\gamma m})}{\eta G(E_\gamma)\,\mathrm{E}(E_\gamma, \Omega_\pi)} \; \frac{\mathrm{cm}^2}{\mathrm{sr}\cdot\mathrm{photon}\cdot\mathrm{nucleus}}, \tag{III.28}$$

where $y(\bar{\theta}_\pi, E_\gamma)$ denotes the yield per unit count of the thin-walled monitor chamber (yield actually measured in the experiment); $\mathscr{E}(E_{\gamma m}) = \int_0^{E_{\gamma m}} \Phi(E_\gamma E_{\gamma m})\,dE_\gamma$ denotes the energy of a single electron in the bremsstrahlung spectrum (obtained from the tables of [76]); $\Phi(E_\gamma, E_{\gamma m})$ denotes the intensity of the bremsstrahlung spectrum; η denotes the number of nuclei per cm^2 of the target; $\mathrm{E}(E_\gamma, \Omega_\pi)$ denotes the probability of detecting a π^0 meson; and $G(E_{\gamma m})$ is the constant (for fixed $E_{\gamma m}$) of the thin-walled monitor chamber; this constant was determined by calibration with the quantum meter.

Table 3 lists the experimental $G(E_{\gamma m})$ values for $\overline{\theta}_\pi$ angles in the lab system. The table indicates only the statistical errors of the results. The inaccuracies (about 6%) due to the absolute calibration of the beam were not taken into account in the figures listed.

TABLE 3

E_γ, MeV	$\frac{d\sigma_\pi}{d\Omega}\cdot 10^{30}$, cm²/sr · nucleus · photon	$\bar{\theta}_\pi$ degrees
214	21±2.0	86
233	26.4±2.5	93

CHAPTER IV

Discussion of the Results

The preceding chapter listed two groups of experimental data. The first group includes the differential cross sections of γ-quanta formation in the decay of π^0 mesons formed at He^4 nuclei by primary photons of the energy interval 160-260 MeV; the data are for the three angles 64°, 96°, and 144° of γ-quanta emission. The second group encompasses the differential cross sections of π^0 meson formation at He^4 for the primary photon energies 215 and 233 MeV and the π^0-meson emission angle ~90° in the center-of-mass system.

Both the differential and the total cross section of reaction (I.12) can be obtained from the data of the first group, and the applicability limit of the theoretical model used for interpreting the reaction can be estimated.

The cross section of elastic photoproduction of π^0 mesons at helium [Eq. (II.20)] can be written in the form

$$\left[\frac{d\sigma_\pi}{d\Omega}\right]_{\text{He}} = (a_0 + a_1\cos\theta_\pi^* + a_2\cos^2\theta_\pi^*\sin^2\theta_\pi^*) = \sum_{n=0}^{4} B_n\cos^n\theta_\pi^*, \tag{IV.1}$$

where

$$B_0 = a_0;\quad B_1 = a_1,\ B_2 = a_2 - a_0;\quad B_3 = -a_1;\quad B_4 = -a_2 \tag{IV.2}$$

and θ_π^* denotes the angle of π^0-meson emission in the center-of-mass system.

The aforementioned expansion of $\left[\frac{d\sigma_\pi}{d\Omega}\right]_{\text{He}}$ up to n = 4 is possible at primary photon energies close to the threshold, because the coefficients before $\sin^2\theta_\pi^*$ in Eq. (II.20) are characterized by a weak angular dependence. The angular dependence of the amplitude $\overset{+}{\mathcal{F}}_2$ follows from Eq. (II.15), with $\mathcal{F}_{21} \ll \mathcal{F}_{20}$ in the energy interval under consideration. The form factor is defined by the momentum transferred to the He^4 nucleus

$$Q^2 = k^2 + q^2 - 2qk\cos\theta_\pi^*, \tag{IV.3}$$

where k and q denote the momenta of photon and meson, respectively, in the center of mass of the meson–nucleon system. In the energy range close to the threshold, we have k = 1 and $q \to 0$ (in units $\hbar = \mu = c = 1$). The term which in Eq. (IV.3) depends upon the angle has in this case no important influence upon the quantity Q^2.

Moreover, the $\overset{+}{\mathcal{F}}_2$ dependencies of θ_π^* and F_{He} are such that the form factor F_{He} decreases at increasing θ_π^* values, whereas $\overset{+}{\mathcal{F}}_2$ increases ($\mathcal{F}_{21} < 0$).

Calculations made with the least-square method have shown that the correlation of the expansion coefficients in Eq. (IV.1) and the errors of these coefficients make it possible to restrict the expansion to terms of the second power of $\cos\theta_\pi^*$.

Equation (IV.1) can be used to determine the cross section of the formation of γ quanta which are emitted under the angle θ_γ and with the energy $\mathcal{E}_\gamma$ in the decay of π^0 mesons (lab system; see Appendix):

$$\frac{d^2\sigma_\gamma}{d\Omega\, d\mathcal{E}_\gamma} = \mathcal{J}\sum_{n=0}^{4} B_n I_n, \tag{IV.4}$$

where $\mathcal{J}$ denotes the Jacobian of the transition from the center-of-mass system to the lab system; the B_n are identical with the expansion coefficients of (IV.2); and I_n depends upon the kinematic parameters of the photoproduction reaction (see Appendix). By multiplying Eq. (IV.4) with the efficiency $\eta(\mathscr{E}_\gamma)$ of detecting γ quanta with the telescope [see Eq. (III.1)] and by integrating over the energy $\mathscr{E}_\gamma$, we obtain the differential cross section of γ-quanta production in the π^0 meson decay for the angle θ_γ in the lab system, with proper account for the detection efficiency:

$$\frac{d\sigma_\gamma}{d\Omega}(\theta_\gamma) = \int_{\mathscr{E}_{\gamma\min}}^{\mathscr{E}_{\gamma\max}} \frac{d^2\sigma}{d\Omega\, d\mathscr{E}_\gamma}\eta(\mathscr{E}_\gamma)\, d\mathscr{E}_\gamma, \tag{IV.5}$$

where $\mathscr{E}_{\gamma\min}$ and $\mathscr{E}_{\gamma\max}$ denote the minimum and maximum energies of the photon resulting from π^0 meson decay, respectively. Integration results in (see Appendix)

$$\frac{d\sigma_\gamma}{d\Omega}(\theta_\gamma) = \sum_{n=0}^{4} B_n C_n(\theta_\gamma). \tag{IV.6}$$

The coefficients $C_n(\theta_\gamma)$ can be calculated for a fixed energy of the primary photons and for a fixed angle at which decay photons are detected. These coefficients were calculated on a computer.

Thus, Eqs. (IV.1) and (IV.6) establish a direct relationship (in two-particle kinematics) between the differential cross section of elastic photoproduction of π^0 mesons at He^4 and the differential cross section of γ-quanta formation at He^4 nuclei.

When the $d\sigma_\pi/d\Omega$ value obtained from Eq. (II.20) and the amplitude $\mathscr{F}_2^+$ of π-meson production at nucleons according to Eq. (II.15) are used, one can calculate $(d\sigma_\gamma/d\Omega)_{\text{theor}}$. We see from a comparison between this cross section figure of γ-quanta formation and the experimental cross section measured in our work for the emission angles θ_γ = 64°, 96°, and 144° in the energy interval E_γ = 160-260 MeV (Fig. 24) that the agreement between the results is satisfactory in the energy interval 160-210 MeV of the primary photons, whereas the experimental curve is above the theoretical curve at energies E_γ > 210 MeV. The discrepancy becomes more pronounced at even higher energies. This discrepancy between theoretical and experimental $d\sigma_\gamma/d\Omega$ values, observed at E_γ > 210 MeV for all angles, seems to be caused by the increasing importance of inelastic π^0-meson photoproduction processes (III.2) which were disregarded in the theoretical calculations.

The agreement between theoretical and experimental $d\sigma_\gamma/d\Omega$ values at energies below 210 MeV indicates that the total cross section of photoproduction can be almost entirely ex-

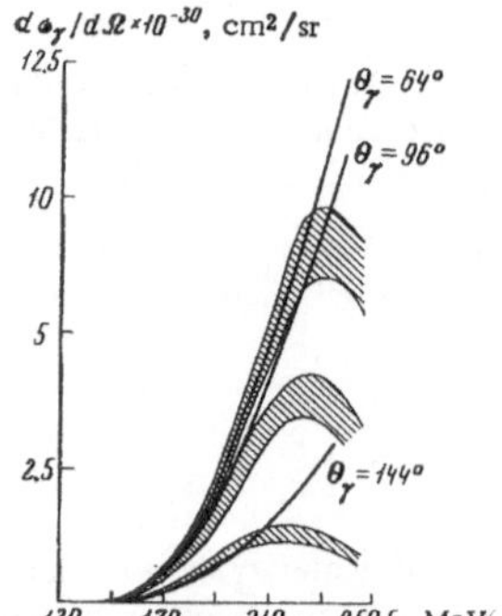

Fig. 24. Energy dependence of the cross section of γ-quanta formation for the emission angles θ_γ = 64°, 96°, and 144°. The shaded area corresponds to the 10% accuracy interval of the theoretical calculations.

TABLE 4. Parameters of the Expansion of the Differential Cross Section in the Form $\left[\frac{d\sigma_\pi}{d\Omega}\right]_{He} = (a_0 + a_1 \cos\theta_\pi{}^* + a_2 \cos^2\theta_\pi{}^*)$ and Values of the Total Cross Section σ_{tot}^{He} Obtained in the Experiment under Consideration

E_γ, MeV	$a_0 \cdot 10^{30}$, cm²/sr	$a_1 \cdot 10^{30}$, cm²/sr	$a_2 \cdot 10^{30}$, cm²/sr	$\sigma_{tot}^{He} \cdot 10^{29}$ cm²
159	0.7±0.5	−0.8±2.8	−0.3±3.2	0.6±0.7
169	3.6±0.5	1.0±2.4	−4.7±2.6	2.3±0.6
179	6.4±0.4	3.0±1.9	−8.6±1.4	3.9±0.4
189	11.6±0.4	5.3±1.6	−12.2±1.7	7.7±0.4
199	16.4±0.4	7.5±1.5	−14.4±2.1	11.4±0.5
209	22.3±0.4	10.2±1.9	−16.8±2.4	15.9±0.7

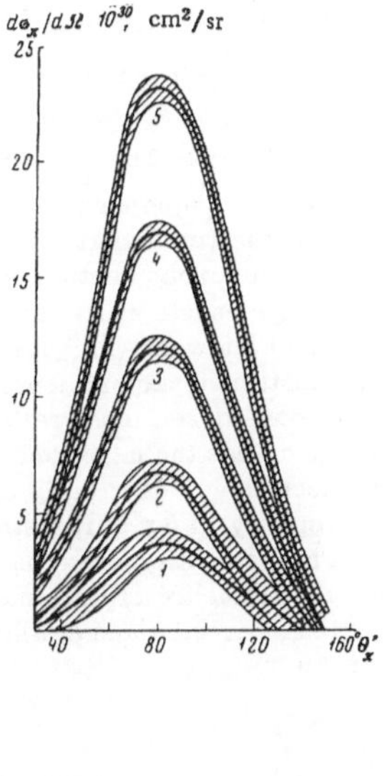

Fig. 25. Differential cross section of the reaction $\gamma + He^4 \rightarrow \pi^0 + He^4$ as a function of the meson-emission angle and the bremsstrahlung energy in the center-of-mass system. Curves 1-5 correspond to E_γ = 169-209 MeV (10 MeV interval). The shaded area corresponds to the interval of experimental errors.

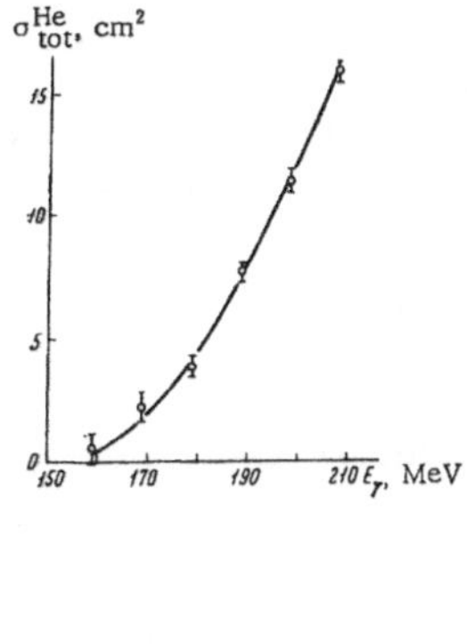

Fig. 26. Energy dependence of the total cross section of π^0 meson photoproduction at He^4 (multiplied by 10^{29}).

plained in this energy interval by the elastic process of (I.12). This is consistent with the assumptions of [24] and [16].

The inverse procedure was used for a more detailed comparison between experimental and theoretical results. The cross sections $d\sigma_\pi/d\Omega$ were determined from the measured $(d\sigma_\gamma/d\Omega)_{exp}$ values. The cross sections of γ-quanta formation from π^0 meson decay and the coefficients $C_n(\theta_\gamma)$ were used to determine the expansion parameters a_0, a_1, and a_2 (IV. 2). The results of these calculations are listed in Table 4. Figure 25 shows the differential cross sections which were calculated with these coefficients in dependence of the angle θ_π^* for the elastic photoproduction of π^0 mesons at He^4. The energy of the bremsstrahlung beam in the lab system was chosen as parameter of the curves. The calculated coefficients a_0 and a_2 can be used to determine the total cross section of the reaction in the energy interval $E_\gamma = 160$–210 MeV of the primary photons. By integrating Eq. (IV.1) over the angle θ_π^* of π^0 meson emission in the center-of-mass system, we obtain the quantity σ_{tot}^{He} in the form

$$\sigma_{tot}^{He}(E_\gamma) = 2\pi\left(\frac{4}{3}a_0 + \frac{4}{15}a_2\right). \tag{IV.7}$$

The resulting $\sigma_{tot}^{He}(E_\gamma)$ values are listed in Table 4 and depicted in Fig. 26.

The experimental $\frac{d\sigma_\pi}{d\Omega}(E_\gamma)$ values for the meson-emission angle $\theta_\pi^* = 90°$ were supplemented by the results of [26] and are shown in Fig. 27 for the three energies 267, 280, and 300 MeV of the primary photons. The $\left[\frac{d\sigma_\pi}{d\Omega}(90°)\right]_{He}$ values which we had obtained in our experiments by detection of one or two γ quanta from the decay of π^0 mesons agree, as put into evidence by the figure, among themselves and are consistent with the results of [26]. Curve 1 depicts the energy dependence of the cross section calculated with Eq. (II.40). The values obtained in [33] were used for the single-nucleon $\mathcal{F}_2^+$ amplitude. The form factor of the He^4 nucleus was calculated with Eq. (II.46) and with $\langle r \rangle = 1.4$ F. The corresponding value of the radius was obtained in electron-scattering experiments [20, 21] and the photofission of He^4 [82] and can at the present time be considered most reliable. It follows from the figure that the calculated curve 1 properly describes the experimental cross section up to energies $E_\gamma \leq 210$ MeV. The theoretical cross section values greatly exceed the experimental cross sections at high energies. This discrepancy is not surprising because the interaction between the meson generated and the nucleus increases with increasing energy (when the $^3/_2$–$^3/_2$ resonance is approached). In order to estimate the influence of this interaction, we calculated the "distorted" form factor of the nucleus at various energies and emission angles. The results of these calculations are shown in Figs. 28-30.

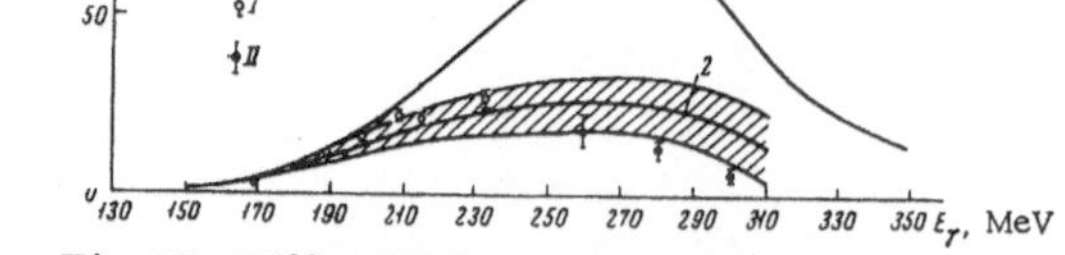

Fig. 27. Differential cross section of the reaction $\gamma + He^4 \rightarrow \pi^0 + He^4$ for the meson-emission angle $\theta_\pi^* \simeq 90°$ in the center-of-mass system. I) Our work; II) results of [26]; 1) result of calculating the cross section with Eq. (II.40), disregarding multiple scattering; 2) calculation including multiple scattering.

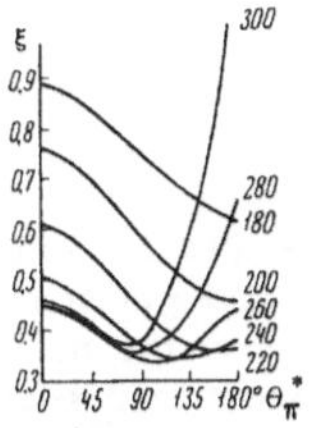

Fig. 28. The function $\xi(\theta_\pi^*)$ for various E_γ values of the interval 180-300 MeV.

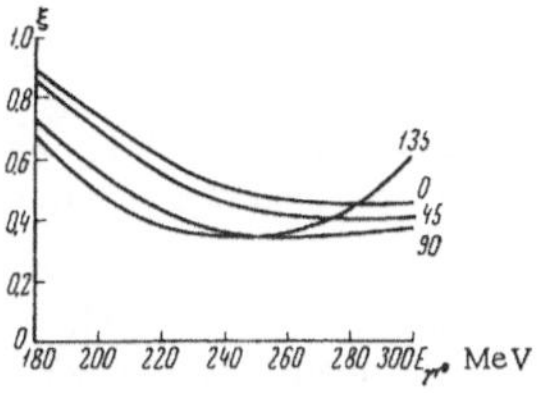

Fig. 29. The function $\xi(E_\gamma)$ for the π meson generation angles $\theta_\pi^* = 0°, 45°, 90°$, and 135°.

According to Eqs. (II.47) and (II.48), a form factor $\tilde{F}$ in which the distortion of the pion wave was taken into account is no longer a function of the transferred momentum only but depends also upon E_γ and θ_π^*. Equation (II.49) is therefore conveniently expressed as a function of $\xi(E_\gamma)$ or $\xi(\theta_\pi^*)$ (at fixed θ_π^* or E_γ, respectively) rather than a function of Q^2. Figures 28 and 29 show these dependencies. Figure 30 shows $F^2_{\text{single}}(\theta_\pi^*)$ and $\tilde{F}^2_{\text{single}}(\theta_\pi^*)$ at $E_\gamma = 180$ MeV as an example how $\xi(\theta_\pi^*)$ manifests itself.

As can be inferred from Figs. 28 and 29, multiple scattering reduces the cross section, and the reduction increases with increasing energies (in the energy interval 180-300 MeV under consideration) and reaches about 50% at 300 MeV. Mainly the production at large angles is suppressed at low energies, whereas the suppression is more isotropic at high energies. We emphasize that in the region of large momenta transferred (near the first diffraction minimum of F_{single}), the results cannot be accepted with great confidence since the form factors of (II.45) and (II.47) differ greatly in that region. This fact seems to explain the strong peak in $\xi(\theta_\pi^*)$ at large angles and at $E_\gamma = 300$ MeV.

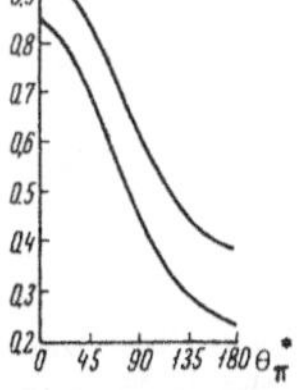

Fig. 30. Squares of the form factors for the model of a homogeneous sphere; various angles; $\langle r \rangle = 1.4$ F, and $E_\gamma = 180$ MeV. The upper curve was calculated with Eq. (II.47), disregarding multiple scattering; the lower curve was calculated with Eq. (II.48), taking into account multiple scattering.

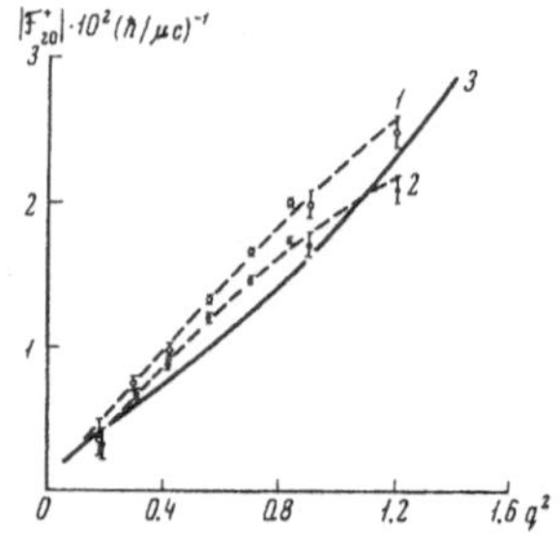

Fig. 31. Dependence of the amplitude $\mathcal{F}^+_{20}$ upon the square of the meson momentum in the center-of-mass of the meson–nucleon system. Curves 1 and 2 correspond to $\langle r \rangle = 1.61$ F and $\langle r \rangle = 1.4$ F, respectively; curve 3 corresponds to the $\mathcal{F}^+_{20}$ calculated in [11].

Unfortunately, the errors of the measurements make it possible to establish an additional angular dependence of the type $\xi(\theta^*_\pi)$ shown in Fig. 30 from the experimental angular distributions.

As far as the cross section at 90° is concerned (for which the measurements are most reliable), multiple scattering was taken into account as shown by curve 2 of Fig. 27. The shaded area, which was constructed under the assumption that the error in the estimates of the effect amounts to 20% at $\theta^*_\pi = 90°$ and $E_\gamma = 180$ MeV, gives an idea of the increasing error at increasing energies. In order to reduce the errors, one must calculate ξ for real form factors (Gauss, Irving). Nevertheless, when we take into account the interaction in the final state, we obviously eliminate the sharp discrepancy which occurs at $E_\gamma > 210$ MeV between the experimental data and curve 1 calculated with Eq. (II.40) in the "pure" momentum approximation. When a comparison with experiments is made, we must bear in mind that, for the sake of simplicity, in the derivation of Eq. (II.40) only the lowest powers of μ/N, q/N, and k/N were kept. Obviously, the error related to this approximation increases with increasing E_γ.

We can estimate the amplitude $\left[\frac{d\sigma_\pi}{d\Omega}\right]_{He}$ π^0 meson photoproduction at nucleons when we use the experimental $\mathcal{F}^+_2$ values in the interval $E_\gamma = 140$–220 MeV. Since the majority of the measurements were made at the angle $\theta^*_\pi = 90°$ in the center-of-mass system, the quantity $\frac{d\sigma_\pi}{d\Omega}(90°)$ was used for the amplitude estimates. The form factor for the He^4 nucleus was determined with Eq. (II.45) in which Irving wave functions were substituted. The mean-square radius used in the calculations was $\langle r \rangle = 1.4$ F.

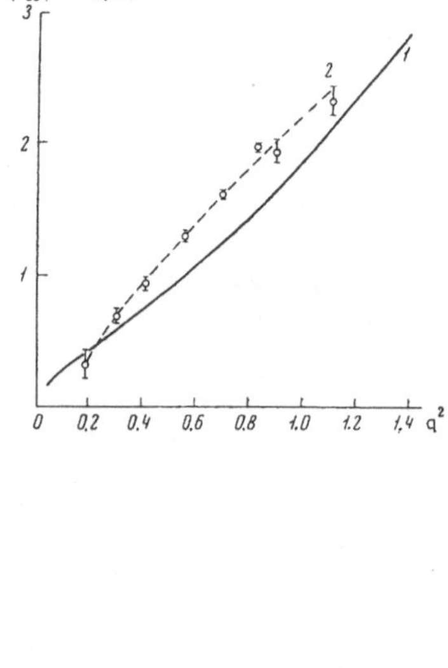

Fig. 32. Dependence of the quantity $\mathcal{F}^+_{20}$ upon the square of the meson momentum in the center of mass of the meson–nucleon system; the dependence was obtained with Eq. (I.10) for $\langle r \rangle$ = 1.4 F and (1) with a Gaussian form factor and (2) from experiments.

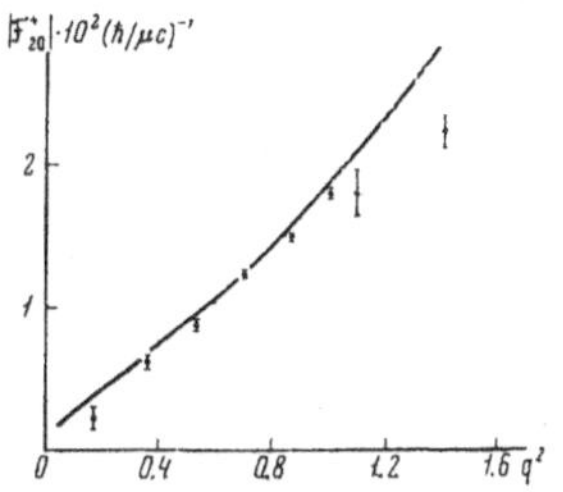

Fig. 33. Dependence of the quantity $\mathcal{F}_{20}^{+}$ upon the square of the meson momentum in the center-of-mass system; the dependence was obtained with Eq. (II.40). The solid curve corresponds to the theory of [4, 11].

The amplitude $\mathcal{F}_2^{+}$ was separated with Eqs. (I.10) and (II.40), respectively, in order to compare the following two versions of calculations:* a) nonrelativistic momentum approximation and b) "relativistic momentum approximation." Figure 31 is a comparison of the $\mathcal{F}_2^{+}$ values obtained with Eq. (I.10) and the theoretical values for $0.4 < q < 1.1$. The $\mathcal{F}_{20}^{+}$ values for $\langle r\rangle = 1.61$ F are also shown to indicate the influence of $\langle r\rangle$. The observed difference between experimental and theoretical $\mathcal{F}_{20}^{+}$ values is obviously unrelated to the type of the nuclear form factor, because an F(Qr) of the type of the Gauss function of Eq. (II.46) even slightly enhances the difference (Fig. 32). It seems that the form factor cannot be related to the mean-square radius, because in order to remove the difference, we have to use a too small $\langle r\rangle$ value which is inconsistent with the available experimental evidence. On the other hand, all calculated versions of $\mathcal{F}_{20\,\mathrm{theor}}^{+}$ agree within a 3% interval (Fig. 20), i.e., the discrepancy cannot be explained by inaccuracies of the theory.

Figure 33 relates to calculations which were made with Eq. (II.40). We believe that the observed satisfactory agreement attests to the validity of the relativistic momentum approximation model.

Calculations which were made with Eq. (II.40) for π^0 meson photoproduction at C^{12} nuclei are in good agreement with the experimental figures of [83]. But in order to confirm that this agreement is not a random result, inelastic effects which were disregarded in the derivation of Eq. (II.40) must be introduced. On the other hand, it is highly desirable to supplement the experimental data, particularly as far as angular distributions are concerned, and to extend the investigations to higher photon energies.

Appendix

Relation between the Differential Cross Section of γ-Quanta Formation and the Differential Cross Section of π^0 Meson Photoproduction

By representing the cross section of π^0 meson photoproduction as a series

$$\frac{d\sigma_\pi}{d\Omega} = \sum_{n=0}^{m} B_n \cos^n \theta_\pi^* \tag{A.1}$$

$d\sigma_\pi/d\Omega$ can be related to the cross section of γ-quanta production in the decay of π^0 mesons in the lab system. In the case of π^0 meson photoproduction, the corresponding expressions were derived in [61-63], but the expansion was restricted to the second power of $\cos\theta_\pi^*$ in all these papers. We obtained in our work relations which allow a conversion of the cross section of π^0 meson photoproduction to the cross section of decay-photon production in the lab system for any number of terms of series (A.1).

* Calculations with Eq. (I.9) result in cross sections which several times differ from the experimental cross section even in the energy range of 200 MeV.

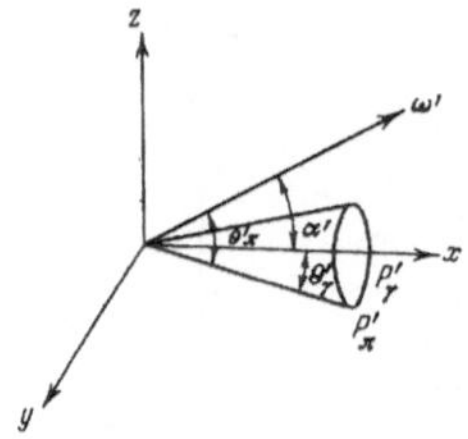

Fig. 34. Relation between the meson exit angle $\theta_\pi^* = \theta_\pi$, the angle α' of decay γ-quantum emission, and the angle of decay γ-quantum emission included by the direction of the γ quantum and the direction of motion of the π^0 meson.

It was shown in [62] that the angular distribution and energy distribution of the γ quanta originating from the decay of π^0 mesons can be expressed as follows for photoproduction in the center-of-mass system:

$$f'(\alpha', \mathscr{E}'_\gamma) = \frac{1}{\pi\sqrt{\mathscr{E}'^2_\pi - 1}} \int_0^{2\pi} \frac{d\sigma_\pi}{d\Omega}\, d\varphi, \tag{A.2}$$

where α' and $\mathscr{E}'_\gamma$ denote the emission angle relative to the accelerator beam and the energy of the decay photon in the center-of-mass system, respectively; $\mathscr{E}'_\pi$ denotes the energy of the π^0 meson; and $d\sigma_\pi/d\Omega$ denotes the cross section of photoproduction. The relation between the angles θ_π^* and α' and the emission angle θ'_γ of the γ quantum relative to the direction of π^0 meson motion in the center-of-mass system (Fig. 34) is given by

$$\cos\theta_\pi^* = \cos\alpha' \cos\theta'_\gamma + \sin\alpha' \sin\theta'_\gamma \cos\varphi. \tag{A.3}$$

Substituting Eq. (A.1) into Eq. (A.2) results in

$$f'(\alpha', \mathscr{E}'_\gamma) = \frac{1}{\pi\sqrt{\mathscr{E}'^2_\pi - 1}} \sum_{n=0}^{m} I_n, \tag{A.4}$$

where

$$I_n = B_n \int_0^{2\pi} \cos^n \theta_\pi^*\, d\varphi. \tag{A.5}$$

By replacing in Eq. (A.5) $\cos^n\theta_\pi^*$ by Eq. (A.3) and by setting $\cos\alpha' \cdot \cos\theta'_\gamma = a$, $\sin\alpha' \cdot \sin\theta'_\gamma = b$, we obtain

$$I_n = B_n \int_0^{2\pi} \Big[a_n + na^{n-1} b\cos\varphi + \frac{n(n-1)}{2!} a^{n-2} b^2 \cos^2\varphi + \ldots +$$

$$+ nab^{n-1}\cos^{n-1}\varphi + b\cos^n\varphi\Big] = B_n\Big[a^n - 2\pi + na^{n-1} b \int_0^{2\pi} \cos\varphi\, d\varphi + \ldots +$$

$$+ \frac{n(n-1)}{2!} a^{n-2} b^2 \int_0^{2\pi} \cos^2\varphi\, d\varphi + b^n \int_0^{2\pi} \cos^n\varphi\, d\varphi; \tag{A.6}$$

$$\int_0^{2\pi} \cos^n\varphi\, d\varphi = \begin{cases} 0 & \text{for n even} \\ 2\pi \dfrac{(n-1)(n-3)(n-5)\ldots 1}{n(n-2)(n-4)\ldots 2} & \text{for n odd} \end{cases} \tag{A.7}$$

The distribution of the γ quanta from the π^0 meson decay over the angles and the energy in the lab system can be stated in the form

$$f(\theta_\gamma, \mathscr{E}_\gamma) = \mathscr{J} f'(\alpha', \mathscr{E}'_\gamma), \tag{A.8}$$

where $\mathscr{J}$ denotes the Jacobian of the transition from the center-of-mass system to the lab system.

By introducing variables which depend upon the emission angle θ_γ, of the decay photons, the photon energy $\mathscr{E}_\gamma$, and the energy ω of the primary photon in the lab system,

$$\Phi = \frac{\cos\theta_\gamma - \beta_c}{1 - \beta_c \cos\theta_\gamma} = \cos\alpha',$$

$$\chi = \frac{1}{2q'\gamma_c(1-\beta_c\cos\theta_\gamma)},$$

$$\psi = \frac{\sin\theta_\gamma}{\gamma_c(1-\beta_c\cos\theta_\gamma)},\qquad (A.9)$$

$$\varkappa = \frac{\varepsilon_\alpha}{q'},\quad \mathscr{E}_\pi = \frac{2M\omega+\mu^2}{2(2M\omega+M^2)^{1/2}},\quad \beta_c = \frac{\omega}{\omega+M},$$

we obtain

$$\gamma_c = \frac{1}{(1-\beta_c^2)^{1/2}},\quad q' = [\mathscr{E}_\pi'^2-\mu^2]^{1/2},\quad \mathscr{J} = 2\chi q', \qquad (A.10)$$

$$a = \Phi\left(\varkappa - \frac{\chi}{\mathscr{E}_\gamma}\right);\quad b = \psi\left[1-\left(\varkappa-\frac{\chi}{\mathscr{E}_\gamma}\right)\right]^{1/2}. \qquad (A.11)$$

By inserting these quantities into Eqs. (A.6) and (A.8) and taking into account Eq. (A.7), we obtain the cross section of the formation of decay photons with the energy $\mathscr{E}_\gamma$, and the emission angle θ_γ with respect to the primary photon beam:

$$\frac{d^2\sigma_\gamma}{d\Omega\, d\mathscr{E}_\gamma} = f(\theta_\gamma, \mathscr{E}_\gamma) = \mathscr{J}\sum_{n=0}^{m} B_n I_n. \qquad (A.12)$$

The counting rate registered by the equipment and reduced to each target nucleus and primary photon in the beam is equal to the measured differential cross section of the formation of γ quanta which are emitted at the angle θ_γ with respect to the primary photon beam:

$$\frac{d\sigma_\gamma}{d\Omega}(\theta_\gamma) = \int_{\mathscr{E}_{\text{min}}}^{\mathscr{E}_{\text{max}}} \frac{d^2\sigma_\gamma}{d\Omega\, d\mathscr{E}_\gamma}\,\eta(\mathscr{E}_\gamma)\, d\mathscr{E}_\gamma, \qquad (A.13)$$

where $\eta(\mathscr{E}_\gamma)$ denotes the efficiency of the equipment for detecting γ quanta with the energy $\mathscr{E}_\gamma$, and

$$\mathscr{E}_{\text{min}} = (1-\bar\beta)\frac{\mathscr{E}_\pi'}{2\gamma_c(1-\beta_c\cos\theta_\gamma)},$$

$$\mathscr{E}_{\text{max}} = (1+\bar\beta)\frac{\mathscr{E}_\pi'}{2\gamma_c(1-\beta_c\cos\theta_\gamma)} \qquad (A.14)$$

denote the minimum and maximum energies, respectively, of decay photons at the angle θ_γ, with $\bar\beta = q'/\mathscr{E}_\pi'$.

We assume that the energy dependence of the efficiency of photon detection can be approximated by the expression

$$\eta(\mathscr{E}_\gamma) = \eta_\infty(1-\xi e^{-k\mathscr{E}_\gamma}), \qquad (A.15)$$

where η_∞, k, and ξ denote constants which must be determined for the actual experimental setup (we have in our case $\eta_\infty = 0.45$, $\xi = 1.105$, and $k = 0.68$). We obtain with Eqs. (A.15) and (A.13) and after integration

$$\frac{d\sigma_\pi}{d\Omega}\sum_{n=0}^{m} B_n C_n(\theta_\gamma), \qquad (A.16)$$

where

$$C_n(\theta_\gamma) = \mathscr{J}\int_{\mathscr{E}_{\text{min}}}^{\mathscr{E}_{\text{max}}} I_n \eta(\mathscr{E}_\gamma)\, d\mathscr{E}_\gamma. \qquad (A.17)$$

We use the notation

$$i_n = \int_{\mathscr{E}_{\text{min}}}^{\mathscr{E}_{\text{max}}} \frac{1-\xi e^{-k\mathscr{E}_\gamma}}{\mathscr{E}_\gamma^n}\, d\mathscr{E}_\gamma. \qquad (A.18)$$

and obtain for n > 1

$$i_n = \frac{1}{n-1}\left(\frac{1}{\mathscr{E}_{\text{min}}^{n-1}} - \frac{1}{\mathscr{E}_{\text{max}}^{n-1}}\right) - \xi\mathscr{J}_n, \qquad (A.19)$$

$$\mathscr{J}_n = \frac{1}{n-1}\left[\left(\frac{e^{-k\mathscr{E}_{\text{min}}}}{\mathscr{E}_{\text{min}}^{n-1}} - \frac{e^{-k\mathscr{E}_{\text{max}}}}{\mathscr{E}_{\text{max}}^{n-1}}\right) - k\mathscr{J}_{n-1}, \right. \qquad (A.20)$$

and for n = 1

$$i_1 = \ln \frac{\mathscr{E}_{\max}}{\mathscr{E}_{\min}} - \xi \mathscr{I}_1, \tag{A.21}$$

$$\mathscr{I}_1 = \int\limits_{\mathscr{E}_{\min}}^{\mathscr{E}_{\max}} \frac{e^{-k\mathscr{E}_\gamma}}{\mathscr{E}_\gamma} d\mathscr{E}_\gamma = \ln \frac{\mathscr{E}_{\max}}{\mathscr{E}_{\min}} - k(\mathscr{E}_{\max} - \mathscr{E}_{\min}) + \frac{k^2}{4}(\mathscr{E}^2_{\max} - \mathscr{E}^2_{\min}) - \frac{k^3}{18}(\mathscr{E}^3_{\max} - \mathscr{E}^3_{\min}) + \ldots, \tag{A.22}$$

$$i_0 = (\mathscr{E}_{\max} - \mathscr{E}_{\min}) + \frac{\xi}{k}(e^{-k\mathscr{E}_{\max}} - e^{-k\mathscr{E}_{\min}}). \tag{A.23}$$

Expressions for the coefficients C_n can be obtained from Eqs. (A.18)-(A.23). We obtain for our formula (IV.6) (m = 4):

$$\begin{aligned}
C_0 &= 4\chi\eta_0 i_0,\\
C_1 &= 4\chi\eta_0\Phi(\varkappa i_0 - \varkappa i_1),\\
C_2 &= 2\chi\eta_0\{\psi^2 i_0 + (2\Phi^2 - \psi^2)[\varkappa^2 i_0 - 2\varkappa\chi i_1 + \varkappa^2 i_2]\},\\
C_3 &= 2\chi\eta_0[(2\Phi^3 - 3\psi^2\Phi)(\varkappa^3 i_0 - 3\varkappa^2\chi i_1 + 3\varkappa\chi^2 i_2 - \chi^3 i_3) + 3\psi^2\Phi(\chi i_0 - \chi i_1)],\\
C_4 &= 2\chi\eta_0\Big[\Big(2\Phi^4 + \frac{3}{4}\psi^4 - 6\Phi^2\psi^2\Big)(\varkappa^4 i_0 - 4\varkappa^3\chi i_1 + 6\varkappa^2\chi^2 i_2 - 4\varkappa\chi^3 i_3 + \chi^4 i^4) +\\
&\quad + \Big(6\Phi^2\psi^2 - \frac{3}{2}\psi^4\Big)(\varkappa^2 i_0 - 2\varkappa\chi i_1 + \chi^2 i_2) + \frac{3}{4}\psi^4 i_0\Big].
\end{aligned} \tag{A.24}$$

Form Factor Approximation for π Meson Scattering at Nuclei

When we assume that in elastic coherent π meson scattering at a nucleus, the scattering amplitude U, which is related to the S matrix by the expression

$$\langle q_2 p_2 | S | q_1 p_1 \rangle = \delta_{fi} + \frac{1}{(2\pi)^2} \frac{\delta(p_1 + q_1 - p_2 - q_2)}{4\sqrt{p_{10} q_{10} p_{20} q_{20}}}, \tag{A.25}$$

is analytic in t [$t = (q_2 - q_1)^2$, $s = (q_1 + p_1)$ denote the usual relativistic invariants] and when we consider in the dispersion relations for U only the first irregular singularity

$$t = t_0 = m^{-2}[(m + N)^2 - M^2][M^2 - (m - N)^2], \tag{A.26}$$

corresponding to the diagram of Fig. 35, we can derive [44, 45] the following approximation for U

$$U = F(t) \oint \frac{d\sigma}{2\pi i} \frac{R}{\sqrt{(\sigma_+ - \sigma)(\sigma_- - \sigma)}}. \tag{A.27}$$

The notation is interpreted as follows: F(t) denotes the form factor of the nucleus [45]; $\sigma = (n_1 + q_1)^2$; m denotes the mass of the nuclear fragment; $m = M - N + \varepsilon$; and ε denotes the binding energy. The contour of integration comprises the points σ_+ and σ_-. As has been mentioned in [44], in contrast to the simplest form factor approximation, Eq. (A.27) takes into account relativistic corrections for the relative motion of the nucleons. The maximum energy σ_+ and the minimum energy σ_- transferred in block T of the diagram (Fig. 35) can be obtained with the dual graph which expresses the conservation laws of the four momenta in each of the apices of the diagram (Fig. 35):

$$\begin{aligned}
\sigma_\pm &= \sigma_0 \pm \Delta,\\
\sigma_0 &= s + m^2 - \frac{2}{4M^2 - t}(M^2 + m^2 - N^2)(s + M^2 - \mu^2),\\
\Delta &= (4M^2 - t)^{-1} 2m\sqrt{t_0 - t}\sqrt{\Lambda(s, t)},\\
\Lambda(s, t) &= st + s^2 - 2s(M^2 + \mu^2) + (M^2 - \mu^2)^2.
\end{aligned} \tag{A.28}$$

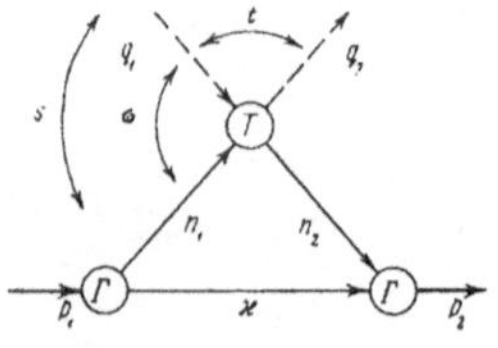

Fig. 35

The equation $t\Lambda(s, t) = 0$ defines the region in which the reaction $\pi + a \to \pi + A$ is feasible in center-of-mass coordinates of the system $A + \pi$

$$\Lambda(s, t) = 2q^2 s(1 + \cos\theta), \tag{A.29}$$

where $q \equiv |\mathbf{q}|$ and θ denote the momentum and the angle of meson scattering in center-of-mass coordinates of the system $A + \pi$.

Equation (A.27) is relativisticly invariant and thus no difficulties are encountered in the selection of the coordinate system [84]. Equations (A.28) establish relations between the kinematics of the nucleon and the nucleus. More particularly, when we assume $t \ll 4M$ and $\varepsilon \ll N$, we can introduce the approximation

$$\sigma_0 \simeq \sigma_0' \equiv 2NE_q^L + (N + \mu)^2, \tag{A.30}$$

where E_q^L denotes the kinetic energy of the meson in the lab system. It is easy to verify from a comparison of Eq. (A.30) and the expression $s = 2ME_q^L + (M + \mu)^2$ that σ_0' is the σ value corresponding to the motion of a nucleon not bound to the nucleus and having the same velocity as the center of mass of the nucleus.

The function R can be expressed by the amplitude T of meson scattering at the nucleon

$$R = \frac{M}{8mN^2}\,\mathrm{Sp}\,\{T(\hat{n}_1 + N)\,a(p_1)(\hat{\varkappa} - m)\,\bar{a}(p_2)(\hat{n}_2 + N)\}. \tag{A.31}$$

We have (see [49])

$$a(p) = \frac{\hat{p} + M}{2M}\,\gamma_5 c \simeq \gamma_5 c, \quad \bar{a} = \beta a^+ \beta \tag{A.32}$$

and the amplitude is given by the relation

$$\langle q_2 n_2 | S | q_1 n_1 \rangle = \delta_{fi} + \frac{N}{(2\pi)^2}\,\frac{\delta(n_1 + q_1 - n_2 - q_2)}{2\sqrt{n_{10} q_{10} n_{20} q_{20}}}\,\bar{u}(n_2)\,Tu(n_1). \tag{A.33}$$

In center-of-mass coordinates of the system $N + \pi$, the notation

$$F = f_1 + i(\sigma\,[\tilde{q}_2' \times \tilde{q}_1'])\,f_2 \equiv \bar{f}_1 + (\sigma\tilde{q}_2')(\sigma\tilde{q}_1')\,\bar{f}_2 \tag{A.34}$$

is used, where

$$f_1 = \bar{f}_1 + \bar{f}_2\cos\theta', \quad f_2 = \bar{f}_2, \quad \tilde{q}' = \frac{\mathbf{q}'}{|q'|}$$

($\mathbf{q}'$ and θ' denote momentum and scattering angle in center-of-mass coordinates of the system $N + \pi$) and

$$\begin{aligned} \bar{f}_1 &= K_1[T_1 + (\omega - N)T_2], \\ \bar{f}_2 &= K_2[-T_1 + (\omega + N)T_2]. \end{aligned} \tag{A.35}$$

We have introduced the notation

$$K_1 = \frac{E + N}{8\pi\omega}; \quad K_2 = \frac{E - N}{8\pi\omega}; \quad E = \sqrt{\mathbf{q}'^2 + N^2}; \tag{A.36}$$

$\omega = \sqrt{\sigma}$ denotes the total energy in center-of-mass coordinates of the system $N + \pi$.

By using Eqs. (A.31)-(A.33) and calculating the spur, we obtain

$$R = \sum_{i=1}^{2} \Lambda_i T_i,$$
$$\Lambda_1 = 2M - \frac{tM}{4N^2}, \quad \Lambda_2 = \frac{M}{N}(\sigma - N^2 - \mu^2) + \frac{tM}{8mN^2}\,[(s - M^2) - (\sigma - N^2) + 4mN]. \tag{A.36}$$

It is convenient to switch from the T_i to amplitudes f_i. We obtain with Eq. (A.35)

$$R = 8\pi\omega \sum \lambda_i f_i = 8\pi\omega \sum \bar{\lambda}_i \bar{f}_i,$$

$$\bar{\lambda}_{1,2} = 1 + \frac{t\,[s-(\omega \pm m)^2]}{16m\,N\omega\,(E \pm N)}. \tag{A.37}$$

It is easy to show with Eq. (A.30) that we have with an accuracy of q/M:

$$\lambda_1 = 1 \text{ and } \lambda_2 = 0, \tag{A.38}$$

so that the amplitude of scattering at a nucleus with spin zero is completely described by the spin-independent part of the single-nucleon amplitude:

$$R = \frac{M}{N}\, 8\pi\omega\, f_1. \tag{A.39}$$

Thus, the differential cross section of elastic coherent π meson scattering in center-of-mass coordinates of the system $A + \pi$ can be written in the form

$$\frac{d\sigma}{d\Omega} = \frac{|U|^2}{(8\pi)^2 s} = A^2 Q F^2 |\Phi|^2, \tag{A.40}$$

where

$$\Phi = \frac{1}{\sqrt{\sigma_0}} \oint \frac{d\sigma}{2\pi i}\, \frac{\sqrt{\sigma}\, f_1(\sigma, t)}{\sqrt{(\sigma_+ - \sigma)(\sigma_- - \sigma)}}. \tag{A.41}$$

The kinematic factor

$$Q = \frac{\sigma_0 M^2}{s N^2} \tag{A.42}$$

can be expressed by the meson energy in the lab system

$$Q = \left(\frac{M}{N}\right)^2 \frac{2NE_q^L + (N+\mu)^2}{2ME_q^L + (M+\mu)^2} \simeq 1 + \frac{2(M-N)}{MN}\, \mathscr{E}_q^L, \tag{A.43}$$

where $\mathscr{E} = E_q^L + \mu$ denotes the total meson energy in the lab system.

The integral in Eq. (A.41) can be easily calculated when we ignore corrections for the relative motion and assume the approximation $\sigma_+ = \sigma_- = \sigma_0$. We obtain the simplest expression for the cross section

$$\left(\frac{d\sigma}{d\Omega}\right)_0 = A^2 Q F^2 |f_1(q', \cos\theta')|^2, \tag{A.44}$$

wherein

$$\cos\theta' = 1 - Q(1 - \cos\theta). \tag{A.45}$$

However, it follows from Eq. (A.29) that generally $\sigma_+ \neq \sigma_+$ and $\sigma_+ - \sigma_- \equiv 2\Delta = 0$ holds only for backward scattering.

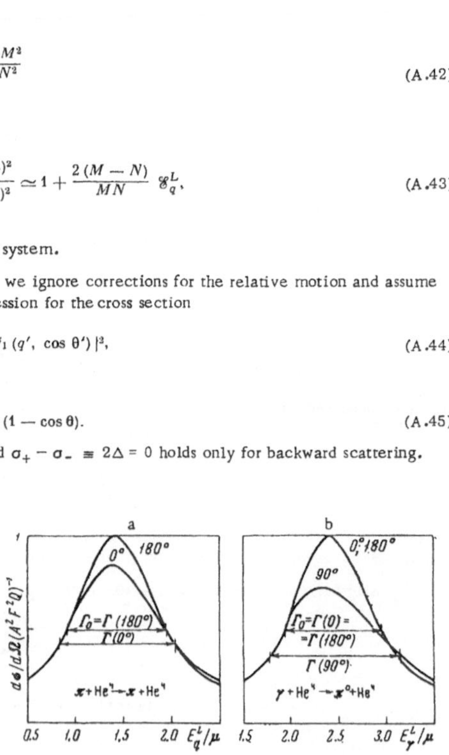

Fig. 36. Resonance curves (a) for the reaction $\pi + He^4 \to \pi + He^4$ and (b) for reaction (I.12).

We have for small energies ($q \ll M$), $\Delta \ll \sigma_0$ and, if the function $f_1(\sigma)$ is a smooth function, we can neglect $\sigma_+ - \sigma_-$. When the energy is increased, particularly near the 3-3 resonance, the approximation is too coarse. We assume for an estimate that $f_1(\sigma)$ has the form of a Breit-Wigner resonance

$$\sqrt{\sigma} f_1 \sim \frac{\gamma}{\sigma - \sigma^* - i\gamma}; \qquad \gamma = N\Gamma. \tag{A.46}$$

By assuming Δ, we would obtain

$$\left(\frac{d\sigma}{d\Omega}\right)_0 \frac{1}{A^2 Q F^2} = \frac{\gamma^2}{(\sigma_0 - \sigma^*)^2 + \gamma^2} \tag{A.47}$$

in place of the exact expression which can be easily obtained by integration of Eq. (A.46):

$$\frac{d\sigma}{d\Omega} \cdot \frac{1}{A^2 Q F^2} = \frac{\gamma^2}{\sqrt{[(\sigma_+ - \sigma^*)^2 + \gamma^2][(\sigma_- - \sigma^*)^2 + \gamma^2]}}. \tag{A.48}$$

Contrary to Eq. (A.47), Eq. (A.48) is also dependent on the angle. As can be inferred from Fig. 36a, the dependence upon the angle reduces the cross section at small angles in the resonance region. The resonance (at small angles) is broadened by 10-20%. The broadening is not described by the simplest form factor approximation in which the relative motion is disregarded.

The same additional angular dependence is also observed in photon scattering at nuclei [the corresponding expressions are easily obtained by assuming $\mu = 0$ in Eq. (A.28)].

This angular dependence assumes a different form in elastic processes such as photoproduction or generation of heavy mesons with mass μ^* by π mesons. We have in this case

$$\sigma_0 = s + m^2 - \frac{2}{4M^2 - t}(M^2 + m^2 - N^2)\left(s + M^2 - \frac{\mu^2 + \mu^{*2}}{2}\right), \tag{A.49}$$

$$t\Lambda(s,t) = st(2M^2 + \mu^2 + \mu^{*2} - s - t) - tM^2\left(M^2 - \mu^2 - \mu^{*2} - \frac{\mu^2\mu^{*2}}{M^2}\right) - M^2(\mu^{*2} - \mu^2)^2, \tag{A.50}$$

and in the center of mass of the A + π system, Δ is proportional to sin θ [45] so that Eq. (A.47) coincides with Eq. (A.48) for $\theta = 0$ and $\theta = 180°$, and that the greatest reduction of the cross section in the resonance region (and, hence, maximum resonance broadening) occurs at $\theta = 90°$ (see Fig. 36b).

However, one must bear in mind that the interaction between the meson and the nucleus in the initial and final states (see Figs. 28-31) can substantially change the angular dependence of the cross section in the resonance region.

References

1. G. Bethe, S. Schweber, and F. Hofman, Mesons and Fields, Vol. 1, Row, Peterson and Co., Evanston, Ill. (1955).
2. G. F. Chew, M. L. Goldberger, F. E. Low, and Y. Nambu, Phys. Rev., 106(6):1345 (1957).
3. J. Ball, Phys. Rev., 124(6):2014 (1961).
4. A. I. Lebedev, Dissertation [in Russian], Physics Institute of the Academy of Sciences of the USSR (1964).
5. R. G. Vasil'ev, B. B. Govorkov, and V. I. Gol'danskii, Zh. Eksp. Teor. Fiz., 37(1):11 (1959); B. B. Govorkov, S. P. Denisov, A. I. Lebedev, and E. V. Minarik, Zh. Eksp. Teor. Fiz., 44(5):1463 (1963). The paper includes an extensive bibliography on the problem.
6. G. Davidson, Thesis, MIT (1959).
7. R. A. Schrack, Thesis, NBS (1960).
8. B. B. Govorkov, S. P. Denisov, and E. V. Minarik, Zh. Eksp. Teor. Fiz., 44(3):878 (1963).
9. A. S. Belousov, A. I. Lebedev, C. V. Rusakov, E. I. Tamm, and L. S. Tatarinskaya, Nucl. Phys., 67:679 (1965); Report of A. I. Lebedev, Physics Institute of the Academy of Sciences of the USSR (1960).
10. Nicola Cabibbo, Phys. Rev. Lett., 7(10):386 (1963).
11. R. G. Vasil'kov, B. B. Govorkov, and V. I. Gol'danskii, Zh. Eksp. Teor. Fiz., 37(4):1149 (1959).
12. B. B. Govorkov, S. P. Denisov, and E. V. Minarik, Zh. Eksp. Teor. Fiz., 42(4):1010 (1962).
13. J. E. Leiss and R. A. Schrack, Rev. Mod. Phys., 30(2):456 (1958).
14. R. S. Schrack, J. E. Leiss, and S. Penner, Phys. Rev., 127(9):1772 (1962).

15. J. Chappelear, Phys. Rev., 99:254 (1955).
16. E. Goldwasser, L. J. Koester, and F. E. Mills, Phys. Rev., 95:1692 (1954).
17. V. V. Balashov, G. Ya. Korenman, and T. S. Macharadze, Yadernaya Fiz., 1(4):668 (1965).
18. L. J. Koester and F. E. Mills, Phys. Rev., 105(6):1900 (1957).
19. R. G. Vasil'kov and B. B. Govorkov, Zh. Eksp. Teor. Fiz., 37:1 (1959).
20. R. Hofstadter, Rev. Mod. Phys., 28:214 (1956).
21. P. Lehmann, Papers of the International Conference on Electromagnetic Interactions at Low and Medium Energies, VINITI, 3:66 (1967).
22. J. Irving, Proc. Phys. Soc. (London), 66A(1):17 (1953).
23. J. Irving, Phil. Mag., 42:338 (1951); J. Gunn and J. Irving, Phil. Mag., 42:1353 (1951).
24. Y. Yamaguchi, Progr. Theoret. Phys. Japan, 13:4 (1955).
25. J. L. Cook, Nucl. Phys., 25:421 (1961).
26. P. Palit and E. H. Bellamy, Proc. Phys. Soc. (London), 72(5):880 (1957).
27. R. W. F. McWhirter, P. Palit, and E.H. Bellamy, Nucl. Instr. and Methods, 3:80 (1958).
28. G. DeSmissure and L. S. Osborn, Phys. Rev., 94:756 (1954).
29. B. Wolfe, A. Silverman, and J. W. DeWire, Phys. Rev., 99:268 (1955).
30. H. L. Davis and D. R. Corson, Phys. Rev., 99:273 (1955).
31. M. Davier, D. Benaksas, D. Drickey, and P. L. Lehman, Phys. Rev., 137:119 (1965).
32. K. D. Stoodly, Ph. D. Thesis, University of Glasgow (1957).
33. S. P. Kharlamov, Dissertation [in Russian], Physics Institute of the Academy of Sciences of the USSR (1966).
34. J. Hamilton and W. S. Woolcock, Rev. Mod. Phys., 35:737 (1963).
35. G. Höhler, Nuovo Cim., 16:585 (1960).
36. K. Berkelman and J. A. Waggoner, Phys. Rev., 117:1364 (1959).
37. L. D. Solov'ev, Zh. Eksp. Teor. Fiz., 33:801 (1957).
38. G. F. Chew and G. C. Wick, Phys. Rev., 85(4):636 (1952).
39. J. Ashkin and G. Wick, Phys. Rev., 85(4):686 (1952).
40. C. A. Engelbrecht, Phys. Rev., 133(4):B988 (1964).
41. A. S. Belousov, S. V. Rusakov, E. I. Tamm, L. S. Tatarinskaya, and P. N. Shareiko, Yadernaya Fiz., 3(3):503 (1966).
42. A. S. Belousov, S. V. Rusakov, E. I. Tamm, and L. S. Tatarinskaya, Yadernaya Fiz., 4(1):110 (1966).
43. J. Nuttall, Nuovo Cim., 29:841 (1963); F. Gross, Phys. Rev., 134:B405 (1964); 136:B140 (1964).
44. R. E. Cutkovsky, Proc. 1960 Annual Intern. Conf. on High-Energy Physics, Rochester, p. 236.
45. V. A. Tsarev, Yadernaya Fiz., 5(1):167 (1967).
46. A. F. Jones, Nuovo Cim., 26:790 (1962).
47. I. S. Shapiro, Theory of Direct Nuclear Reactions, Gosatomizdat (1963).
48. L. S. Dul'kova, I. B. Sokolova, and N. G. Shafranova, Zh. Eksp. Teor. Fiz., 35:313 (1958).
49. G. M. Vagradov and I. B. Sokolova, Zh. Eksp. Teor. Fiz., 36:948 (1959).
50. G. Brunhavt, G. S. Faughn, and V. P. Kenney, Nuovo Cim., 29:1162 (1963).
51. H. N. Pendleton, Phys. Rev., 131:1833 (1963).
52. Electromagnetic Structure of Nuclei and Nucleons [Russian translation], Foreign Literature Press (1958).
53. Yu. A. Budakov, P. F. Ermolov, E. A. Kushnirenko, and V. I. Moskalev, Zh. Eksp. Teor. Fiz., 42:1191 (1962).
54. E. Goldwasser and L. J. Koester, Nuovo Cim. Suppl., 2:951 (1956).
55. Elementary Particle Accelerators [in Russian], Atomizdat (1957).
56. L. I. Slovokhotov, Report of the Physics Institute of the Academy of Sciences of the USSR (1962).
57. Problems of Low-Temperature Cooling [Russian translation], Foreign Literature Press (1961).
58. V. V. Matveev and A. D. Sokolov, Photomultipliers and Scintillation Counters [in Russian], Gosatomizdat (1962).
59. V. P. Agafonov, B. B. Govorkov, S. P. Denisov, and E. V. Minarik, Pribory Tekhn. Eksp., No. 5, 47 (1962).
60. G. M. Shklyarevskii, Zh. Eksp. Teor. Fiz., 46(2):690 (1964).
61. E. Fermi, H. L. Anderson, A. Lundby, D. E. Nagle, and G. B. Yodn, Phys. Rev., 85:935 (1952).
62. G. Cocconi and A. Silverman, Phys. Rev., 88:1230 (1952).
63. R. G. Vasil'kov, B. B. Govorkov, and V. I. Gol'danskii, Zh. Eksp. Teor. Fiz., 37(1):11 (1959).
64. M. F. Kaplon and T. Yamahuchi, Nuovo Cim., 15(4):519 (1961).
65. A. S. Belousov, S. V. Rusakov, E. I. Tamm, and L. S. Tatarinskaya, Zh. Eksp. Teor. Fiz., 41:6 (1961).
66. R. G. Vasil'kov, B. B. Govorkov, and A. V. Kutsenko, Zh. Eksp. Teor. Fiz., 2:23 (1960).
67. A. A. Rudenko, Pribory Tekhn. Eksp., No. 6, 60 (1958).
68. N. F. Moody, G. J. R. Maclusky, and M. O. Deighton, Millimicrosecond Pulse Techniques, CREL, 463 (1950).
69. P. N. Shareiko, Pribory Tekhn. Eksp., No. 4, 134 (1964).
70. P. N. Shareiko, Pribory Tekhn. Eksp., No. 5, 149 (1965); FIAN Preprint A-70 (1964).
71. A. S. Belousov, S. V. Rusakov, E. M. Tamm, and P. N. Shareiko, Report of the Physics Institute of the Academy of Sciences of the USSR (1961).

72. P. N. Shareiko, A. S. Belousov, S. V. Rusakov, E. I. Tamm, and L. S. Tatarinskaya, Report of the Physics Institute of the Academy of Sciences of the USSR (1963).
73. R. R. Wilson, Nucl. Instr., 1:101 (1957).
74. N. P. Klepikov and S. N. Sokolov, Analysis and Planning of Experiments with the Maximum Justification Technique [in Russian], Nauka Press (1964).
75. N. V. Smirnov and I. V. Dunin-Barkovskii, Short Course of Mathematical Statistics for Engineering [in Russian], Fizmatgiz (1959).
76. A. S. Penfold and J. E. Leiss, Analysis of Photo Cross Section, University of Illinois Press (1958); O. V. Bogdankevich and F. A. Nikolaev, Work with a Bremsstrahlung Beam [in Russian], Atomizdat (1964).
77. G. J. Schiff, Phys. Rev., 83:252 (1951).
78. E. I. Tamm, Dissertation [in Russian], Physics Institute of the Academy of Sciences of the USSR (1963).
79. G. Bethe and F. Hofman, Mesons and Fields, Vol. 2, Row, Peterson and Co., Evanston, Ill. (1956).
80. S. P. Denisov, FIAN Preprint A-154 (1962).
81. Y. Goldschmidt, H. S. Osborn, and M. Scott, Phys. Rev., 97:188 (1955); R. L. Walker, D. C. Oakley, and A. V. Tollestrup, Phys. Rev., 97:1279 (1955); W. John and G. Stoppini, Nuovo Cim., 6:1206 (1957); D. B. Miller and E. H. Bellamy, Proc. Conf. on High-Energy Phys., CERN, 205 (1962); A. S. Belousov, S. V. Rusakov, E. I. Tamm, and L. S. Tatarinskaya, Zh. Eksp. Teor. Fiz., 43(4):1550 (1962).
82. A. N. Gorbunov, Trudy Fiz. Inst. Akad. Nauk, Vol. 13, p. 174 (1960).
83. B. B. Govorkov, S. P. Dénisov, and E. V. Minarik, Yadernaya Fiz., 5:190 (1967).
84. M. Gourdin, Nuovo Cim., 33:1374 (1964).

OPPOSITELY DIRECTED ELECTRON–POSITRON BEAMS IN THE SYNCHROTRON

Yu. M. Ado, K.A. Belovintsev, E.G. Bessonov, and P. A. Cherenkov

The possibility of producing oppositely directed electron–positron beams in synchrotrons was studied from the theoretical and experimental viewpoints. Problems related to the influence of both the low injection energy and the variation in the magnetic-field strength upon the principal parameters of the particle-storage process are discussed in detail. The article outlines results of experiments to generate oppositely directed electron–positron beams and describes studies of the storage conditions in the 280-MeV synchrotron of the Physics Institute of the Academy of Sciences of the USSR. Soviet and foreign schemes using the method under consideration in synchrotrons with energies of the order of 1 GeV and more are briefly discussed.

Introduction

The technique of oppositely directed beams, which makes it possible to increase considerably the exploitation efficiency of the energy of accelerated particles and which increases the number of elementary particle interactions to be examined, has now been generally recognized. Oppositely directed beams allow well-organized investigations of interactions in which momenta of the order of 1.0 GeV and higher are transferred. Oppositely directed beams could be generated once high-current accelerators and special storage systems had become available. Soviet, Italian, American, and French scientists deserve appreciation for their work in this field [1-6]. Modern storage devices are complicated, expensive machines whose production has required the solution of numerous scientific, technological, and economic problems. The expenses for the construction of storage systems exceed frequently the expenses for large accelerators. It is therefore of particular interest to investigate the possibility of producing oppositely directed beams in the accelerator proper.

The first device of this type, a symmetric ring phasotron, was designed about ten years ago at the Physics Institute of the Academy of Sciences of the USSR [7]. Ado published a paper [8] in which the accumulation of oppositely directed electron–positron beams in synchrotrons was proposed. Contrary to conventional storage devices, Ado's design makes use of a variable magnetic field which changes in the course of time but maintains a single polarity. More particularly, when a sinusoidal function of time is used as the variable magnetic field component, the guiding magnetic field has the form shown in Fig. 1.

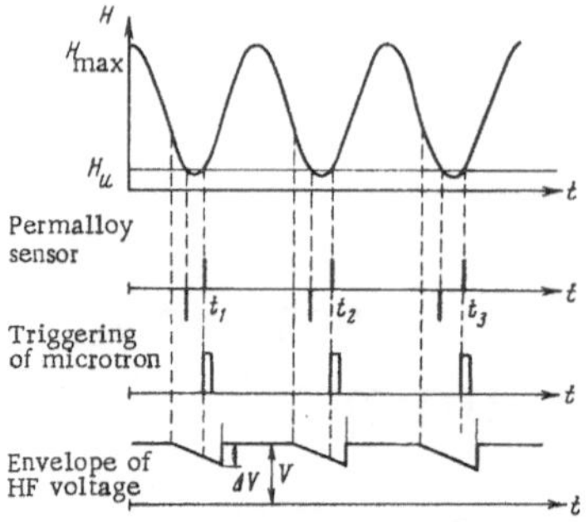

Fig. 1. Time dependence of the guiding magnetic field H and the amplitude of the hf voltage V when the synchrotron is operated as a storage device.

In each variation cycle of the guiding field, radiation damping of the betatron and synchrotron oscillations occurs near H_{max}. Accordingly, the phase volume filled by the particles increases. It is therefore possible to inject additional particles in the depleted part of the phase space at times t_1, t_2, t_3, ..., near the minimum of the magnetic field strength.

The first experiments of accumulating electrons and positrons in a synchrotron were made in the 280-MeV synchrotron of the Physics Institute of the Academy of Sciences of the USSR [8-10]. These experiments confirmed the high efficiency of the storage technique, which, in turn, stimulated further developments for other accelerators with higher energies [11-13].

The present article is a generalization of the principal results of the research on a technique of obtaining oppositely directed electron–positron beams in synchrotrons.

CHAPTER I

Physical Processes in Synchrotron Storage Amplitudes

A time-dependent guiding magnetic field (variable energy) and a low injection energy are, essentially, the main distinguishing features of a synchrotron-storage device, which have deciding influence upon the principal characteristics of the particle-accumulation process, e.g., upon the radiation damping of the oscillation amplitudes, the lifetime of the beam, and the accumulation rates.

Radiation Damping of Oscillation Amplitudes of Particles

In the case of a storage device with a variable, sinusoidal component of the guiding magnetic field, the expression for the particle energy can be written in the form

$$\gamma = \gamma_0 - \gamma_{\sim} \cos \frac{2\pi t}{T}, \tag{1}$$

where $\gamma = E/mc^2$ denotes the relative energy, T denotes the period during which the magnetic field changes, $\gamma_{\sim}$ denotes the amplitude of the variable energy component, and γ_0 denotes the constant component of the particle energy. The particle energy changes periodically in this storage device from the minimum value $E_{min} = E_0 - E_{\sim}$ to the maximum value $E_{max} = E_0 + E_{\sim}$.

Due to radiation damping, the changes in the amplitude of the betatron and phase oscillations of the particles are qualitatively given by damping decrements $\langle\zeta_z\rangle$, $\langle\zeta_x\rangle$, and $\langle\zeta_\theta\rangle$

$$A(\vartheta) = A(0)\exp\left(-\int\langle\zeta\rangle\, d\vartheta\right),$$

where $\vartheta = 2\pi\sigma/\Pi$ denotes the generalized azimuth; Π denotes the perimeter of the orbit; the length of the arc is $\sigma = vt$; v denotes the particle velocity; and the symbol $\langle\ \rangle$ denotes averaging over an orbit.

During each period T of magnetic field strength variations, the oscillation amplitudes of the particles are attenuated by $e^{-T/\tau}$, where

$$\tau = \frac{1}{1/T \int\limits_0^T \langle\zeta\rangle\, dt}$$

denotes the attenuation of the oscillations averaged over one period. When the minimum energy of the particles is much smaller than the maximum energy, Eq. (1) can be approximated by the following expression:

$$\gamma \simeq \gamma_{max} \cos^2 \frac{\pi t}{T}.$$

Thus, the expression for τ assumes the form

$$\tau_{z,x,\eta} = \frac{16\Pi}{5\pi\beta c \Gamma_0(\gamma_{max}) G_{z,x,\eta}}, \tag{2}$$

where $\beta = v/c$, $\Gamma_0 = \frac{2}{3} r_e K_0 \gamma^3$; r_e denotes the classical electron radius; $K_0 = \langle K \rangle$; and K denotes the curvature of the orbit, whereas the coefficients $G_{z,x,\eta}$ are defined by the structure of the magnetic system.

In the case of a weakly focusing synchrotron, we have

$$K_0 = \frac{1}{R(1+\lambda)},\ \lambda = \frac{ml}{2\pi R},\ G_z = 1 + \lambda,\ G_x = (1+\lambda)\frac{n}{1-n},$$
$$G_\eta = (1+\lambda)\frac{3-4n}{1-n},$$

where m denotes the number of magnetic sectors, l the length of the straight gaps, R the radius of curvature, and n the exponent of the magnetic field.

The expressions for calculating the attenuation time of the amplitudes of vertical and radial oscillations and of phase oscillations of the particles can be written in the form

$$\tau_z = \frac{48}{5}\frac{R^2(1+\lambda)}{r_e c \gamma_{max}^3},$$
$$\tau_x = \frac{1-n}{n}\tau_z, \qquad \tau_\eta = \frac{1-n}{3-4n}\tau_z. \tag{3}$$

It follows from a comparison of expressions (3) with the corresponding expressions for a storage device with constant energy [14] that in the case of a variable magnetic field, the attenuation times of the particle oscillations are approximately three times increased.

Lifetime of the Particles

A magnetic field which varies in the course of time has a strong influence upon the loss mechanism of the particles in the synchrotron-storage device. The effect manifests itself particularly in that particles, whose oscillation amplitudes increase due to various disturbing interactions during a cycle of magnetic field changes, may not be lost immediately, but only at the end of the cycle when their energy has decreased to values close to the minimum value. Particles are then lost due to an adiabatic increase in the oscillation amplitude reaching values which are limited only by the dimensions of the effective magnetic field area, or reaching the greatest possible amplitude of the phase oscillations.

Coulomb scattering and bremsstrahlung of the particles at atoms of the residual gas are among single-event processes which have great influence upon the lifetimes τ of the particles in the synchrotron-storage device. The particle lifetime which results from single-event Coulomb scattering of particles in the synchrotron-storage device is given by a formula of [9]:

$$\tau_p \simeq \frac{2\pi\gamma_{\min}\sqrt{\gamma_{\min}\gamma_{\max}}}{r_e^2 c Z^2 N_0\left(\frac{\lambda_x^2}{x_k^2}+\frac{\lambda_z^2}{z_k^2}\right)}, \tag{4}$$

where N_0 denotes the number of atoms of the residual gas (per cm^3); Z denotes the charge (Z = 7.2 for air); $2x_k$ and $2z_k$ denote the radial and vertical dimensions of the effective area; and λ_x and λ_z denote the wavelengths of the radial and axial betatron oscillations. Equation (4) was derived under the assumption that $\tau_z \gg T$. When $\tau_z \sim T$, the right side of Eq. (4) increases by a factor of about 2.

The partial lifetime due to bremsstrahlung emission is almost independent of the particle energy [14] and is entirely determined by the pressure in the chamber of the storage device:

$$\tau_\gamma \simeq \frac{2.7\cdot 10^{-4}}{p_{\text{torr}}}\ (\text{sec}). \tag{5}$$

It is a typical feature of a synchrotron-storage device that multiple processes originating from the build-up of betatron and phase oscillations have a very strong influence upon particle losses. The particles experience in this case the interaction of quantum-type fluctuations of the synchrotron radiation, mainly at energies close to the maximum energy. Certain beam dimensions with a Gaussian distribution of the oscillation amplitudes of the particles become stationary at these energies. When the energy decreases, the dimensions of the beam increase adiabatically and a part of the particles with large oscillation amplitudes is necessarily lost as soon as these amplitudes exceed the limits given by the size of the magnetic field area or the size of the region with stable phase oscillations. The partial lifetime resulting from quantum-type build-up of betatron and phase oscillations is given by a formula of [9]

$$\tau_q = Te^{\frac{u_m^2}{\overline{A^2(\gamma_{\min})}}}, \tag{6}$$

where $\overline{A^2(\gamma_{\min})}$ denotes the mean-square of the stationary oscillation amplitude and u_m^2 the square of the admissible oscillation amplitude of the particles (for phase oscillations, $u_m = 4$, and for betatron oscillations, $u_m = x_k^2, z_k^2$).

The beam size in radial direction and the beam dimensions in phase space (dimensions resulting from the build-up of oscillations by quantum-mechanical fluctuations of the synchrotron radiation) depend upon the mean-square $\overline{A^2}$ of the oscillation amplitude which can be obtained from the solution of the Einstein–Focker equation [15]. We assume that oscillation damping is weak during a period of the magnetic field variation ($\tau > T$) and that the amplitude of the accelerating voltage of the hf field changes in each cycle from $V_{\min}$ to $V_{\max}$. We obtain under these assumptions in the case of a weakly focusing synchrotron [13]:

$$\overline{A^2(\gamma_{\min})} = \frac{847\sqrt{3}}{1024}\,\frac{\Lambda R\gamma_{\max}^3}{n(1-n)\gamma_{\min}},$$
$$\overline{A_n^2(\gamma_{\min})} = \frac{45056}{3003\sqrt{3}r_e}\,\frac{e\Lambda q}{R(1+\lambda)(3-4n)}\,\frac{\gamma_{\max}^{3,5}}{(\gamma_{\min}V_{\min}V_{\max})^{0.5}}, \tag{7}$$

where $\Lambda = \hbar/mc$ and q denotes the multiplying factor of the high frequency.

Along with the excitation of phase oscillations of the particles, there are excited radial synchrotron oscillations with amplitudes which are related to $\overline{A_\eta^2}$ by the following expression of [16] in the case of weakly focusing synchrotrons:

$$\overline{A_{x\eta}^2} = \frac{R^2(1+\lambda)\, eV \sin\varphi_s}{2\pi q E_s (1-n)} \overline{A_\eta^2}, \tag{8}$$

where E_s and φ_s denote the equilibrium values of both the energy and the phase of a particle, respectively.

By substituting Eq. (7) into Eq. (6), we can easily obtain an expression which can be used to calculate the partial lifetime of particles in a storage device with variable energy. This partial lifetime results from multiple processes in quantum-mechanical fluctuations of the synchrotron radiation

$$\tau_{q,x} = T \exp\left[4.7 \cdot 10^7 \frac{x_k^2 n (1-n) E_{min}}{R E_{max}^3}\right],$$
$$\tau_{q,\eta} = T \exp\left[3.1 \cdot 10^8 \frac{(3-4n)(1+\lambda) R \sqrt{E_{min} V_{min} V_{max}}}{q E_{max}^{3.5}}\right], \tag{9}$$

where E is expressed in MeV, V in kilovolts, and R in meters. Since τ_q depends strongly upon E_{max} in Eq. (9), the limit of the maximum energy of the synchrotron-storage device is almost entirely determined by quantum-mechanical fluctuation effects. The quantity E_{max} is usually chosen so that τ_q exceeds the partial lifetime of the particles which is given by single-event scattering at the atoms of the residual gas. The restriction imposed on the E_{max} value implies a decrease in the characteristic attenuation time of the oscillation amplitudes of the particles [see Eq. (3)]. When in this case τ_{att} remains greater than the period T of magnetic field variation, the oscillation amplitude reached at the time at which the next particle bunch is injected could not yet drop to the stationary value and a portion of the phase volume is not used for particle capture. This, in turn, reduces the storage rate of particles.

In general, the specific influence of quantum fluctuations in the synchrotron radiation upon the operation of a synchrotron-storage device necessitates careful selection of the synchrotron parameters in order to maintain high particle-accumulation rates. Actual measures which can be used in a successful solution of this problem are discussed below.

Particle Injection

A low energy of the particles injected into the synchrotron-storage device is important for two reasons. On the one hand, as has been mentioned above, the particle lifetime in the synchrotron-storage device decreases at low injection energies, whereas the attenuation time of the particle-oscillation amplitudes increases. Furthermore, the entire injection process is more easily performed at low energies. As a matter of fact, the injection systems in conventional constant-energy storage devices are custom-built, expensive, and complicated and include an accelerator for an energy which is at least equal to the storage energy. In addition, the injection systems comprise a complicated deflector configuration. In the case under consideration, any high-current accelerator for low energy and with beam parameters corresponding to conditions or effective capture can be used as injector.

However, in a synchrotron-storage device which is designed for generating oppositely directed electron–positron beams, type and parameters of the injection device are primarily dictated by the need for obtaining a sufficiently intensive positron beam.

Positrons are usually produced by conversion of an electron beam at a target situated either in an intermediate point or at the output of a high-current electron accelerator for injection purposes. When the acceptance of the storage device is high enough, the conversion can take place at a target situated in the storage device proper, at the edge of the effective magnetic field region [11]. It is then not necessary to employ a complicated deflector, and both stability and reliability of operation of the injection system are greatly improved.

In storage devices with a variable magnetic field, a low injection energy produces favorable conditions for application of peculiar devices producing positron beams, the devices being distinguished by simplicity and low cost. Among these devices, there are the positron microtron of [17] and various positron traps proposed by foreign and Soviet researchers [18, 19].

There is reason to believe that careful examination of these possibilities will result in highly efficient low-cost positron sources.

Since in every injection mode, the efficiency of particle capture in the storage device depends upon the degree to which the emittance of the injected beam is matched to the acceptance of the storage device, calculations of these parameters are of great importance. The acceptance is given by the magnetic structure of the system, the size of the effective magnetic field region, and the dimensions of the storage device proper. In order to estimate the order of magnitude of these quantities, one can use the following relations which render the limits of the phase region of capture:

for radial acceptance

$$(\rho_x)_{\max} \simeq 2x_k, \quad (\Delta\gamma_x)_{\max} \simeq \frac{2K_0 x_k}{|f_x(\vartheta_u)|^2}; \tag{10}$$

for vertical acceptance

$$(\rho_z)_{\max} \simeq 2z_k, \quad (\Delta\gamma_z)_{\max} \simeq \frac{2K_0 z_k}{|f_z(\vartheta_u)|^2}; \tag{10a}$$

for longitudinal acceptance

$$l_{\max} = \Pi, \quad \left(\frac{\Delta E}{E}\right)_{\max} \simeq \frac{2K_0 x_k}{\psi(\vartheta_u)}, \tag{10b}$$

where $|f(\vartheta_u)|$ denotes the absolute value of the Floquet function at the azimuth of the deflector, and $\psi(\vartheta)$ denotes the function describing orbital expansion. We have for a weakly focusing racetrack of the particles

$$\psi(\vartheta) = \frac{1}{(1-n)(1+\lambda)}, \quad |f_z|^4 \simeq \frac{1+\lambda}{n}, \quad |f_x|^4 \simeq \frac{1+\lambda}{n-1}.$$

Contrary to accelerators for which expressions (10)-(10b) are valid without important restrictions, the injection of each ensuing particle bunch is in storage devices possible only into that part of phase space which is depleted by radiation damping of the oscillation amplitudes. This means that in the case of storage devices, the right sides of expressions (10)-(10b) must be reduced by the factor

$$\delta = 1 - e^{-T/\tau}, \tag{11}$$

which characterizes the extent of radiation damping during the time T between two neighboring injection cycles.

Depending upon the parameters of the injected beam, the phase space of the storage device can be filled during a single orbital revolution or during several orbital revolutions. Since the radial emittance and, more particularly, the vertical emittance of the injected beams is usually much lower than the corresponding acceptances of the storage system, multiorbit injection is preferable.

One can use several almost equivalent modifications of multiorbit injection into the storage device. The most widely used version is as follows. A perturbation, which shifts the equilibrium orbit toward a deflector, is produced on the equilibrium orbit before injection takes place. During injection of the particles, the perturbation is gradually removed, and the orbit returns to the original position. Capture of the multienergy particle beam with the width

$$h \simeq a_e k_{\text{eff}} \tag{12}$$

and the average angular spread

$$\Delta\gamma \simeq \frac{4\sqrt{2}}{3} K_0 \frac{\sqrt{x_k h}}{|f(\vartheta)|^2}, \tag{13}$$

can be obtained in this case, where a_e denotes the shift of the equilibrium orbit during the orbital revolution, and k_{eff}, the number of effective orbital runs.

Injection may take place during a time

$$\tau_i \simeq T_s k_{\text{eff}} \frac{x_k \delta}{h}, \tag{14}$$

where T_s denotes the period of rotational motion of a particle in equilibrium. Obviously, optimum capture corresponds to the condition that the parameter h is equal to the radial width of the injected beam. When the angular spread of the particles in the beam can be ignored, the main advantage of multiorbit injection over single-orbit injection is expressed by the factor $k_{\text{eff}}(x_k/2h)$. When the angular spread is large, the improvement is reduced to $k_{\text{eff}}(x_k/2h)^{1/2}$.

In many cases of actual importance, multiorbit injection systems can increase the efficiency of particle capture in a storage device by one order of magnitude or even more.

A more detailed analysis reveals several specific properties which are typical for particle capture in storage devices. Let us briefly discuss the main features.

1. Particles are injected into a storage device while an accelerating high-frequency field is applied. Consequently, radial oscillations and phase oscillations of the particles occur along with betatron oscillations during the capture process. This means that a proper analysis of the capture efficiency must take into account this point which manifests itself particularly in the fact that particles are removed from the deflector with different velocities, depending upon the phase of the particles relative to the phase of the accelerating voltage. The quantities h and $\Delta\gamma$ [see Eqs. (12) and (13)] depend in this case also upon the initial phases, and h may remain negative, whereas $\Delta\gamma$ may be imaginary. This means that a typical form of losses occurs in these storage devices. These losses are an analog to particle losses in synchrotrons during passage of the particles from quasi-betatron conditions to synchrotron conditions.

2. Each new particle bunch must be injected under conditions such that no previously accumulated particles are lost. This requirement results in two important limitations for the injection process. First, when the attenuation of the oscillation amplitudes of the particles during the time between two successive injection cycles is small ($\tau > T$), the shift of the equilibrium orbit toward the deflector must not exceed $x_k\delta$ in order to eliminate possible losses of particles which are captured in the preceding injection cycle.

The second limitation relates to the extent to which the equilibrium orbit is shifted and results from quantum-mechanical fluctuations in the synchrotron radiation. The second limitation implies that the equilibrium orbit must not be shifted toward the deflector over a distance of less than $(3-4)\sqrt{A_x^2(\gamma_{\min})}$, in order to avoid a sharp reduction of τ_{qx}. In synchrotron-storage devices in which τ_{qx} imposes a limitation upon the maximum particle energy, it is not advantageous to use the above-described multiorbit injection.

These features of particle injection into storage devices necessitate very careful selection of both the injection mode proper and the quantities of the injected particles, their amplitude, and the multiplication factor of the accelerating voltage.

When we denote the capture coefficient by η, the expression for the accumulation rate can be written in the form

$$\frac{dN}{dt} = \frac{N\eta}{T}, \tag{15}$$

where $N = I\tau_i/e$ denotes the number of injected particles. When the injected beam is characterized by an angular distribution and energy distribution over a wide range of values and by a certain distribution function $\Phi(\gamma, E)$ (MeV · sr)$^{-1}$ (e.g., when the conversion of positrons occurs at the edge of the effective region of the storage device), the following relation must be used for calculating the acceleration rate:

$$\frac{dN}{dt} = \frac{N\Phi\eta}{T}.$$

The phase volume which corresponds to the acceptance of the storage device is given by the product of the quantities of Eqs. (10) and some factor which is smaller than unity. The extent to which this volume is filled by particles depends upon the particular injection scheme. Equations (10), (12), and (13) facilitate estimates of the capture coefficient of particles for an injection mode to be selected. For example, if the beam is characterized by a large angular spread in vertical direction and in radial direction, we have $\eta = \eta_z \eta_x$, where

$$\eta_z = \frac{2K_0 z_k}{|f_z(\vartheta_d)|^2} \text{ (r)},$$

and the coefficient η_x is given by

$$\eta_x = \frac{4K_0^2 x_k^2 \delta E_{inj}}{3|f_x(\vartheta_d)|^2 \psi(\vartheta_d)} \text{ (MeV · r)}$$

provided that single-orbit injection is used and the beam has a large energy spread:

$$\eta_x = \frac{4\sqrt{2}}{3} \frac{K_0 \sqrt{x_k h}}{|f_x(\vartheta_d)|^2} \text{ (r)},$$

provided that multiorbit injection is used and the beam has a small energy spread;

$$\eta_x = \frac{8\sqrt{2}}{3} \frac{K_0^2 x_k \sqrt{x_k h}\, E_d}{|f_x(\vartheta_d)|^2 \psi(\vartheta_d)} \text{ (MeV · r)}$$

provided that multiorbit injection is used and the beam has a large energy spread.

CHAPTER II

Investigation of the Particle Accumulation Process and Generation of Oppositely Directed Electron–Positron Beams with the 280-MeV Synchrotron

First experimental studies of the particle accumulation process and the generation of oppositely directed electron–positron beams were made with the 280-MeV synchrotron of the Physics Institute of the Academy of Sciences of the USSR.

The principal parameters of the synchrotron are as follows: radius of the equilibrium orbit R = 82 cm; dimensions of the effective magnetic field region $2z_k \times 2x_k$ = 8 cm × 16 cm; exponent of magnetic field reduction n = 0.6; accelerating voltage V = 2.0 kV; multiplying factor of the high frequency 1; and frequency with which the guiding magnetic field varies f = 50 Hz.

Electrons were for the first time stored in a synchrotron in March 1963 [9]. The goal of that stage in the investigations was mainly to check whether it is possible, in principle, to accumulate particles in a synchrotron with a variable magnetic field of a single polarity.

A microtron with an energy of 7 MeV, a pulse current of about 30 mA, and an injection-pulse duration of 2 msec was used for electron injection (see Fig. 2). About 80% of the particles of the microtron beam were fed through an injection duct with a length of about 4 m and focused with the aid of a double quadrupole lens L_1 and L_2 upon the scattering target P of the synchrotron (about 0.2 mm Ta), i.e., onto a spot with a diameter of about 0.8 cm. The scattered magnetic field caused the electrons to impinge upon the target (which was placed at the edge of the effective magnetic field region of the synchrotron) at an angle of 15° relative to the tangent to the instantaneous orbit. In order to capture particles in the storage mode, amplitude modulation of the accelerating hf voltage was introduced; the modulation was $\Delta V = V_1 - V_2$, as shown in Fig. 1. The modulation ΔV was experimentally determined and amounted to approximately 20%. The injection scheme used made it possible to capture electrons from the external area of the radial oscillation and phase oscillation spaces of the particles (see Appendix). The lifetime of the particles amounted to fractions of a second in the first stage of the investigations (about 20 periods of variation in the guiding magnetic field). The maximum number of accumulated particles did not exceed 10^5.

After stabilization of the minimum magnetic field H_{min} with an accuracy of $3 \cdot 10^{-2}$ and the degree of modulation ΔV of the accelerating high-frequency voltage with an accuracy of $5 \cdot 10^{-2}$ and after very careful adjustment of the electron transfer system between the microtron and the scatterer, the second stage of the investigations was initiated. The second stage was directed toward the explanation of factors which strongly influence the accumulation rate and the lifetime of the particles [10].

Particular attention was paid to investigations of the dependence of the particle lifetime τ upon the following quantities: accelerating voltage V; pressure p in the synchrotron chamber; maximum energy of the particles; and degree of modulation $\Delta V/V$ of the accelerating voltage.

An analysis of the experimental data revealed that single Coulomb scattering of particles at the atoms of the residual gas and quantum-mechanical fluctuations of the synchrotron radiation are the principal effects influencing τ. The experimental data are in good agreement with the theoretical values calculated with formulas (4) and (6). Multiple processes, which are related to quantum-mechanical fluctuations of the radiation make it impossible to increase the maximum particle energy above 200 MeV with the available parameters and operation conditions of the synchrotron.

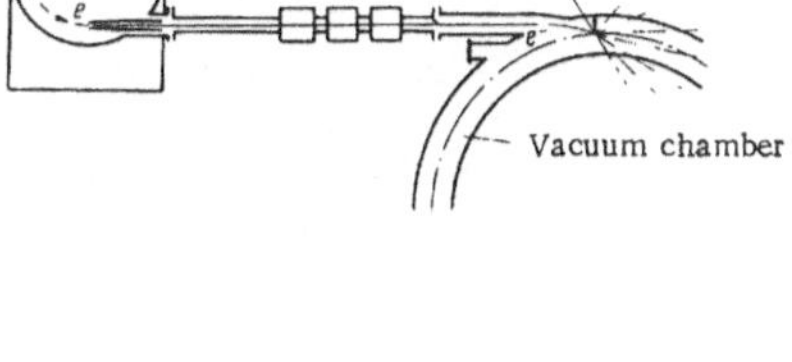

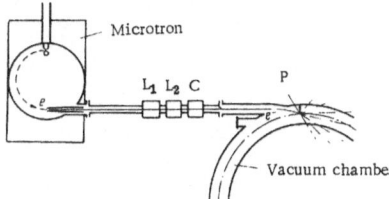

Fig. 2. Electron injection into the synchrotron.

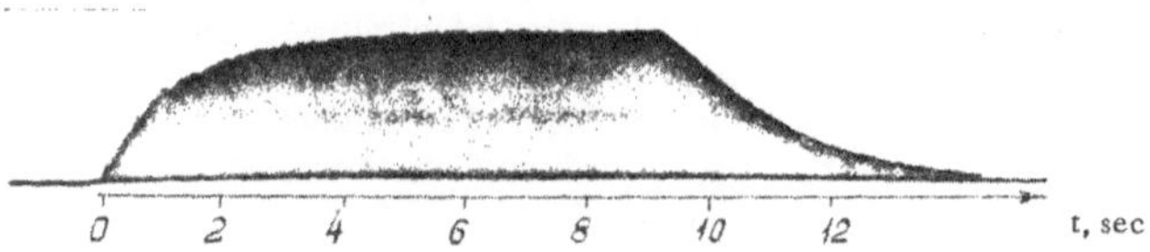

Fig. 3. Oscillogram of the electron accumulation process.

The number of stored particles was derived from measurements of the intensity of the synchrotron radiation and recorded with a loop oscilloscope. Figure 3 is one of the oscillograms. Particles are accumulated during the rising portion of the oscillogram waveform. The exponentially decreasing portion characterizes the lifetime of the particles. The beginning of the decreasing portion coincides with the moment at which the microtron is switched off.

The results which were obtained in the first two stages of the investigations confirmed that it is possible, in practice, to accumulate particles in a synchrotron. These results were the basis for the ensuing development of the technique.

The third stage of the research work comprised storage of positrons simultaneously with the storage of electrons and, hence, made it possible to generate oppositely directed electron–positron beams. Apart from this, several features of the particle storage in a variable magnetic field were considered. The possibility of operating the synchrotron-storage device with constant energy was considered [11].

As before, electrons were injected with an energy of 7.5 MeV from the microtron onto the scattering target. In order to obtain positrons, an electron beam with an energy of 14 MeV from a second microtron was used. Electrons were converted into positrons at a tantalum target (thickness about 2 mm) which was placed inside the vacuum chamber of the synchrotron-storage device (Fig. 4).

In this series of experiments, which were completed in 1966, the accumulation time amounted to 2 sec. During this time, 10^4 positrons and 10^8 electrons were accumulated in oppositely directed electron–positron beams.

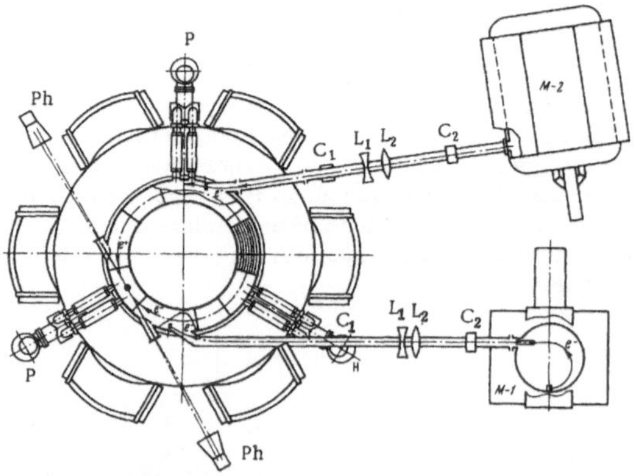

Fig. 4. Injection of electrons and positrons into the synchrotron. P) Pumps; C_1, C_2) adjustment devices; L_1, L_2) lenses; Ph) photomultiplier.

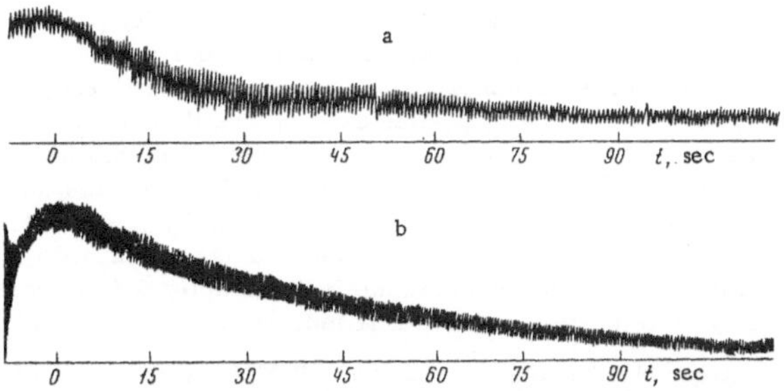

Fig. 5. Oscillograms of the signals derived from photomultipliers; the oscillograms illustrate the decay of (a) the positron beam and (b) the electron beam at 130 MeV.

Since particle capture during synchrotron operation of the storage device involves many orbits, it was of interest to determine in experiments the quantities h and K_{eff} [see Eq. (12)]. To do this, the dependence of the intensity of the beam of accumulated electrons upon the radial position of the converter was measured. When the magnetic field increased at a rate of $1.5 \cdot 10^5$ Oe/sec at the time of particle injection, the measured h value amounted to 3 mm and the number of effective orbits was approximately 20. The fact that K_{eff} was increased about four times over the usual value can be explained by particles bypassing the scatterer in vertical direction.

The synchrotron-storage device was transferred into operation with constant energy after accumulation of particles in operation with variable energy. This variable magnetic field component was switched off and the ensuing motion of the particles took place in a constant magnetic field. The field strength of the constant magnetic component could be fixed at some value in agreement with the type of the experiment and the capacity of the supply system of the accelerator. Transition to operation with constant energy increased the lifetime of the stored particles to 1 min (Fig. 5).

The lifetimes of the particles at various energies of stored particles were measured during operation with constant energy. An analysis of the observed dependence made it possible to determine the actual dimensions of the effective area and the effective pressure of the residual gas in the vacuum chamber. The dimensions of the effective area, which were determined in this way, agreed with the geometrical dimensions, whereas the effective pressure of the residual gas reached $2 \cdot 10^{-6}$ torr, which exceeds by one order of magnitude the pressure measured with an ionization manometer. The discrepancy can be qualitatively explained by oil vapors and by products of oil cracking which penetrate into the vacuum chamber when it is evacuated with diffusion pumps. Actually, all experiments on accumulation and generation

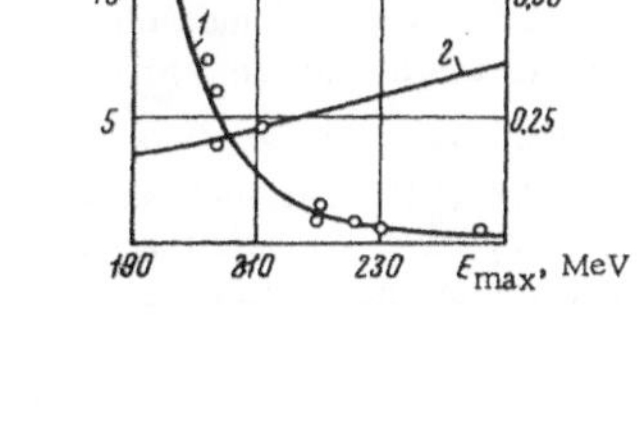

Fig. 6. Dependence of $\tau_{q\eta}$ and δ upon E_{max} (curves 1 and 2, respectively).

of oppositely directed beams were made with a vacuum chamber of porcelain whose inner surface was coated with a conductive aquadag layer. The chamber was evacuated with three diffusion pumps, as shown in Fig. 4. Each of these pumps had an efficiency of about 500 liters/sec. Moreover, the dependence of τ_q upon E_{max} was experimentally determined. The importance of E_{max} for the operation of the synchrotron-storage device was emphasized above. The results obtained are denoted by circles in Fig. 6. Solid line 1 represents the calculated relation which is normalized to the experimental point for the energy $E_{max} = 220$ MeV. As can be inferred from the figure, the experimental data are in good agreement with the calculated values. The figure includes the E_{max} dependence of damping the amplitudes of phase oscillations of the particles δ during a cycle of magnetic field variation.

The future program of experiments with the 280-MeV synchrotron of the Physics Institute of the Academy of Sciences of the USSR envisages, first of all, storage of the greatest number of electrons which the device can hold (about 10^{11} particles) and studies of possible beam instabilities. Apart from this, it is proposed to increase the number of stored particles in the oppositely directed beams to 10^{10}-10^{11} electrons and 10^{7}-10^{8} positrons, mainly by improving the vacuum, increasing the number of injected particles, and rising the efficiency of capture. Interaction between electron and positron beams will also be studied in detail in experiments.

CHAPTER III

Synchrotron-Storage Devices for High Energies

Even the first considerations of the possible generation of oppositely directed electron–positron beams in operating synchrotrons (the possibility was immediately confirmed by experiments with the 280-MeV synchrotron of the Physics Institute of the Academy of Sciences of the USSR) attracted the attention of numerous physicists. Of particular interest were the projects of synchrotron-storage devices for energies of about 1 GeV and more. At these energies, however, the effects caused by quantum-mechanical fluctuations of the synchrotron radiation are particularly noticeable in synchrotrons operated as storage devices. These effects can unfavorably influence both the particle accumulation rates and accumulation times. Apart from this, superhigh vacuum (about 10^{-9} torr) which is required for experiments with oppositely directed beams is hard to attain because the magnetic field of the synchrotron is variable and therefore implies various requirements which the vacuum chamber must satisfy.

To overcome these difficulties is one of the most important tasks which are encountered in the development and testing of synchrotron-storage devices.

Foreign Projects

The research workers of the Italian ADON project for a 1.5-GeV electron–positron storage device made for the first time attempts to accumulate particles in a variable magnetic field [7]. The calculations proved that the characteristic attenuation time of the oscillation amplitudes of the particles must be of the order of 1 sec, when the range of energy variation extends from 100 MeV to 440 MeV. This time decreases to 1/8 sec, when the energy varies from 100 to 900 MeV. The difficulties encountered in this manufacture of the vacuum chamber and the electromagnet of the storage device forced the researchers to abandon this project.

In 1964, a group at Harvard University began work on the project to generate oppositely directed beams in the 6-GeV electron synchrotron at Cambridge [12]. The method proposed by scientists of the Physics Institute of the Academy of Sciences of the USSR was the basis of the project. An injection energy of 130 MeV was used. The maximum energy in operation as a storage device amounted to about 3.0 GeV. The guiding magnetic field varied with a frequency of 60 Hz. A special linear accelerator was supposed to be used as the injector. According to estimates, the accumulation time at currents of about 0.1 A amounted to 90 sec in each of the oppositely directed electron-positron beams, provided that the average residual pressure in the chamber of the accelerator amounted to about $5 \cdot 10^{-8}$ torr. A so-called bypass system is the characteristic feature of the Cambridge project. The principle of this system is as follows: After accumulating the maximum current in the synchrotron chamber, the equilibrium orbit is distorted so that the particles which are at a certain azimuth deviate from the circular trajectory, pass through a magnetic bypass, and return to the circular trajectory. The bypass section includes a long, straight gap in which recording equipment with a large solid angle of observation can be conveniently located. The magnetic design features of the bypass are chosen so that the cross section is minimal at the point at which the beams intersect. The originators of the project hope that this design will yield a high brightness of $5 \cdot 10^{31}$ cm^2 $\cdot$ sec^{-1}. An additional advantage of the bypass results from the fact that it is possible to produce a high vacuum of about 10^{-9} torr at the point at which the beams intersect, whereas the pressure in the rest of the chamber can be kept at 10^{-7}–$5 \cdot 10^{-8}$ torr.

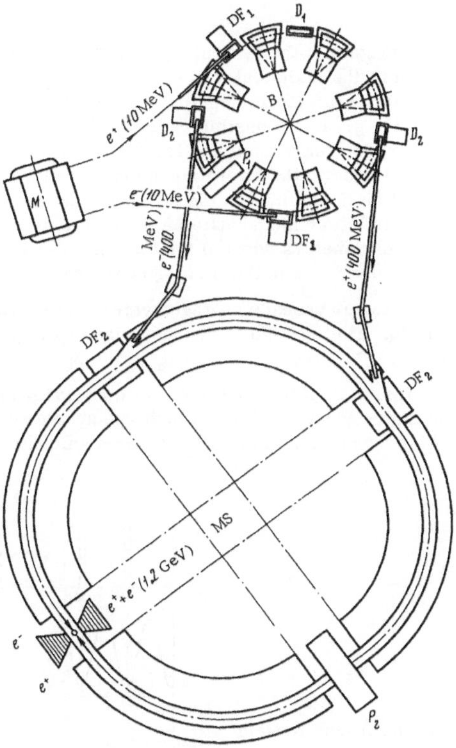

Fig. 7. Schematic view of the cascade storage device (B, booster; M, microtron; MS, main synchrotron; DF, deflector).

Cascade Accumulation System

Attempts to find efficient methods of generating oppositely directed electron–positron beams in synchrotrons at an energy of 1 GeV or more were made by the group of the Laboratory of Photomesonic Processes and the Laboratory of Accelerators of the Physics Institute of the Academy of Sciences of the USSR, when they worked on the development of a cascade accumulation system [14].

Figure 7 shows the principle of the cascade device for generating oppositely directed beams with an energy of about 1.2 GeV in a synchrotron. The particle accumulation process is in this case divided into two stages. In the first stage (see Fig. 8), particles are accumulated in an intermediate (booster) synchrotron. After intervals, which are equal to the accumulation times of particles in the booster, the electron and positron beams must be transferred into the main synchrotron (second stage of the storage device). The magnetic field in the main synchrotron is kept almost constant. Only when the particles are transferred, the magnetic field is reduced to values corresponding to the energy of the booster. The principal function of the booster synchrotron is therefore to reduce the frequency of the injection cycles and to increase the energy with which the particles are injected into the main synchrotron. These two measures result in an overall increase in particle lifetime at high accumulation rates.

In designing the electromagnet of the system, synchrotron operation in the accumulation mode and ease of operation were the main features considered. For example, in order to obtain a high particle-accumulation rate, the dimensions of the effective region and the guiding magnetic field must be given the greatest possible values. Eight straight sections provide space for the beam inlet and outlet components and for the accelerating hf systems. It is proposed that the particles produced in the microtron be captured in a large number of orbital runs. The vacuum which must be maintained in the booster must not be very high. Estimates show that a pressure of about 10^{-7} torr is adequate as far as the accumulation time of particles is concerned. With the parameters adopted for the cascade system (Fig. 7), the time required for accumulating 10^{11} positrons in the main synchrotron amounts to about 20 min.

Figures 9 and 10 show another cascade storage system for particles and the time dependence of the magnetic field. This system is characterized by high particle accumulation rates in the main synchrotron.

The system allows an increase in the coefficient of converting the electron beam into the positron beam by means of high-energy electrons which are accelerated in the main synchrotron. The time calculated for accumulation of 10^{11} positrons amounts only to several seconds

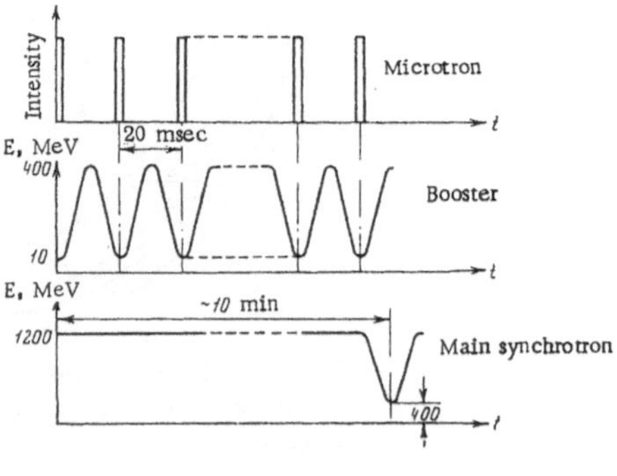

Fig. 8. Time dependence of the magnetic field in the booster and in the main synchrotron.

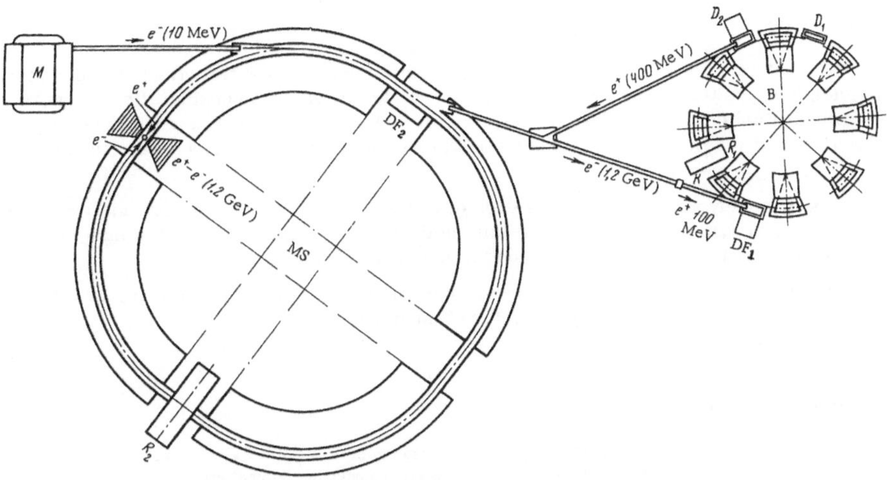

Fig. 9. Cascade storage device in which the positrons are generated in the booster (M, microtron; MS, main synchrotron; B, booster; DF, deflector).

in this system. The actual construction of this cascade storage system depends mainly upon the successful generation of a superhigh vacuum in the chamber of the main synchrotron.

The booster synchrotron of the cascade storage device may be an independent system, because the high intensity of the beam and the rather high energy of the particles (about 400 MeV) allow experiments on photomesonic and photonuclear reactions. Moreover, additional possibilities result from the generation of monochromatic gamma quanta during in-flight annihilation of positrons. Research with intensive positron beams is possible with this booster synchrotron. Of equal importance is the fact that when the parameters are properly chosen and the necessary conditions are satisfied, the booster synchrotron makes it possible to increase both the intensity and the energy of the particle beam which is accelerated in the main synchrotron.

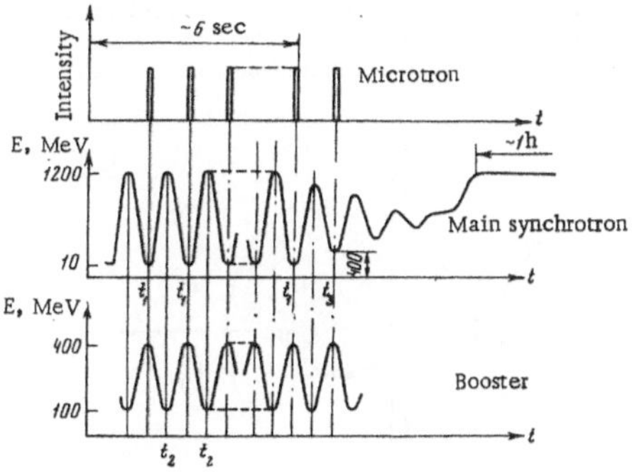

Fig. 10. Time dependence of the magnetic field in the booster and in the main synchrotron; operation as schematically shown in Fig. 9.

Conclusions

The selection of a particular storage system for studies of interactions of elementary particles in oppositely directed beams may be dictated by various considerations. Usually, the technological conditions, cost considerations, and the problems to be considered in research work are the main points. The final goal can be reached in various ways in each real case.

The main advantage of the generation of oppositely directed beams considered in the present work results from the fact that this method can be easily employed in many research centers having synchrotrons in operation for medium and high energies.

It can be assumed that the future development of the method will involve mainly improvements of cascade storage systems for elementary particles.

Appendix

We outline the calculation of the efficiency of particle capture in the storage device when a multiorbit injection system of the type considered in the text is employed. We denote the deviation, which betatron oscillations produce, from the instantaneous orbit by x, the deviation of the instantaneous orbit from the equilibrium orbit, by x_η, the shift of the equilibrium orbit from the undisturbed state, by x_s, and the distance between the deflector and the medium orbit by x_d. The condition of particle capture on the instantaneous orbit can be written in the form $x_k + x_{\eta k} + x_{sk} < x_d$, where the subscript k denotes the number of runs of the particles, and the deviation is taken at the azimuth of the deflector. This condition, which can be expressed by the initial deviations of the particle, has the form

$$x_0 \cos 2\pi \nu_x k + \alpha_0 \sin 2\pi \nu_x k < x_d - x_{\eta 0} - x_{s_0} + D(k), \qquad (A.1)$$

where $\alpha_0 = K_0^{-1} | f(\vartheta_d)|^2(\gamma_0 - \gamma^1)$, $\gamma' = x_0 K_0(| f|' / | f |)$; ν_x denotes the frequency of the betatron oscillations; $D(k) = (x_{\eta 0} - x_{\eta k}) + (x_{s_0} - x_{sk})$ denotes the distance over which the instantaneous particle orbit is shifted from the deflector during k orbital revolutions.

The group of limit lines defined by Eq. (A.1) and the straight line $x_d - x_{\eta 0} - x_{s0}$, which corresponds to the edge of the deflector on the x_0, α_0 plane form a certain region [20-22] in which the capture conditions are satisfied. This region is determined by the variation of D(k), i.e., by the initial phase φ_0 of the particle and by the initial energy of the particle (or by $x_{\eta 0}$).

Figures 11 and 12 depict the φ, x_η plane which corresponds to the radial motion and phase motion of the particles. The particles can be captured either immediately in the region of stable phase oscillations (Fig. 11) or can be captured from beyond this region (Fig. 12). Particles can be captured from the outside area if at the time at which the particles pass along the separating line, the accelerating voltage is rapidly increased (Fig. 12). When this is done, particles which are incident between the new and the old separating lines are captured.

The calculation of the efficiency with which particles are captured is reduced to the investigation of the function D(k) for various initial conditions. A straight line on the x_0, α_0 plane corresponds to each k value. The capture region on this plane is defined either by the first orbital revolution (at small k values) or by the orbital revolutions at which D(k)

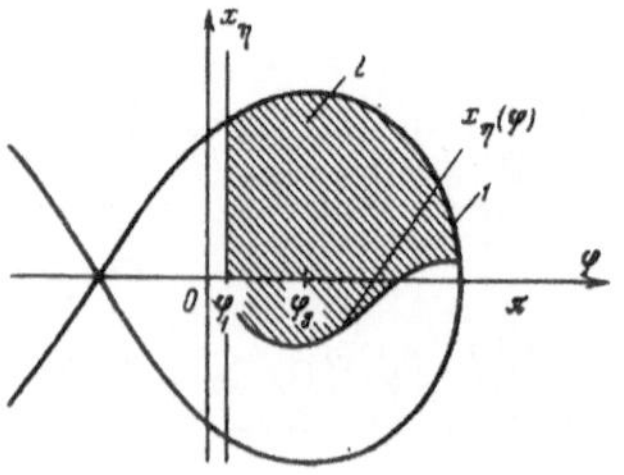

Fig. 11. Plane of radial oscillations and phase oscillations. 1) Separating line; 2) region of favorable initial conditions.

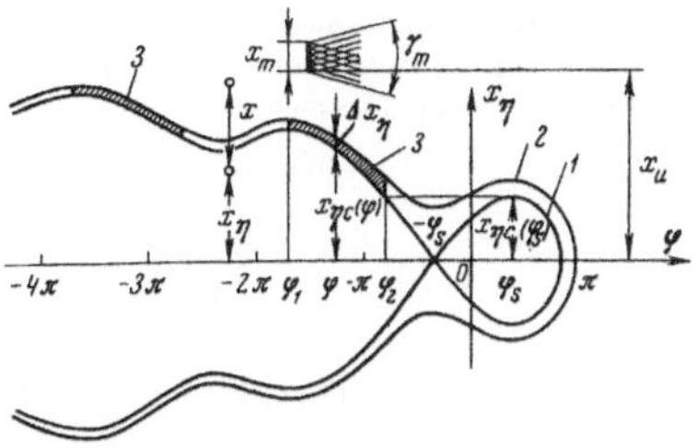

Fig. 12. Plane of radial oscillations and phase oscillations. 1) Separation line; 2) limit trajectory; 3) region of favorable initial conditions.

passes through the minimum. When $a_e > 0$ holds for all initial phase values, the function D(k) cannot have a minimum. When during its motion along a phase trajectory, the particle passes through the phase φ_s (or when at this time $\dot{x}_s = 0$ holds), or through a phase which is defined by the condition $a_e(\varphi) = 0$, ($\dot{x}_s \neq 0$), or through a phase which corresponds to the beginning of the equilibrium-orbit shift, the function D(k) has a minimum at these phase values.

In order to determine D_{min}, we must calculate $x_\eta(\varphi)$ and $x_\eta(\varphi_0)$ and the time $\int_{\varphi_0}^{\varphi} \frac{d\varphi}{\dot{\varphi}}$ during which the particle moves along the phase trajectory between the time of the particle's exit and the time at which the particle passes through the corresponding phase φ.

When the capture region is defined by the first orbital runs, the capture region is conveniently replaced by the area of a segment with height $h_1 = a_e k_{eff}$ and the radius $r = x_d - x_{\eta 0} - x_{s0} + h_1$ in the calculations. We have

$$a_e \simeq \dot{x}_s T_s + \psi(\vartheta_d)\, \frac{eV}{K_0 E}\,(\cos\varphi_0 - \cos\varphi). \tag{A.2}$$

When the capture region is determined by the second condition, we must take a segment with the height $h_2 = D_{min}$. One must assume in this case that only particles which are characterized by a region limited by the curve $h(\varphi_0, x_{\eta 0}) = 0$ on the plane of the radial oscillations and phase oscillations can be captured in storage operation, where

$$h = \min(h_1, h_2). \tag{A.3}$$

In this case a segment with radius r and a height $h > 0$ corresponds to captured particles on the plane of the betatron oscillations. This segment has the chord $\Delta\alpha \simeq 2\sqrt{2rh}$ and an area of about $\simeq \frac{4\sqrt{2}}{3} h \sqrt{rh}$.

The dimensions of the segment determine the capture coefficient η' of the particles to which a region of initial conditions close to φ_0 and $x_{\eta 0}$ corresponds to the phase plane. For example, if the beam is wide ($x_{beam} > h$) and has a large angular spread ($\alpha_{beam} > \Delta\alpha$), we have

$$\eta' = \frac{4\sqrt{2}}{3}\,\frac{K_0 h^{3/2}\sqrt{x_d - x_{\eta 0} - x_{s0}}}{|f(\vartheta_d)|^2 (x_{beam}\gamma_{beam}}.$$

The total capture coefficient of the particles is obtained by averaging η' over the time and the initial conditions φ_0, and $x_{\eta 0}$:

$$\eta_x = \int_0^t dt \iint \eta'\, d\varphi_0\, dx_{\eta 0}. \tag{A.4}$$

The coefficient η can be approximately calculated for each accelerator. To do this, the injection time Δt must be divided into several equal intervals and the capture efficiency η_i must be averaged over φ_0 and $x_{\eta 0}$ in each interval (Figs. 11 and 12). The total efficiency η is in this case $\frac{1}{n}\sum_1^n n_i$.

The high-frequency field deforms the phase volume which is filled by particles injected into the storage device. In the general case, this may result in additional losses during the capture process. The conditions under which the particles are injected into the storage device must therefore be carefully selected. When the injected beam is characterized by a large energy spread, it may be convenient to capture particles at the same time from the outside region of the phase-oscillation area and from the region defined by the separating line proper.

Obviously, particles can be directly injected without application of an accelerating voltage. To do this, the high-frequency voltage must be adiabatically slowly switched off at the time at which cos $\varphi_s \simeq 0$ (minimum of the guiding magnetic field), particles must be injected, and the voltage must be slowly increased. This injection mode is sensible when it is possible to store 100% of the previously accumulated particles.

References

1. E.A. Abromyan, V. L. Auslender, V. N. Baier, et al., Transactions of the International Conference on Accelerators (Dubna) [in Russian], 274 (1963).
2. C. Bernardini, U. Bizzarri, G. F. Corazza, G. Ghigo, R. Querzoli, and B. Touschek, Nuovo Cim., 23:202 (1962).
3. F. Amman, R. Andreani, M. Basseti, M. Bernardini, et al., Transactions of the International Conference on Accelerators (Dubna) [in Russian], 249 (1963).
4. H. Bruck, Transactions of the International Conference on Accelerators (Dubna) [in Russian], 288 (1963).
5. W. C. Barber, B. O. Gittelman, G. K. Neil, and B. Richter, Transactions of the International Conference on Accelerators (Frascati), 266 (1965).
6. F. Amman, Labor. Naz. (Frascati) Nota Intern., 79 (1961).
7. A. A. Kolomenskii, V. A. Petukhov, and M. S. Rabinovich, Pribory Tekhn. Eksp., No. 2, 26 (1956).
8. Yu. M. Ado, Atomnaya Energiya, 12:54 (1962).
9. Yu. M. Ado, K. A. Belovintsev, A. Ya. Belyak, et al., Transactions of the International Conference on Accelerators (Dubna) [in Russian], 355 (1963).
10. Yu. M. Ado, E. G. Bessonov, and P. A. Cherenkov, Atomnaya Energiya, 18:104 (1965).
11. Yu. M. Ado, K. A. Belovintsev, E. G. Bessonov, P. A. Cherenkov, and V. S. Shirchenko, Atomnaya Energiya, 23:3 (1967).
12. K. W. Robinson and G. A. Voss, CEAL-TM-149 (1965).
13. K. W. Robinson, T. R. Sherwood, and G. A. Voss, Nucl. Sci., NS-14:670 (1967).
14. K. A. Belovintsev, A. Ya. Belyak, E. G. Bessonov, et al., Translations of the International Conference on Accelerators (Cambridge), A66 (1967).
15. A. I. Alikhanyan, Usp. Fiz. Nauk, 81:7 (1963).
16. A. A. Kolomenskii and A. N. Lebedev, Theory of Cyclic Accelerators [in Russian], Fizmatgiz (1962).
17. K. A. Belovintsev and P. A. Cherenkov, Transactions of the International Conference on Accelerators (Dubna) [in Russian] (1963).
18. G. I. Budker, Atomnaya Energiya, 19:505 (1965).
19. K. W. Robinson, CEAL-1016 (1965).
20. I. S. Danilkin and M. S. Rabinovich, Zh. Eksp. Teor. Fiz., 28:351 (1958).
21. Yu. M. Ado and E. G. Bessonov, Transactions of the 6th International Conference on Accelerators [in Russian], 65 (1966).
22. M. S. Rabinovich, Trudy Fiz. Inst. Akad. Nauk, Vol. 10, p. 23 (1958).

OPTICAL ANISOTROPY OF ATOMIC NUCLEI *

S.F. Semenko

The article deals with the optical anisotropy (tensor form) of the polarizability of nuclei, which manifests itself in photonuclear reactions in the range of the giant dipole resonance. The kinematics of the interaction between radiation and an optically anisotropic system is examined. The various experiments for determining the parameters of optical anisotropy are discussed and their efficiencies are compared. (Photon scattering by oriented nuclei is discussed in detail.) Summation rules and microscopic models (various "particle–hole" approximations) are used to estimate the parameters of the optical anisotropy by strongly deformed nuclei with axial symmetry. A schematic model of the relation between dipole and quadrupole oscillations is formulated. The model is used to estimate optical anisotropy effects which result from various deformation forms of nuclei.

Introduction

Optical anisotropy is one of the most remarkable properties which are exhibited by atomic nuclei upon excitation of a giant resonance.

The sharp increase in the cross section of photonuclear reactions at energies of 10-20 MeV is termed "giant resonance" (see, e.g., [1-4] and the references therein). The interaction of the nucleus with the photons in the range of a giant resonance is mainly an electric dipole interaction. This is proved, first of all, by the agreement between the most important experimental characteristics and the theoretical estimates derived from the assumption of a dipole interaction.

The Thomas–Reiche–Kuhn summation rule is the most general of these estimates for the cross section of the dipole-like photoabsorption. The summation rule was applied to a nucleus having the Levinger–Bethe form of [1]:

$$\int_0^\infty \sigma \, d\hbar\omega = \frac{2\pi^2 e^2 \hbar}{Mc} \frac{NZ}{A}(1+\varkappa) \approx 0.06 \frac{NZ}{A}(1+\varkappa) \text{ (MeV} \cdot \text{barn)} \tag{1}$$

where $\varkappa$ denotes an additional term resulting from Majorana and Heisenberg exchange forces.

The cross section of photoabsorption (i.e., the sum of the cross sections of all photoreactions) for medium and heavy nuclei (integration over the range of the giant resonance) makes up the dipole sum for $\varkappa \leq 0.4$.

---

* Dissertation submitted in partial fulfillment of requirements for the degree of Candidate of Physicomathematical Science and defended on June 20, 1966. Scientific advisor: Professor A. M. Baldin.

The principal details of a giant resonance are described in the collective models of [5-7] in which a giant resonance is treated as an excitation of dipole oscillations of the totality of nuclear protons and neutrons.

The frequency of these oscillations can be related to the symmetry energy $k[(N-Z)/A]^2$ which appears in the Weizsäcker formula and which, in the case under consideration, functions as the potential of the exciting forces.

This relation was established in its most general form by Migdahl [5] when he estimated the static dipole-type polarizability of a nucleus and the integral $\int_0^\infty (\hbar\omega)^{-2}\sigma\, d\hbar\omega$

$$(0|c(\omega)=0)|0)=e^2R^2A/(40k)=\hbar c 2^{-1}\pi^{-2}\int_0^\infty (\hbar\omega)^{-2}\sigma\, d\hbar\omega, \tag{2}$$

where k = 23 MeV, and R denotes the radius of the nucleus ($R = A^{1/3}\cdot 1.1\cdot 10^{-13}$ cm). An estimate for the average energy of the dipole maximum (for $\varkappa = 0$) is obtained from Eqs. (1) and (2):

$$\hbar\omega_d = 82\, A^{-1/3}\ \text{MeV}, \tag{3}$$

and is in good agreement with the experimental results obtained with heavy nuclei ($A > 100$).

Afterwards, considerable progress was made in designing the microscopic picture of a dipole resonance [8-11] (see also [12-15]). It was shown on several occasions [8-11] that it is possible to explain the energy of the dipole maximum with the shell model.

The interpretation of finer details of a giant resonance (e.g., the width of the photoabsorption curve and the properties resulting from the magnetization of the nucleus, etc.) requires more complicated microscopic models ("particle-hole" approximation) and, hence, cumbersome formulas which are much more complicated than those resulting from the "particle-hole" concept (see [16, 17] and also [18, 19]).

Thus, as in other areas of nuclear physics, the theory of giant resonances describes certain properties of the nucleus with the aid of models which are based on averaged, overall characteristics.

Optical anisotropy is a property through which the predominantly collective properties of dipole excitation are put into evidence. In order to describe optical anisotropy, collective models and various averaging calculations proved useful.

The term "optical anisotropy" of a nucleus was introduced by Baldin [20-22] in analogy to molecular optics. The importance of this analogy is based on the adiabatically slow motion of the nuclear surface, compared to the dipole oscillations of that surface. It is therefore possible to compare the dipole degrees of freedom of a nucleus to the electron's degrees of freedom in a molecule and to establish a relation between the motion of the nuclear surface and the rotations and oscillations of the nuclear core of a molecule [23].

The dipole polarizability c_{ij} of an optically anisotropic system includes, in addition to the scalar part, a vector part (antisymmetric) c^1_{ij} and a tensor part (symmetric with spur zero) c^2_{ij} [24].

We can speak of two aspects of the theory of optical anisotropy: the first aspect is mainly kinematic and refers to the description of observation methods. The other aspect relates to the explanation of the parameters of optical anisotropy with the aid of a model.

Several important results concerning these two aspects were obtained by Baldin [22, 23]. In order to measure the tensor component of the polarizability, Baldin proposed an experiment on the photoabsorption by oriented nuclei and predicted the magnitude of the effect to be expected for strongly deformed nuclei.

In an experiment which was afterwards made by Ambler, Fuller, and Marshak [25] and which concerned measurements of the photoinduced neutron yield of composite Ho^{165} nuclei, a tensor-type polarizability was found. Within the experimental error limits, the magnitude of this tensor-type polarizability coincided with the theoretical predictions.

A detailed investigation of the kinematics of photon scattering at oriented, optically anisotropic nuclei was one of the ensuing theoretical problems.

Strongly deformed nuclei are the most conspicuous example of optically anisotropic nuclei. The cross section of photoabsorption of heavy, strongly deformed nuclei is divided into two maxima. The difference between the corresponding resonance energies is proportional to the extent of the deformation:

$$\omega_b - \omega_a = \omega_d (a - b) a^{-1/3} b^{-2/3}, \tag{4}$$

where a and b denote the lengths of the longitudinal and transverse semiaxes of the nuclear ellipsoid.

The existence of two frequencies characterizing oscillations in the directions of the two principal directions of a spheroidal nucleus was for the first time assumed by Okamoto [26], who emphasized the strong broadening of the giant resonance in the case of strongly deformed nuclei (in comparison with other nuclei).

Okamoto used the various collective models to estimate the dependence of the longitudinal and transverse oscillation frequencies upon the deformation in the case of spheroidal nuclei. Okamoto [26, 27] and Danos [28] obtained the following relation from the hydrodynamic model of Steinwedel and Jensen [7]:

$$\omega_b / \omega_a \approx b/a. \tag{5}$$

This relation is consistent with experimental results.

Since the hydrodynamic model is a very specialized concept, it was of interest to interpret Eq. (5) with more realistic concepts [22], particularly with the "particle—hole" model. This problem was also of interest for the explanation of the reasons for the 1.5-2 MeV broadening of the transverse maximum relative to the longitudinal maximum [29].

A certain progress in the theory of the optical anisotropy of spheroidal nuclei [22, 23] made it possible to look for methods of generalizing the theory to all other deformation forms.

An important step in this direction was made by Inopyn [30] who proposed triple splitting of giant resonances in the case of a static, nonaxial deformation.

Inclusion of the nonstatic nuclear deformation was the next problem to be solved, as has been mentioned by numerous researchers [31, 32].

The above problems determined the direction of the investigations which the authors made in 1960-1965 and which are reviewed in the present article.

Chapter 1 is a description of the kinematics of absorption and photon scattering by an optically anisotropic system.

Chapter 2 outlines estimates of the effects of tensor-type polarizability of heavy spheroidal nuclei and discusses several models of vector-type polarizability. Chapter 2 estimates, in addition, the contribution of the tensor-type polarizability to the constant of hyperfine splitting of atomic and molecular spectra.

Chapter 3 uses the summation rules as well as the "particle–hole" model to estimate the parameters of the two maxima of a giant resonance in the case of heavy, spheroidal nuclei.

Chapter 4 describes the formulation and application of a schematic model for the coupling between dipole oscillations and surface oscillations of a nucleus for the solution of several concrete problems.

CHAPTER I

Kinematic Relations in the Theory of Optical Anisotropy

Scalar, Vector, and Tensor Components of the Polarizability

We restrict our considerations to processes which result from dipole interactions of electromagnetic radiation with matter. The Hamiltonian of this type of interactions is

$$H_\gamma = -\mathbf{d}\mathbf{E}\exp(i\omega t), \tag{I.1}$$

where $\mathbf{d}$ denotes the operator of the electric dipole moment of the system; ω denotes the frequency of the electromagnetic radiation; and $\mathbf{E}$ denotes the electric field strength.

The scattering of γ quanta by the system is described with the aid of the S matrix

$$\hat{S} = \hat{1} + i\hat{R}. \tag{I.2}$$

In the first nonvanishing order of perturbation theory on interactions of the type defined in Eq. (I.1), the $\hat{R}$ matrix is [33]

$$\hat{R} = R\delta(E_i + \hbar\omega - E_f - \hbar\omega'), \tag{I.3}$$

$$(f\omega'\lambda' \mid R \mid i\omega\lambda) = \frac{1}{2\pi}\left(\frac{\omega'\omega}{c^2}\right)^{3/2}\sum_r\left[\frac{(f\mid d_i\mid r)(r\mid d_j\mid i)}{F_r - E_i - \hbar\omega - i\hbar\Gamma i/2} + \frac{(f\mid d_j\mid r)(r\mid d_i\mid i)}{E_r - E_f + \hbar\omega + i\hbar\Gamma_r/2}\right]\lambda_i'^{*}\lambda_j - \frac{\omega}{2\pi}\frac{e^2}{Mc^3}\frac{Z^2}{AM}\delta_{ij}\lambda_i'^{*}\lambda_j, \tag{I.4}$$

where $\left.\begin{matrix}\omega, \lambda\\ \omega', \lambda'\end{matrix}\right\}$ denotes the frequency and the vector of photon polarization before and after scattering, respectively; $|i)$ and $|f)$ denote the state of the system before and after scattering, respectively; E_i and E_f denote the corresponding energies of the system; E_r and $\hbar\Gamma_r$ denote the energy and the width of level r, respectively; and M denotes the mass of the nucleon. The second term in Eq. (I.4) describes the Thomson scattering at the charge of the nucleus.

In our ensuing discussion, we use the notation

$$R^T \equiv -(2\pi)^{-1}(\omega/c)^3,\quad t \equiv -(2\pi)^{-1}\omega e^2 Z^2 A^{-1} M^{-1} c^{-3} \tag{I.5}$$

for the amplitude of Thomson scattering.

The quantity

$$(f\mid c_{ij}\mid i) = \sum_r \frac{(f\mid d_i\mid r)(r\mid d_j\mid i)}{E_r - E - \hbar\omega - i\hbar\Gamma_r/2} + \frac{(f\mid d_j\mid r)(r\mid d_i\mid i)}{E_r - E + \hbar\omega + i\hbar\Gamma_r/2} \tag{I.6}$$

is the matrix element of the dipole polarizability of the nucleus. This quantity is the coefficient of the proportionality between the external electric field and the induced electric dipole moment $\mathbf{d}'$

$$d_i' = c_{ij}E_j. \tag{I.7}$$

The dipole component of the polarizability can be resolved into three parts which represent a scalar, an axial vector, and a symmetric tensor with zero trace:

$$c_{ij} = c^0\delta_{ij} + c^1_{ij} + c^2_{ij}, \tag{I.8}$$

$$c^1_{ij} = -c^1_{ji}, \tag{I.9}$$

$$c^2_{ji} = c^2_{ij}; \qquad c^2_{ii} = 0. \tag{I.10}$$

These quantities will be termed scalar, vector, and tensor polarizability.

Vector and tensor polarizabilities characterize the optical anisotropy of the system.

The following expressions for the scalar, vector, and tensor polarizabilities result from Eqs. (I.6) and (I.8)-(I.10):

$$(f\,|\,c^0\,|\,i) = 2\,(3\hbar)^{-1}\sum_r (f\,|\,\mathbf{d}\,|\,r)(r\,|\,\mathbf{d}\,|\,i)\,\omega_r(\omega_r^2 - \omega^2 - i\omega\Gamma_r)^{-1}, \tag{I.11}$$

$$(f\,|\,c^1_{ij}\,|\,i) = \hbar^{-1}\sum_r [(f\,|\,d_i\,|\,r)(r\,|\,d_j\,|\,i) - (f\,|\,d_j\,|\,r)(r\,|\,d_i\,|\,i)]\,(\omega + i\Gamma_r/2)(\omega_r^2 - \omega^2 - i\omega\Gamma_r)^{-1}, \tag{I.12}$$

$$(f\,|\,c^2_{ij}\,|\,i) = 2\hbar^{-1}\sum_r \left\{\frac{1}{2}[(f\,|\,d_i\,|\,r)(r\,|\,d_j\,|\,i) + (f\,|\,d_j\,|\,r)(r\,|\,d_i\,|\,i)] - \frac{1}{3}(f\,|\,\mathbf{d}\,|\,r)(r\,|\,\mathbf{d}\,|\,i)\right\}\omega_r(\omega_r^2 - \omega^2 - i\omega^2\Gamma_r)^{-1}, \tag{I.13}$$

where $\hbar\omega_r = E_r - E_i$. (Here, and in our ensuing discussion, we assume that $E_r - E_f \approx E_r - E_i$.)

It is convenient for our ensuing calculations to switch to circular coordinates. These are defined by relations of [33]:

$$\mathbf{e}_0 = \mathbf{e}_z,\ \mathbf{e}_{\pm 1} = (i\mathbf{e}_y \pm \mathbf{e}_x)/\sqrt{2},\quad \mathbf{e}^{\mu} = (-1)^{\mu}\,\mathbf{e}_{-\mu}. \tag{I.14}$$

The circular covariant coordinates of the various components of polarizability are

$$(f\,|\,c^g_\sigma\,|\,i) = \exp\{i\pi[1 - (-1)^g]/4\}\,(1, 1, g)\,\hbar^{-1}\sum_{\mu r}(11\mu\nu|g\sigma)\,(f\,|\,d_\mu\,|\,r)\,\times$$
$$\times\,(r\,|\,d_\nu\,|\,i)\,[(\omega_r - \omega - i\Gamma_r/2)^{-1} + (-1)^g(\omega_r + \omega + i\Gamma_r/2)^{-1}]; \qquad g = 0, 1, 2; \tag{I.15}$$

the symbol (e, f, g) denotes a Clebsh–Gordan coefficient of the following form:

$$(e, f, g) \equiv \left(e, f, \frac{1 - (-1)^{e+f-g}}{2}, \frac{(-1)^{e+f-g} - 1}{2}\,\middle|\, g0\right). \tag{I.16}$$

Thus, the R matrix assumes the form

$$R = \frac{1}{2\pi}\left(\frac{\omega'\omega}{c^2}\right)^{3/2}\sum_{g\sigma}(c^g_\sigma - t\delta_{g0})\,(-1)^\sigma\,T^g_{-\sigma}(\boldsymbol{\lambda}'^*, \boldsymbol{\lambda}), \tag{I.17}$$

$$T^g_\sigma(\boldsymbol{\lambda}'^*, \boldsymbol{\lambda}) \equiv \exp\{i\pi[1 - (-1)^g]/4\}\,(1, 1, g)^{-1}\sum_\mu (11\mu\nu|g\sigma)\,\lambda'^*_\mu\lambda_\nu. \tag{I.18}$$

Principal Effects of Optical Anisotropy

Let us consider the kinematic properties of scattering and absorption of γ quanta, since the kinematic properties make it possible to determine the parameters of the optical anisotropy of nuclei.

Photoabsorption [22, 29]

According to the optical theorem of [34], the cross section of total photoabsorption is

$$\sigma = 8\pi^2 (c/\omega)^2 \operatorname{Im} (i\lambda|\mathscr{R}|i\lambda). \qquad \text{(I.19)}$$

It follows from Eqs. (I.17)-(I.19)

$$\sigma = 4\pi (\omega/c) \operatorname{Im} \left\{ (i|c^0|i) - c^1_{JJ}\overline{\mathbf{J}\,[\lambda^*\lambda]/J} + c^2_{JJ}\overline{\left[\frac{3}{2}(\mathbf{J}\lambda^*)(\mathbf{J}\lambda) + \frac{3}{2}(\mathbf{J}\lambda)(\mathbf{J}\lambda^*) - J(J+1)\right]} \Big/ J(2J-1) \right\}, \qquad \text{(I.20)}$$

where $c^{1,2}_{JJ} \equiv (i, M = J|c^{1,2}_0|i, M = J)$ denotes the spectroscopic values of the vector and tensor polarizabilities, i.e., the average value of the ground state at the maximum projection of the moment.

The spin factors of Eq. (I.20) must be averaged over the density matrix ρ_{M_1, M_2} which describes the orientation of the nuclei:

$$\overline{\varphi(\mathbf{J})} \equiv \sum_{M_1, M_2} (M_1|\varphi(\mathbf{J})|M_2)\, \rho_{M_2M_1}. \qquad \text{(I.21)}$$

As can be inferred from Eq. (I.21), the vector and tensor polarizabilities make the cross section of photoproduction depend upon the orientation of the electric vector of the photon relative to the spin vector of the nucleus.

The vector polarizability provides a contribution to the photoabsorption only when the electric vector is complex, i.e., when the emission is characterized by circular or elliptic polarization.

The absorption cross section of an unpolarized photon beam is

$$\sigma^J = 4\pi \frac{\omega}{c} \operatorname{Im} \{(i|c^0|i) - c^2_{JJ}[3(\mathbf{Jk})^2 - J(J+1)]/[2J(2J-1)]\}. \qquad \text{(I.22)}$$

The tensor polarizability in this case is only half as large as in the case of linearly polarized photons.

The dependence of an effect upon the polarization of the photon is conveniently described by expressing the cross section with the Stokes parameters [33, 35]. The Stokes parameters are related to the density matrix of the photon by the relation

$$\Pi_{i', j'} = \overline{\lambda^*_{i'}\lambda_{j'}} = \frac{1}{2}\begin{pmatrix} 1+\zeta_3 & \zeta_1 - i\zeta_2 \\ \zeta_1 + i\zeta_2 & 1-\zeta_3 \end{pmatrix}, \qquad \text{(I.23)}$$

where i' and j' = 1 or 2; the vectors $\mathbf{e}'_1$, $\mathbf{e}'_2$, and **k** (wave vector of the photon) form an orthonormal right-hand system. By definition, ζ_3 is the difference between the probabilities of polarization in the directions $\mathbf{e}'_1$ and $\mathbf{e}'_2$; ζ_1 is the difference between the probabilities of polarization in the directions $(\mathbf{e}'_1 + \mathbf{e}'_2)$ and $(\mathbf{e}'_1 - \mathbf{e}'_2)$; and ζ_2 is the difference between the probabilities of right-hand and left-hand circular polarization.

When the density matrix of the spin states of the nuclei is diagonal

$$\rho_{M_1M_2} = a_{M_1}\delta_{M_1M_2} \qquad \text{(I.24)}$$

we obtain for the absorption cross section of the photons with arbitrary polarization:

$$\sigma^J = 4\pi \frac{\omega}{c}\left\{ (i|c^0|i) + if_1\zeta_2 \cos\theta^A c^1_{JJ} + f_2\left[-\frac{1}{2}\left(\frac{3}{2}\cos^2\theta^A - \frac{1}{2}\right) + \frac{3}{4}\sin^2\theta^A (\zeta_3\cos 2\psi^A - \zeta_1 \sin 2\psi^A)\, c^2_{JJ}\right]\right\}, \qquad \text{(I.25)}$$

where φ^A, ψ^A, and θ^A denote the Euler angles which define the position of the coordinate system e_1', e_2', and k (in which the density matrix of the incident photon beam is expressed) relative to a system in which the matrix of the nuclear orientation is given; and f_1 and f_2 denote the first and second parameters of orientation.

Here, as well as in our ensuing discussion, we use the notation of [36] for the orientation parameter of rank g:

$$f_g \equiv \sum_M \rho_{MM}(JgM0|JM)/(JgJ0|JJ) \tag{I.26}$$

Scattering at Nonoriented Systems

Scattering is the main experimental device for investigating the optical anisotropy by molecules and, hence, scattering has been discussed in detail in the literature [24].

Since the experimental methods of experimental and nuclear physics differ, it is indicated to bring now an additional discussion of the effect "scattering."

The scattering cross section in the transition from a state $|i, \lambda)$ into a state $|f, \lambda')$ is

$$\frac{d\sigma}{d\Omega}(f\lambda' \leftarrow i\lambda) = 4\pi^2 (c/\omega)^2 (2J+1)^{-1} \sum_{M, M'} |(f, \lambda' | R | i, \lambda)|^2. \tag{I.27}$$

After using the Wigner–Eckhart theorem [37], the corresponding matrix elements of the polarizability are stated in the form

$$(fM' | c_\sigma^g | iM) = (gJ\sigma M | J'M')(f \| c^g \| i). \tag{I.28}$$

By substituting Eq. (I.28) into Eq. (I.27), we obtain

$$\frac{d\sigma}{d\Omega}(f\lambda' \leftarrow i\lambda) = \omega'^3 \omega c^{-4} (2J'+1)(2J+1)^{-1} \sum_g (2g+1)^{-1} |(f \| c^g - t\delta_{g_0} \| i)|^2 \times$$
$$\times \sum_\sigma |T_\sigma^g(\lambda'^*\lambda)|^2 = \omega'^3 \omega c^{-4} (2J+1)(2J+1)^{-1} \Big[|(f \| c^0 - t \| i)|^2 \ |(\lambda'^*\lambda)|^2 + \frac{1}{3} |(f \| c^1 \| i)|^2 (1 - |\lambda'\lambda|^2) +$$
$$+ \frac{1}{20} |(f \| c^2 \| i)|^2 (3 + 3|\lambda'\lambda|^2 - 2|\lambda'^*\lambda|^2) \Big]. \tag{I.29}$$

With the transformations

$$|\lambda'^*\lambda|^2 = \sum_{g\sigma} (-1)^\sigma \sum_\alpha (11\alpha\beta | g\sigma) \lambda_\alpha'^* \lambda_\beta' \sum_\gamma (11\gamma\varepsilon | g, -\sigma) \lambda_\gamma^* \lambda_\varepsilon, \tag{I.30}$$

$$|\lambda'\lambda|^2 = \sum_{g\sigma} (-1)^{g+\sigma} \sum_\alpha (11\alpha\beta | g\sigma) \lambda_\alpha'^* \lambda_\beta' \sum_\gamma (11\gamma\varepsilon | g, -\sigma) \lambda_\gamma^* \lambda_\varepsilon, \tag{I.31}$$

and, in addition,

$$\sum_\alpha (11\alpha\beta | g\sigma) \lambda_\alpha'^* \lambda_\beta' = \sum_\mu D_{\sigma\mu}^g \sum_{\alpha, \beta = -1, 1} (11\alpha\beta | g\mu) \Pi_{\alpha\beta}', \tag{I.32}$$

where $\Pi_{\alpha\beta}'$ denotes the density matrix of the scattered photon, we obtain from Eq. (I.29) formulas for the scattering cross section in the direction of k' and for the Stokes parameters ζ_1', ζ_2', ζ_3' which describe the photons scattered in this direction:

$$\frac{d\sigma}{d\Omega}(\lambda, \mathbf{k}') = \sum_{\lambda'} \frac{d\sigma}{d\Omega}(f\lambda' \leftarrow i\lambda) = \frac{\omega'^3\omega}{2c^4} \frac{2J'+1}{2J+1} \Big\{ |(f \| c^0 - t \| i)|^2 (1 + \cos^2\theta^S) + \frac{1}{3} |(f \| c^1 \| i)|^2 (3 - \cos^2\theta^S) +$$
$$+ \frac{1}{20} |(f \| c^2 \| i)|^2 (13 + \cos^2\theta^S) - \Big[|(f \| c^0 - t \| i)|^2 - \frac{1}{3} |(f \| c^1 \| i)|^2 +$$
$$+ \frac{1}{20} |(f \| c^2 \| i)|^2 \Big] \sin^2\theta^S (\zeta_3 \cos 2\varphi^S + \zeta_1 \sin 2\varphi^S) \Big\}, \tag{I.33}$$

$$\zeta_3' \frac{d\sigma}{d\Omega}(\lambda, \mathbf{k}') = \frac{\omega'^3\omega}{c^4}\frac{2J'+1}{2J+1}\left[|(f\|c^0 - t\|i)|^2 - \frac{1}{3}|(f\|c^1\|i)|^2 + \frac{1}{20}|(f\|c^2\|i)|^2\right]\left\{-\frac{\sin^2\theta^S}{2}\cos 2\psi^S + \right.$$

$$\left. + \cos^4\frac{\theta^S}{2}[\zeta_3\cos 2(\varphi^S + \psi^S) + \zeta_1\sin 2(\varphi^S + \psi^S)] + \sin^4\frac{\theta^S}{2}[\zeta_3\cos 2(\varphi^S - \psi^S) + \zeta_1\sin 2(\varphi^S - \psi^S)]\right\}, \quad \text{(I.34)}$$

$$\zeta_1' \frac{d\sigma}{d\Omega}(\lambda, \mathbf{k}') = \frac{\omega'^3\omega}{c^4}\frac{2J'+1}{2J+1}\left[|(f\|c^0 - t\|i)|^2 - \frac{1}{3}|(f\|c^1\|i)|^2 + \frac{1}{20}|(f\|c^2\|i)|^2\right]\left\{\frac{\sin^2\theta^S}{2}\sin 2\psi^S + \right.$$

$$\left. + \cos^4\frac{\theta^S}{2}[\zeta_1\cos 2(\varphi^S + \psi^S) - \zeta_3\sin 2(\varphi^S + \psi^S)] - \sin^4\frac{\theta^S}{2}[\zeta_1\cos 2(\varphi^S - \psi^S) - \zeta_3\sin 2(\varphi^S - \psi^S)]\right\}, \quad \text{(I.35)}$$

$$\zeta_2' \frac{d\sigma}{d\Omega}(\lambda, \mathbf{k}') = \frac{\omega'^3\omega}{c^4}\frac{2J'+1}{2J+1}\left[|(f\|c^0 - t\|i)|^2 - \frac{1}{3}|(f\|c^1\|i)|^2 - \frac{1}{4}|(f\|c^2\|i)|^2\right]\zeta_2\cos\theta^S. \quad \text{(I.36)}$$

φ^S, θ^S, and ψ^S denote the Euler angles which describe the position of the system $\mathbf{e}_1''$, $\mathbf{e}_2''$, and $\mathbf{e}_3'' \equiv \mathbf{k}'$ (the density matrix of the scattered photon is expressed in this system) relative to the system $\mathbf{e}_1'$, $\mathbf{e}_2'$, and $\mathbf{e}_3' = \mathbf{k}$ in which the density matrix of the incident photon was given. The scattering angle is denoted by θ^S.

Thus, experiments on photon scattering allow determinations of the absolute value of the matrix elements c^0, c^1, and c^2.

In molecular optics, the polarization of scattered radiation is measured for this purpose (e.g., $d\sigma/d\Omega$, ξ_1' and ξ_2' at $\xi_2 \neq 0$ are measured, in order to determine the absolute values of the matrix elements of the three polarizabilities).

The experimental data on the kinematics of photon scattering by nuclei in the range of a giant resonance are at the present time limited to the dependence on the scattering angle θ^S (see [38, 39]).

Information which could be deduced from these experimental results on optical anisotropy parameters is generally ambiguous, because the measurements of the cross section in dependence of θ^S yield two equations for the determination of three unknowns $|(f\|c^{0,1,2}\|i)|^2$. Additional information is therefore needed. (In the case of heavy nuclei, the rather self-evident assumption that the polarizability is small is made; see Chapter 2.)

Polarization of an incident photon beam makes it possible to immediately determine three parameters: $|(f\|c^{0,1,2}\|i)|^2$ from the dependence of the cross section upon the angles φ^S and θ^S. Moreover, the optical anisotropy effects are enhanced in this case.

Modern experimental techniques provided intensive beams of polarized photons [40]. Research on the optical anisotropy of nuclei with the aid of polarized photons is therefore a matter of the near future.

Scattering at Oriented Nuclei [41, 42]

These experiments allow measurements of the "interference terms" which are products of the various polarizabilities and vanish in the averaging over the orientation of the nuclei.

We will not expound the effects of photon polarization, because, as will be shown below, even without these effects our experiments provide a large number of measurable quantities.

The scattering cross section is in this case

$$\frac{d\sigma^J}{d\Omega}(f, \mathbf{k}' \leftarrow i, \mathbf{k}) = 2\pi^2(c/\omega)^2 \sum_{\lambda', \lambda, M_1, M_2 M'} (fM'\lambda'|R|iM_1\lambda)\,\rho_{M_1M_2}\,(fM'\lambda'|R|M_2 i\lambda)^*. \quad \text{(I.37)}$$

With Eqs. (I.17), (I.18), and (I.28), and with the equalities

$$T_\sigma^{g^*}(\lambda'^*\lambda) = (-1)^\sigma T_{-\sigma}^g(\lambda'\lambda^*), \quad \text{(I.38)}$$

$$\sum_{\lambda} \lambda_\alpha \lambda_\gamma = (-1)^\alpha \delta_{-\alpha\gamma} - k_\alpha k_\gamma, \tag{I.39}$$

$$(eJ\rho M_1|J'M')(fJ\sigma M_2|J'M') = (-1)^{J'-J-\sigma}(2J'+1)(2J+1)^{-1/2} \times$$
$$\times \sum_g (-1)^g (JgM_1\tau|JM_2)(2g+1)^{1/2}(ef, -\rho\sigma|gM_1-M_2)\, W(JJef; gJ') \tag{I.40}$$

we can rewrite Eq. (I.37) in the form

$$\frac{d\sigma^{\mathbf{J}}}{d\Omega}(f, \mathbf{k}' \leftarrow i, \mathbf{k}) = \frac{d\sigma}{d\Omega}(f, \mathbf{k}' \leftarrow i, \mathbf{k}) + \frac{\omega'^3\omega}{2c^4} \sum_{e, f, g>0} (f\|c^e - t\delta_{e0}\|i)(f\|c^f - t\delta_{f0}\|i)^*(-1)^{J-J'+g} \times$$
$$\times (2J'+1)(2J+1)^{-1/2}(2g+1)^{1/2}\, W(JJef; gJ), \tag{I.41}$$

where

$$\frac{d\sigma}{d\Omega}(f, \mathbf{k}' \leftarrow i, \mathbf{k}) = \frac{\omega'^3\omega}{2c^4}\, \frac{2J'+1}{2J+1} \Big\{ |(f\|c^0 - t\|i)|^2 [1 + (\mathbf{k}'\mathbf{k})^2] +$$
$$+ \frac{1}{3} |(f\|c^1\|i)|^2 [3 - (\mathbf{k}'\mathbf{k})^2] + \frac{1}{20} |(f\|c^2\|i)|^2 [13 + (\mathbf{k}'\mathbf{k})^2] \Big\},$$

$$B_g^{ef}(\mathbf{k}', \mathbf{k}', \mathbf{J}) = \sum_{M_1 M_2} \rho_{M_1 M_2} (JgM_1\tau|JM_2)\, \mathcal{T}_\tau^{(ef)\,g}(\mathbf{k}'\mathbf{k}); \tag{I.42}$$

$$\mathcal{T}_\tau^{(ef)\,g}(\mathbf{k}'\mathbf{k}) = \exp\{i\pi[2 - (-1)^e - (-1)^f]/4\}\,(1, 1, e)^{-1}(1, 1, f)^{-1} \times$$
$$\times (e, f, g)^{-1} \sum_{\rho\alpha\gamma} (ef\rho\sigma|g\tau)(11\alpha\beta|e\rho)(11\gamma\varepsilon|f\sigma)\,[\delta_{-\alpha\gamma}(-1)^\alpha - k'_\alpha k'_\gamma] \times$$
$$\times [\delta_{\beta,\,\varepsilon}(-1)^\beta - k_\beta k_\varepsilon] = \exp\{i\pi[2 - (-1)^e - (-1)^f]/4\}\,(1, 1, e)^{-1} \times$$
$$\times (1, 1, f)^{-1}(e, f, g)^{-1} \Big\{ (2f+1)^{1/2}(2e+1)^{1/2}\, W(11ef; 21) \sum_\beta (11\beta\varepsilon|2\tau) \times$$
$$\times [(-1)^f k_\beta k_\varepsilon + (-1)^e k'_\beta k'_\varepsilon]\,\delta_{g2} + \sum_{\rho\alpha\gamma} (ef\rho\sigma|g\tau)(11\alpha\beta|e\rho)(11\gamma\varepsilon|f\sigma)\, k'_\alpha k'_\gamma k_\beta k_\varepsilon \Big\}. \tag{I.43}$$

We obtain immediately from Eq. (I.43) that

$$\mathcal{T}_\tau^{(01)\,1} = \mathcal{T}_\tau^{(10)\,1} = (\mathbf{k}'\mathbf{k})\,[\mathbf{k}'\mathbf{k}]_\tau, \tag{I.44}$$

$$\mathcal{T}_\tau^{(12)\,1} = \mathcal{T}_\tau^{(21)\,1} = -\frac{1}{2}(\mathbf{k}'\mathbf{k})\,[\mathbf{k}'\mathbf{k}]_\tau, \tag{I.45}$$

and also that

$$\mathcal{T}_\tau^{(ef)\,2} = \mathcal{T}_\tau^{(fe)\,2} = a^{(ef)}[D_{0\tau}^2(\mathbf{k}') + D_{0\tau}^2(\mathbf{k})] + b^{(ef)} D_{0\tau}^2([\mathbf{k}'\mathbf{k}]) \tag{I.46}$$

for $e + f = 2$ or 4, and

$$\mathcal{T}_\tau^{(ef)\,2} = -\mathcal{T}_\tau^{(fe)\,2} = c\,[D_{0\tau}^2(\mathbf{k}') - D_{0\tau}^2(\mathbf{k})] \tag{I.47}$$

for $e + f = 3$.

The coefficients a, b, and c can be easily obtained by equating Eqs. (I.46) and (I.47) to Eqs. (I.43) for concrete $\mathbf{k}'$ and $\mathbf{k}$ values. (For example, $\mathbf{k}' = \mathbf{k} = \mathbf{e}_z$, $\mathbf{k}' = \mathbf{e}_y$, $\mathbf{k} = \mathbf{e}_z$, $\mathbf{k}' = \mathbf{e}_z$, and $\mathbf{k} = \mathbf{e}_x$.) Accordingly, we obtain

$$\mathcal{T}_\tau^{(02)\,2} = \mathcal{T}_\tau^{(20)\,2} = -\frac{1}{2}[D_{0\tau}^2(\mathbf{k}') + D_{0\tau}^2(\mathbf{k})] + \frac{1}{2} D_{0\tau}^2([\mathbf{k}'\mathbf{k}]), \tag{I.48}$$

$$\mathcal{T}_\tau^{(11)\,2} = D_{0\tau}^2(\mathbf{k}') + D_{0\tau}^2(\mathbf{k}) + D_{0\tau}^2([\mathbf{k}'\mathbf{k}]), \tag{I.49}$$

$$\mathcal{T}_\tau^{(22)\,2} = -\frac{5}{4}[D_{0\tau}^2(\mathbf{k}') + D_{0\tau}^2(\mathbf{k})] - \frac{1}{4} D_{0\tau}^2([\mathbf{k}'\mathbf{k}]), \tag{I.50}$$

$$\mathcal{T}_\tau^{(12)\,2} = -\mathcal{T}_\tau^{(21)\,2} = i\sqrt{\frac{5}{3}}\,[D_{0\tau}^2(\mathbf{k}') - D_{0\tau}^2(\mathbf{k})]. \tag{I.51}$$

The quantity $\mathcal{T}_{\tau}^{(12)3}$ is a symmetric sum of products of the type $k'_\alpha k_\beta [k'k]_\gamma$, which transform into $D^3_{0\tau}(\alpha)$ for $\mathbf{k}' = \mathbf{k} = [\mathbf{k}'\mathbf{k}] = \boldsymbol{\alpha}$. $\mathcal{T}^{(22)4}$ is the symmetric sum of the quantities $k'_\alpha k_\beta k'_\gamma k_\varepsilon$ and transforms into $D^4_{0\tau}(\alpha)$ at $\mathbf{k}' = \mathbf{k} = \boldsymbol{\alpha}$.

These definitions can be used to uniquely determine the actual form of given functions.

More particularly, we have

$$\mathcal{T}_{\tau=0}^{(12)3} = \mathcal{T}_{\tau=0}^{(21)3} = \frac{5}{2} k'_z k_z [k'k]_z - \frac{1}{2} (\mathbf{k}'\mathbf{k}) [\mathbf{k}'\mathbf{k}]_z; \tag{I.52}$$

$$\mathcal{T}_{\tau=0}^{(22)4} = \frac{35}{8} k'^2_z k^2_z - \frac{5}{8} [k'^2_z + k^2_z + 4k'_z k_z (\mathbf{k}'\mathbf{k})] + \frac{1}{8} [1 + 2 (\mathbf{k}'\mathbf{k})^2]. \tag{I.53}$$

Obviously, in the case of a diagonal density matrix,

$$B_g^{ef} = (JgJ0|JJ) f_g \mathcal{T}_{\tau=0}^{(ef)g}, \tag{I.54}$$

where f_g denotes an orientation parameter of order g [see Eq. (I.26].

In the general case of a nondiagonal density matrix, the above considerations and the fact that the quantity $\sum_{M_1M_2} (JgM_1\tau|JM_2) \rho_{M_1M_2} D^g_{0\tau}(\alpha)$ is the zero component of the spin tensor of rank g in a coordinate system whose z axis is identical with vector $\boldsymbol{\alpha}$ lead to a reformulation of the quantities $B_g^{ef}(\mathbf{k}', \mathbf{k}, \mathbf{J})$ in the form

$$B_1^{01} = B_1^{10} = (\mathbf{k}'\mathbf{k}) \overline{\mathbf{J}[\mathbf{k}'\mathbf{k}]} [J(J+1)]^{-1/2}, \tag{I.55}$$

$$B_1^{12} = B_1^{21} = -\frac{1}{2} (\mathbf{k}'\mathbf{k}) (\overline{\mathbf{J}[\mathbf{k}'\mathbf{k}]}) [J(J+1)]^{-1/2}, \tag{I.56}$$

$$B_2^{02} = B_2^{20} = \frac{1}{2} \{-[3\overline{(\mathbf{J}\mathbf{k}')^2} + 3\overline{(\mathbf{J}\mathbf{k})^2} - 2J(J+1)] + 3\overline{(\mathbf{J}[\mathbf{k}'\mathbf{k}])^2} - J(J+1)[\mathbf{k}'\mathbf{k}]^2\} [(2J-1)J(J+1)(2J+3)]^{-1/2}, \tag{I.57}$$

$$B_2^{11} = \{3\overline{(\mathbf{J}\mathbf{k})^2} + 3\overline{(\mathbf{J}\mathbf{k})^2} - 2J(J+1) + 3\overline{(\mathbf{J}[\mathbf{k}'\mathbf{k}])^2} - J(J+1)[\overline{\mathbf{k}'\mathbf{k}}]^2\} [(2J-1)J(J+1)(2J+3)]^{-1/2}, \tag{I.58}$$

$$B_2^{22} = -\frac{1}{4} \{5[3(\mathbf{J}\mathbf{k}')^2 + 3(\mathbf{J}\mathbf{k})^2 - 2J(J+1)] + 3(\mathbf{J}[\mathbf{k}'\mathbf{k}])^2 - J(J+1) \times [\mathbf{k}'\mathbf{k}]^2\} [(2J-1)J(J+1)(2J+3)]^{-1/2}, \tag{I.59}$$

$$B_2^{12} = -B_2^{21} = i\sqrt{\frac{5}{3}} [3\overline{(\mathbf{J}\mathbf{k}')^2} - 3\overline{(\mathbf{J}\mathbf{k})^2}] [(2J-1)J(J+1)(2J+3)]^{-1/2}, \tag{I.60}$$

$$B_3^{12} = B_3^{21} = \left\{5S\overline{(\mathbf{J}\mathbf{k}')(\mathbf{J}\mathbf{k})(\mathbf{J}[\mathbf{k}'\mathbf{k}])} - \left[J(J+1) - \frac{1}{3}\right] (\mathbf{k}'\mathbf{k}) (\overline{\mathbf{J}[\mathbf{k}'\mathbf{k}]})\right\} \times [(J-1)J(2J-1)J(J+1)(2J+3)(J+2)]^{-1/2}, \tag{I.61}$$

$$B_4^{22} = \left\{\frac{35}{2} \overline{S(\mathbf{J}\mathbf{k}')(\mathbf{J}\mathbf{k}')(\mathbf{J}\mathbf{k})(\mathbf{J}\mathbf{k})} - \frac{5}{2}\left(J^2 + J - \frac{5}{6}\right) \overline{[(\mathbf{J}\mathbf{k}')^2 + (\mathbf{J}\mathbf{k})^2 + 2(\mathbf{J}\mathbf{k}')(\mathbf{J}\mathbf{k})(\mathbf{k}'\mathbf{k}) + 2(\mathbf{J}\mathbf{k})(\mathbf{J}\mathbf{k}')(\mathbf{k}'\mathbf{k})]} + \frac{1}{2}(J-1)J(J+1)(J+2)[1 + 2(\mathbf{k}'\mathbf{k})^2]\right\} [(2J-3)(J-1)(2J-1)J(J+1)(2J+3)(J+2)(2J+5)]^{-1/2}, \tag{I.62}$$

where S(Ja)(Jb)... denotes a product which was made symmetric with respect to vectors **a**, **b**,... .

Due to the optical anisotropy, the scattering cross section depends upon the angles between the wave vectors of the incident and scattered photons and the direction in which the nuclei are aligned. This implies, in particular, an "azimuthal asymmetry" of the scattering, i.e., an angular dependence of the scattering in the plane perpendicular to the beam. Measurements of this effect allow determinations of the matrix elements of the tensor-type polarizability.

Differences in the scattering intensity to the right and to the left of the plane of the incident beam and of the nuclear polarization z are a specific effect of the vector-type polarizability. The vector polarizability can be detected by measuring the difference

$$\frac{d\sigma^J}{d\Omega}\left(\mathbf{k}'\mathbf{k} = -\frac{1}{\sqrt{2}}, [\mathbf{k}'\mathbf{k}]_z = \frac{1}{\sqrt{2}}\right) - \frac{d\sigma^J}{d\Omega}\left(\mathbf{k}'\mathbf{k} = -\frac{1}{\sqrt{2}}, [\mathbf{k}'\mathbf{k}]_z \equiv \frac{-1}{\sqrt{2}}\right).$$

In order to determine the tensor-type polarizability, the nuclei must be aligned ($f_{2,4} \neq 0$). The vector-type polarizability can be measured on polarized nuclei ($f_{1,3} \neq 0$).

As follows from Eqs. (I.41)-(I.54), the above-described experiment facilitates determinations of the six parameters of optical anisotropy: $|(f\|c^1\|i)|^2$, $|(f\|c^2\|i)|^2$, $\mathrm{Re}\,[(f\|c^0\|i)^* (f\|c^1\|i)]$, $\mathrm{Re}\,[(f\|c^0\|i)^* (f\|c^2\|i)]$, $\mathrm{Re}\,[(f\|c^1\|i)^* (f\|c^2\|i)]$, and $\mathrm{Im}\,[(f\|c^1\|i)^* (f\|c^2\|i)]$. Experiments on photon scattering at oriented nuclei are at the present time possible because nuclei can be aligned and uncontaminated samples are available.

The main advantage of this experiment results from the fact that the experiment provides directly information on the optical anisotropy, on a large number of measurable quantities, and a direct relation between the cross section and the polarizability.* The experiment is therefore very important (at least among the experiments discussed) for quantitative research on the optical anisotropy of nuclei.

CHAPTER II

Some Very Simple Calculations of the Effects of Optical Anisotropy

Effects of Tensor-Type Polarizability of Heavy Spheroidal Nuclei

Let us consider the effects of tensor polarizability of heavy nuclei with a large axially symmetric deformation, as encountered in the majority of rare-earth and transuranium elements.

We assume the form of the nucleus to be stationary and disregard weak surface oscillations.

Since the deformation β_0 is large and the rotation rate of the core is small ($\hbar^2/I \approx 20$-30 keV) so that $h\omega_d\beta_0 \gg h^2/I$, the forces resulting from the nonspherical field exceed greatly the forces acting in the system of main deformation axes of the Coriolis force. The orientation of the dipole-type oscillations relative to the main deformation axes is constant in the course of time, i.e., the dipole-type oscillations are internal excitations of the core.

In order to unify the ensuing calculations, the dipole resonance of odd nuclei is conveniently assumed to result only from excitations of the core, i.e., $\mathbf{d} = \mathbf{d}_{\mathrm{core}}$.

The error which is made under these assumptions is immaterial because it is generally accepted that an unpaired nucleon has practically no influence upon the photoabsorption curve of heavy nuclei.

* For example, the cross section of photoabsorption in heavy nuclei is defined as the sum of the cross sections of reactions (γ, n) and (γ, 2n). An experimental discrimination between these processes was obtained in 1962 by the group of Fultz [43]. Before, that time, the value of the cross section above the threshold of the reaction (γ, 2n) was practically unknown (see, e.g., [3]). The resulting ambiguity affects the results of Ambler, Fuller, and Marshak [25] who measured the photo-induced neutron yield from oriented nuclei.

The dipole polarizability can then be expressed by two parameters, namely the "internal" scalar and tensor polarizabilities

$$c^S \equiv (0|c'^0|0) = \frac{2}{3\hbar}\sum_r\left[\frac{|(0|d_{z'}|r)|^2\omega_r}{\omega_r^2-\omega^2-i\omega\Gamma_r}+\frac{|(0|d_{x'}|r)|^2\omega_r+|(0|d_{y'}|r)|^2\omega_r}{\omega_r^2-\omega^2-i\omega\Gamma_r}\right], \quad \text{(II.1)}$$

$$c^T \equiv (0|c_0'^2|0) = \frac{2}{3\hbar}\sum_r\left[\frac{2|(0|d_{z'}|r)|^2\omega_r}{\omega_r^2-\omega^2-i\omega\Gamma_r}-\frac{|(0|d_{x'}|r)|^2\omega_r+|(0|d_{y'}|r)|^2\omega_r}{\omega_r^2-\omega^2-i\omega\Gamma_r}\right] \quad \text{(II.2)}$$

(0 and r denote the quantum numbers which characterize the ground state and the excited state of the core; $d_{x',\,y',\,z'} = \mathbf{d}\mathbf{e}'_{1,2,3}$).

The vector-type polarizability of an even – even core is zero.

After replacing the set of levels of the longitudinal and transverse excitation by two "longitudinal" and transverse" resonances with frequencies ω_a and ω_b and widths Γ_a and Γ_b, we can rewrite Eqs. (II.1) and (II.2) in the following form:

$$c^S = \frac{2}{3\hbar}\left[\frac{(0|d_{z'}^2|0)\,\omega_a}{\omega_a^2-\omega^2-i\omega\Gamma_a}+\frac{(0|d_{x'}^2+d_{y'}^2|0)\,\omega_b}{\omega_b^2-\omega^2-i\omega\Gamma_b}\right], \quad \text{(II.3)}$$

$$c^T = \frac{2}{3\hbar}\left[\frac{2\,(0|d_{z'}^2|0)\,\omega_a}{\omega_a^2-\omega^2-i\omega\Gamma_a}-\frac{(0|d_{x'}^2+d_{y'}^2|0)\,\omega_b}{\omega_b^2-\omega^2-i\omega\Gamma_b}\right]. \quad \text{(II.4)}$$

These formulas allow calculations of the effects of tensor-type polarizability on the basis of experimental data on photoabsorption. Since $(\omega_b - \omega_a) \gtrsim \Gamma_a,\ \Gamma_b$, we have

$$c^T(\omega - \omega_a) \approx 2c^S(\omega = \omega_a); \qquad c^T(\omega = \omega_b) \approx -c^S(\omega = \omega_b).$$

The operator of scalar polarizability is simply a number in this approximation.

$$(f|c^0|i) = c^S(f|i). \quad \text{(II.5)}$$

The matrix elements of tensor-type polarizability are calculated in full analogy to the matrix elements of the quadrupole moment:

$$(f|c_\mu^2|i) = c^T(f|D_{\mu 0}^2(\varphi\theta\psi)|i). \quad \text{(II.6)}$$

The wave functions $|f)$ and $|i)$ of the low-energy states of an even – even spheroidal nucleus are (see [44]):

$$\sqrt{\frac{2J+1}{8\pi^2}}\,D_{MK}^J(\varphi\theta\psi)\,\varphi_0, \qquad K = 0; \quad J = 0, 2, 4, \ldots, \quad \text{(II.7)}$$

where K denotes the projection of the total moment upon the symmetry axis of the nucleus, and φ_0 denotes the wave function of the internal state.

The wave functions of odd spheroidal nuclei are less reliably known.

The approximation of a strong bond between an odd nucleon and the core is usually used [45, 46],

$$\Psi_{JMK} = \sqrt{\frac{2J+1}{16\pi^2}}\,[D_{MK}^J\varphi_{K_e=K} + D_{M,-K}^J(-1)_J^K R_1\varphi_{K_e=K}] \quad \text{(II.8)}$$

(the subscript e refers to the outer nucleon and R_1 denotes an operator effecting 180° rotation around the e_1' axis).

The universal validity of this approximation is disputed [47]. We have in the general case

$$\Psi_{JM\tau} = \sum_K A_\tau^K \Psi_{JMK}. \qquad (II.9)$$

As will be shown below, the selection of the functions of an odd nucleus is not crucial for calculating observed quantities.

The matrix elements of the tensor-type polarizability for wave functions of Eqs. (II.7) and (II.9) are, respectively,

$$(J'M'K\,|\,c_\mu^2\,|\,JMK) = c^T\,(2J\mu M\,|\,J'M')\,(2J0K\,|\,J'K)\,(2J+1)^{1/2}(2J'+1)^{-1/2}, \qquad (II.10)$$

$$(J'M'\tau'\,|\,c_\mu^2\,|\,JM\tau) = c^T\,(2J\mu M\,|\,J'M')\,(2J+1)^{1/2}(2J'+1)^{-1/2}\sum_K A_{\tau'}^K A_\tau^K\,(2J0K\,|\,J'K). \qquad (II.11)$$

Accordingly, we have for the matrix elements of Eq. (I.28):

$$(f\,\|\,c^0\,\|\,i) = c^S\,(f\,|\,i), \qquad (II.12)$$

$$(f\,\|\,c^1\,\|\,i) \approx 0, \qquad (II.13)$$

$$(f\,\|\,c^2\,\|\,i) = c^T\,(2J+1)^{1/2}(2J'+1)^{-1/2}\sum_K A_{\tau'}^{J'K} A_\tau^{JK}\,(2J0K\,|\,J'K). \qquad (II.14)$$

We will give some numerical estimates of the effects of tensor polarizability in various experiments.

Photoabsorption by Oriented Nuclei [22]

The following expression for the absorption cross section of oriented, spheroidal nuclei is obtained from Eqs. (I.22), (II.3)-(II.5), and (II.11) for photons with the frequency ω which are incident at an angle θ^A relative to the direction of nuclear alignment:

$$\sigma(\theta^A) = \frac{\sigma_a}{[(\omega_a^2-\omega^2)/(\omega\Gamma_a)]^2+1} + \frac{\sigma_b}{[(\omega_b^2-\omega^2)/(\omega\Gamma_b)]^2+1} - \left(\frac{3}{4}\cos^2\theta^A - \frac{1}{4}\right) f_2 P_2 \times$$
$$\times\left[\frac{2\sigma_a}{[(\omega_a^2-\omega^2)/(\omega\Gamma_a)]^2+1} - \frac{\sigma_b}{[(\omega_b^2-\omega^2)/(\omega\Gamma_b)]^2+1}\right], \qquad (II.15)$$

where

$$\sigma_a = 8\pi\omega_a\,(0\,|\,d_{z'}^2\,|0)(3\hbar c\Gamma_a)^{-1}, \qquad (II.16)$$

$$\sigma_b = 8\pi\omega_b\,(0\,|\,d_{x'}^2 + d_{y'}^2\,|\,0)\,(3\hbar c\Gamma_b)^{-1}, \qquad (II.17)$$

and P_2 denotes the projection operator which establishes a relation between the internal quadrupole moment and the spectroscopic moment:

$$P_2 \equiv (M=J\,|\,Q_\mu\,|\,M=J)/Q_0 = \sum_K |\,A_\tau^{JK}\,|^2\,(2J0K\,|\,JK)\,(2J0J\,|\,JJ). \qquad (II.18)$$

We obtain for K = J in the limit of a strong bond

$$P_2 = J\,(2J-1)/[J+1]\,[2J+3]. \qquad (II.19)$$

The maximum effect of tensor polarizability can be characterized by the quantity:

$$T^A(\hbar\omega) = \left[\sigma^J\left(\theta^A = \frac{\pi}{2}\right) - \sigma^J(\theta^A = 0)\right] \Big/ \sigma = \frac{3}{4} P_2 f_2 \left\{ \frac{2\sigma_a}{[(\omega_a^2 - \omega^2)/(\omega\Gamma_a)]^2 + 1} - \frac{\sigma_b}{[(\omega_b^2 - \omega^2)/(\omega\Gamma_b)]^2 + 1} \right\} \times$$
$$\times \left\{ \frac{\sigma_a}{[(\omega_a^2 - \omega^2)/(\omega\Gamma_a)]^2 + 1} + \frac{\sigma_b}{[(\omega_b^2 - \omega^2)/(\omega\Gamma_b)]^2 + 1} \right\}^{-1}. \quad (\text{II}.20)$$

When we use the parameters of the photoabsorption curve* measured in [48] and the value $P_2 = 7/15$, which follows from Eq. (II.19), we obtain for the Ho^{165} nucleus:

$$T^A(\omega = \omega_a) = 0.48 f_2; \qquad T^A(\omega = \omega_b) = -0.24 f_2.$$

The results of [25] are, within the experimental error limits, in good agreement with the estimates which the authors made with relations similar to those stated above.

Scattering of Unpolarized Photons by Nonoriented Nuclei

The total scattering with excitation of levels having energies $E_f \le 2$ MeV ("quasi-elastic" scattering) is of greatest interest, because this effect is measured in the experiment under consideration due to the poor energy resolution (about 15%) of the scattered γ rays.

The quasi-elastic scattering cross section is

$$\frac{d\sigma}{d\Omega}(\mathbf{k}' \leftarrow \mathbf{k}) = \sum_{J'\tau'} \frac{d\sigma}{d\Omega} J'\tau' \leftarrow J\tau. \quad (\text{II}.21)$$

We obtain from Eqs. (II.21), (I.33), (II.5), and (II.11):

$$\frac{d\sigma}{d\Omega}(\mathbf{k}' \leftarrow \mathbf{k}) = \frac{1}{2}\left(\frac{\omega}{c}\right)^4 |c^S - t|^2 [1 + (\mathbf{k}'\mathbf{k})^2] +$$
$$+ \frac{1}{40}\left(\frac{\omega}{c}\right)^4 |c^T|^2 [13 + (\mathbf{k}'\mathbf{k})^2] \sum_{J'\tau'} (\omega - E_{J'\tau'}/\hbar)^3 \Big/ \sum_K A_{\tau'}^{J'K} A_{\tau}^{JK} (2J0K|J'K)/^2 =$$
$$= \frac{1}{2}\left(\frac{\omega}{c}\right)^4 \left\{ |c^S - t|^2 [1 + (\mathbf{k}'\mathbf{k})^2] + \frac{1}{20} |c^T|^2 \sum_{J'\tau'} \Big/ \sum_K A_{\tau'}^{J'K} A_{\tau}^{LK} (2J0K|J'K)^2/(1 - \delta) \right\}, \quad (\text{II}.22)$$

where δ denotes a small correction.

With the orthogonality relations

$$\sum_{\tau'} A_{\tau'}^{J'K'} A_{\tau'}^{J'K} = \delta_{K'K} \quad [47], \quad (\text{II}.23)$$

$$\sum_{J'} (2J0K|J'K)^2 = 1, \quad (\text{II}.24)$$

and ignoring the correction δ which in the first order of magnitude in $E_f/h\omega$ has the value

$$\delta = 3(\hbar\omega)^{-1} \sum_{J'\tau'} \left| \sum_K A_{\tau'}^{J'K} A_{\tau}^{JK} (2J0K|J'K) \right|^2 E_{J'\tau'} = \frac{9\hbar^2/I}{\hbar\omega} \simeq 2\%, \quad (\text{II}.25)$$

we obtain for the cross section of quasi-elastic scattering

$$\frac{d\sigma}{d\Omega}(\mathbf{k}' \leftarrow \mathbf{k}) = \frac{1}{2}\left(\frac{\omega}{c}\right)^4 \left\{ |c^S - t|^2 [1 + (\mathbf{k}'\mathbf{k})^2] + \frac{1}{20} |c^T|^2 [13 + (\mathbf{k}'\mathbf{k})^2] \right\}. \quad (\text{II}.26)$$

* See Table 1.

This expression, which was obtained for the first time with the approximation of a strong bond between the surface and the outer nucleon [3] is, as put into evidence by the above considerations, generally independent of structural details originating from the odd nucleon.

The relative effect of tensor-type polarizability (for $\theta^S = 90°$) is

$$t^S = \frac{13}{20}|c^T|^2 \Big/ \left[|c^S - t|^2 + \frac{13}{20}|c^T|^2\right] \tag{II.27}$$

amounts to $t^s(\hbar\omega = \hbar\omega_a) = 0.46$; and $t^S(\hbar\omega = \hbar\omega_b) = 0.32$. for the parameters measured in [48] for the giant resonance of Ho^{165}.

The results of numerous experiments on quasi-elastic scattering [3, 32, 38, 39, 49] are consistent with the estimates obtained with Eqs. (II.3), (II.4), and (II.26).

As has been mentioned above, the tensor-type polarizability can in these experiments be deduced from the dependence of the cross section upon the scattering angle [38, 39]. Tensor-type polarizability can also be determined by subtracting from the cross section the contribution resulting from scalar polarizability [3, 32, 49] (the imaginary part of scalar polarizability is proportional to the cross section of photoabsorption; the real part is obtained from the imaginary part with the aid of dispersion relations).

The accuracy of experiments on scattering and photoabsorption amounts to 10%. This inaccuracy, which increases in the ensuing analysis, results in errors which in the final results are comparable to the magnitude of the effect under consideration. The existing experimental results on photon scattering cannot be considered undisputable proof of tensor polarizability in nuclei.

Scattering at Oriented Nuclei [42]

We will show below that the relative effect of tensor-type polarizability is in this case greater than in the two above-described experiments.

The cross section of quasi-elastic scattering at oriented spheroidal nuclei is

$$\frac{d\sigma^J}{d\Omega}(\mathbf{k}' \leftarrow \mathbf{k}) = \sum_{J'\tau'} \frac{d\sigma^J}{d\Omega}\, J'\tau'\mathbf{k}' \leftarrow J\tau\mathbf{k}\,. \tag{II.28}$$

As in research on scattering at nonoriented nuclei, we make the substitution $(\omega - E_{J'\tau'}/\hbar)^3 \to \omega^3$. The error thus incurred is only a few per cent, as can be inferred from the above estimates [see Eq. (II.25)].

By substituting Eqs. (II.12)-(II.14) into Eq. (I.14), we obtain

$$\begin{aligned}\frac{d\sigma^J}{d\Omega}(\mathbf{k}' \leftarrow \mathbf{k}) - \frac{d\sigma}{d\Omega} = \frac{1}{2}\left(\frac{\omega}{c}\right)^2 \sum_{e,\, f=0,\,2;\; g=2,4} [(c^S - t)\,\delta_{e0} + c^T\delta_{e2}]\,[(c^S - t)\,\delta_{f0} + \\ + c^T\delta_{f2}]^* \sum_{K'K} A_\tau^{JK} A_\tau^{JK'} A_{\tau'}^{J'K} A_{\tau'}^{J'K'} (eJ0K|J'K)(fJ0K'|J'K')(-1)^{J'-J-g} \times \\ \times (2J+1)^1 (2g+1)^{1/2} W(JJef;\, gJ')\, B_g^{ef}(\mathbf{k}', \mathbf{k}, \mathbf{J}).\end{aligned} \tag{II.29}$$

Furthermore, when we use Eqs. (II.23) and (I.40) and the orthogonality relation of the Racah coefficients

$$\sum_{J'} (2J'+1)(2g+1)^{1/2}(2g'+1)^{1/2} W(JJef;\, gJ')\cdot W(JJef;\, g'J') = \delta_{gg'} \tag{II.30}$$

we obtain the following simple expression for the cross section of quasi-elastic scattering at oriented spheroidal nuclei

$$\frac{d\sigma^J}{d\Omega}(\mathbf{k}' \leftarrow \mathbf{k}) - \frac{d\sigma}{d\Omega}(\mathbf{k}' \leftarrow \mathbf{k}) =$$

$$= \frac{1}{2}\left(\frac{\omega}{c}\right)^4 \sum_{e,\, f=0,\, 2;\; g=2,\, 4} [(c^S - t)\,\delta_{e0} + c^T \delta_{f2}]^* (e, f, g)^2 (JgJ0 \mid JJ)^{-1} P_g B_g^{ef}(\mathbf{k}', \mathbf{k}, J), \qquad (\text{II}.31)$$

where P_g denotes the projection operator of the following form:

$$P_g = (JgJ0|JJ) \sum_K |A_\tau^{JK}|^2 (JgJK|JK) \qquad (\text{II}.32)$$

[see also Eqs. (II.18) and (II.19)]. In the limit of a nucleon strongly bound to the surface we obtain for J = K:

$$P_g = (JgJ0|JJ)^2. \qquad (\text{II}.33)$$

We include the formula for the case of a diagonal density matrix, i.e., for the case in which the cross section can be expressed by the orientation parameters f_g of Eq. (I.26):

$$\left(\frac{d\sigma^J}{d\Omega} - \frac{d\sigma}{d\Omega}\right) 2 \left(\frac{c}{\omega}\right)^4 = -\left\{\mathrm{Re}\,[(c^S - t)^* c^T] + \frac{5}{14}|c^T|^2\right\}\left(\frac{3}{2}k_z'^2 + \right.$$

$$\left. + \frac{3}{2}k_z^2 - 1\right) P_2 f_2 + \left\{\mathrm{Re}\,[(c^S - t)^* c^T] - \frac{1}{14}|c^T|^2\right\}\left(\frac{3}{2}[k'k]_z^2 - \frac{[\mathbf{k}'\mathbf{k}]^2}{2}\right) P_2 f_2 +$$

$$+ \frac{18}{35}|c^T|^2 \left\{\frac{35}{8} k_z'^2 k_z^2 - \frac{5}{8} k_z' k_z \mathbf{k}'\mathbf{k} + \frac{1}{2}[1 + 2(\mathbf{k}'\mathbf{k})^2] P_4 f_4\right\} \qquad (\text{II}.34)$$

[when the degree of alignment is not very low, the last term in Eq. (II.34) can be ignored].

An azimuthal scattering asymmetry, which results from tensor-type polarizability, can be described by the quantity

$$T^S = \left[\frac{d\sigma^J}{d\Omega}(k_x' = 1, k_y = 1) - \frac{d\sigma^J}{d\Omega}(k_z' = 1, k_y = 1)\right] \Big/ \frac{d\sigma}{d\Omega}(\mathbf{k}'\mathbf{k} = 0) =$$

$$= \left[3\mathrm{Re}\,[(c^S - t)^* c^T] + \frac{2}{7}|c^T|^2\right\} P_2 f_2 \Big/ \left[|(c^S - t)|^2 + \frac{13}{20}|c^T|^2\right]. \qquad (\text{II}.35)$$

For $P_2 = 7/15$ and the giant-resonance parameters of [48] for Ho^{165} we obtain

$$T^S(\omega = \omega_a) = 0.61 f_2; \quad T^S = (\omega = \omega_b) = -0.58 f_2.$$

Values of $f_2 \approx 0.5$ can be attained at the present time, i.e., even when the cross section is measured with an accuracy of 10%, the effect of interest can be safely established.

The results of the calculations can be considered valid for nuclei with J = 7/2 and $\beta_0 = 0.3$, i.e., for Ho^{165}, Er^{167}, Lu^{175}, Hf^{177}, and Ta^{181}.

Figure 1 shows the function $T^S(\hbar\omega)/f_2$ and includes for comparison the curve $T^A(\hbar\omega)/f_2$ representing the tensor polarizability effect in photoabsorption [see Eq. (II.20)]. As can be in-

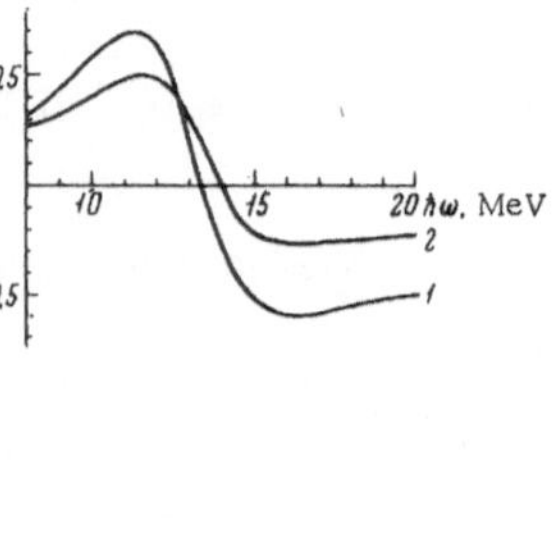

Fig. 1. Characteristics of the tensor-type polarizability in scattering and absorption by oriented nuclei (T^S/f_2 and T^A/f_2, curves 1 and 2, respectively); the calculations refer to the Ho^{165} nucleus.

ferred from the figure, the relative effect of tensor polarizability in scattering is 1.5-2 times greater than the corresponding effect in photoabsorption.

Vector Polarizability

It was mentioned above that the vector polarizability of heavy nuclei must be extremely small because the vector polarizability results from an odd nucleon which in the case of heavy nuclei has practically no influence upon the photoabsorption curve. We will illustrate this conclusion with the aid of actual calculations and will try to establish in which cases the effects of vector polarizability may be important.

It follows from the equation for the part of the induced dipole moment due to vector polarizability

$$\mathbf{d}'_v = [\mathbf{c}^1\mathbf{E}], \tag{II.36}$$

where **E** denotes the external electric field, that this characteristic reflects rotations or a magnetic field in the system. Let us compare, as an example, Eq. (II.36) and the equation of electron "drift" in perpendicular electric and magnetic fields:

$$\mathbf{v} = \frac{c}{H^2}[\mathbf{HE}]. \tag{II.37}$$

The Faraday effect [50], i.e., the rotation of the plane of polarization when light passes through a medium located in a magnetic field, is the classical example of vector polarizability.

Let us return to our model calculations.

The Two-Dimensional Harmonic Oscillator ($\omega_x = \omega_y$) in a Magnetic Field Perpendicular to the xy Plane

We obtain in this simple example estimates of the vector polarizability, which, in general, are valid as far as their order of magnitude is concerned.

The self-energy of this system is

$$E_{n_\perp,\Lambda} = \hbar(\omega_x^2 + \omega_L^2)^{1/2}(n_\perp + 1 - \hbar\omega_L\Lambda), \tag{II.38}$$

where $\omega_L = eH/(2Mc)$ denotes the Larmor frequency; $n_\perp$ denotes the principal quantum number; and Λ denotes the projection of the momentum upon the direction z of the magnetic field. The eigenfunctions are the oscillator wave functions having the frequency $\omega_\perp \equiv (\omega_x^2 + \omega_L^2)^{1/2}$.

The polarizability components in the xy plane are

$$(n_\perp\Lambda|c^0_{xx} + c^2_{xx}|n_\perp\Lambda) = e^2(M\omega_\perp)^{-1}\{(\omega_\perp + \omega_L)[(\omega_\perp + \omega_L)^2 - \omega^2 - i\omega\Gamma]^{-1} + \\ + (\omega_\perp - \omega_L)[(\omega_\perp - \omega_L)^2 - \omega^2 - i\omega\Gamma]^{-1}\}, \tag{II.39}$$

$$(n_\perp\Lambda|c^1_{xy}|n_\perp\Lambda) = -ie^2\omega(M\omega_\perp)^{-1}\{[(\omega_\perp + \omega_L)^2 - \omega^2 - i\omega\Gamma]^{-1} - [(\omega_\perp - \omega_L)^2 - \omega^2 - i\omega\Gamma]^{-1}\}. \tag{II.40}$$

It follows from Eqs. (II.39) and (II.40) that when the "rotational" structure of the dipole spectrum can be resolved, i.e., when $\Gamma \ll \omega_L$, the vector polarizability of the region of the dipole peaks is comparable to the scalar polarizability:

$$c^1_{xy}(\omega_\perp \pm \omega_L)/c_{xx}(\omega_\perp \pm \omega_L) = \pm i. \tag{II.41}$$

When this is not the case (i.e., if $\Gamma \gg \omega_d$), we have

$$|c^1_{xy}|/|c_{xx}| \simeq \omega_L\omega/|\omega_\perp^2 - \omega^2 - i\omega\Gamma| \lesssim \omega_L/\Gamma. \tag{II.42}$$

Interestingly enough, when we switch to the static limit in Eq. (II.40), we obtain

$$\lim_{\omega\to 0} i\omega c^{1}_{xy} = -ec/H. \tag{II.43}$$

After inserting this expression into Eq. (II.36) and assuming $e\mathbf{v} = i\omega\mathbf{d}$, we obtain the equation for electron drift [Eq. (II.37)].

Vector-Type Polarizability of Heavy Spheroidal Nuclei

In the approximation of a strong bond between the outer nucleon and an even–even core, the vector polarizability of a spheroidal nucleus is characterized by a single parameter which is the zeroth component of this quantity in the system of internal axes:

$$(0\,|\,c^{1}_{x'y'}\,|\,0) \equiv c^{V} = \sum\left[\frac{|(0\,|\,d_{-1}\,|\,r)|^{2}}{\omega_r^2-\omega^2-i\omega\Gamma_r} - \frac{|(0\,|\,d_{1}\,|\,r)|^{2}}{\omega_r^2-\omega^2-i\omega\Gamma_r}\right]. \tag{II.44}$$

In the case under consideration, the "rotation," which leads to vector-type polarizability, originate from a single external nucleon.

We replace the set of transitions $K \to K+1$ and $K \to K-1$ by two levels with the effective energies $\hbar\omega_{\pm 1}$, respectively, and the widths $\Gamma_b' \equiv \Gamma_b - |\omega_{+1} - \omega_{-1}|$, where Γ_b denotes the width of the transverse maximum of a giant resonance.

The frequencies $\omega_{\pm 1}$ are the "centers of gravity" of the transitions $\Delta K = \pm 1$. The difference between the frequencies can be defined in the following way:

$$\omega_1 - \omega_{-1} = \left\{\sum_r \omega_r\,|(0\,|\,d_{-1}\,|\,r)|^2 - \sum_r \omega_r\,|(0\,|\,d_1\,|\,r)|^2\right\}\Big/\,(0\,|\,d^2_{x'}\,|\,0). \tag{II.45}$$

We can evaluate Eq. (II.45) with the model of independent particles:

$$\omega_1 - \omega_{-1} = \left\{\sum_n (\varepsilon_n - \varepsilon_{j_e})\,[\,|(j_e\,|\,x' - iy'\,|\,n)|^2 - |(j_e\,|\,x' + iy'\,|\,n)|^2] - \right.$$
$$\left. - \sum_j (\varepsilon_{j_e} - \varepsilon_j)\,[\,|(j_e\,|\,x' - iy'\,|\,j)|^2 - |(j_e\,|\,x' + iy'\,|\,j)|^2\right\} \sum_{nj}[\,|(n\,|\,x\,|\,j)|^2 + |(n\,|\,y\,|\,j)|^2], \tag{II.46}$$

where j_e denotes the quantum numbers of the outer nucleon; and j and n denote the quantum numbers of the filled and unfilled states, respectively.

For the numerical calculation with Eq. (II.46), we use the asymptotic Nilsson wave functions and energies of [51]. These energies are characterized by the intrinsic quantum numbers $n_{\parallel}$, $n_{\perp}$, and Λ of the spheroidal oscillator and the projection σ of the spin upon the symmetry axis. The asymptotic value of Nilsson's single-particle energy is

$$\varepsilon_{n_z n_\perp \Lambda\sigma} = \hbar\omega_{\parallel}\left(n_{\parallel} + \frac{1}{2}\right) + \hbar\omega_{\perp}(n_{\perp}+1) - \alpha\hbar\omega_0\Lambda\sigma - \alpha\varkappa\hbar\omega_0\,[\Lambda^2 + 2n_{\parallel}n_{\perp} + 2n_{\parallel} + n_{\perp}]. \tag{II.47}$$

We have in this case

$$\hbar\omega_1 - \hbar\omega_{-1} = -\left[2\alpha\sigma_e + 4\varkappa\alpha\left(\Lambda_e + \frac{1}{4}\right)\right]\hbar\omega_0/A. \tag{II.48}$$

In the case of the nuclei of the rare-earth series, $\alpha\hbar\omega_0 \approx 1$ MeV and $\alpha\varkappa\hbar\omega_0 \approx 0.2$ MeV, which means that $\hbar\omega_1 - \hbar\omega_{-1} \lesssim 4$ MeV/A ≈ 20 keV.

The difference $\hbar(\omega_1 - \omega_{-1})$ is much smaller than the width of the transverse maximum $\hbar\Gamma_b \approx 4$ MeV. The ratio of vector polarizability to scalar polarizability can therefore be estimated with Eq. (II.42)

$$|c^V(\omega = \omega_b)| : |c^S(\omega = \omega_b)| \leqslant \frac{3}{2}\frac{\omega_1 - \omega_{-1}}{\Gamma_b} \approx 1\ \%. \tag{II.49}$$

We must bear in mind that, since $\omega_1 - \omega_{-1}$ is of the same order of magnitude as the rotation frequency of the nucleus ($\hbar^2/I \approx$ 20-30 keV), the vector polarizability can, generally speaking, not be taken as characteristic of only the internal motion and, strictly speaking, the rotation of the nucleus must be taken into consideration. However, the order of magnitude of the estimates must remain unchanged.

Isolated Dipole Level with a Certain Spin Value

The above estimate is acceptable when the giant resonance is free of any fine structure.

In the case of light nuclei ($A \leq 30$), the giant resonance is characterized by a large number of maxima. The width of several of the peaks is very small ($\hbar\Gamma \approx$ 0.5-1 MeV) [52, 53].

It is generally accepted that in the low-energy spectrum of odd nuclei such as F^{19}, Mg^{25}, and Al^{27}, there exist several levels which can be interpreted as results of collective rotations [54]. The parameter $\hbar^2/I$ of the collective rotation is 200-300 keV. Levels with a certain spin value, which are separated by intervals of $(\hbar^2/I)J \approx$ 0.5-1 MeV, can therefore appear in a giant resonance of light nuclei. If the width of this level is small enough [$\hbar\Gamma_r <$ (1-0.5) MeV], the level can be assumed isolated, i.e., for $\omega = \omega_r$, all other resonances can be ignored.

In order to estimate the ratio of the vector and tensor polarizabilities to scalar polarizability, we express the polarizabilities in this case by the reduced matrix elements of the dipole moment

$$(fJ'M'|d_\mu|rJ''M'') = (1\mu J''M''|J'M')(f\|d\|r), \tag{II.50}$$

$$(rJ''M''|d_\mu|iJM) = (1J\mu M|J''M'')(r\|d\|i) \tag{II.51}$$

and, with the equality

$$\sum_{\mu\nu}(11\mu\nu|g\sigma)(1\mu J''M''|J'M')(1\nu JM|J''M'') = (2J+1)(2J''+1)^{-1/2}\sum_g(2g+1)^{1/2}\times$$
$$\times(-1)^{J-J'+g}(JgM\sigma|J'M')\,W(JJ'11;gJ'') \tag{II.52}$$

we transform Eqs. (I.15)-(I.17) to the following form

$$(f|c_\sigma^g|i) = \exp\{i\pi[1-(-i)^g]/4\}(1,1,g)(2J+1)(2g+1)^{1/2}\times$$
$$\times(-1)^{J-J'-g}(JgM\sigma|J'M')\sum_{r,J''}(f\|d\|r)(r\|d\|1)(2J''+1)^{-1/2}W(JJ'11;gJ'')\times$$
$$\times[(\omega_{rJ''}-\omega-i\Gamma_{rJ''}/2)^{-1}+(-1)^g(\omega_{rJ''}+\omega+i\Gamma_{rJ''}/2)^{-1}], \tag{II.53}$$

where the symbol (1, 1, g) is defined by Eq. (I.16).

This means that in the region of an isolated level $\hbar\omega_{J''}$ with a certain spin value $J^\pi = J_r$, the various polarizabilities behave like Racah coefficients which are multiplied by a constant:

$$\sum_\sigma|(f|c_\sigma^g|i)|^2_{\omega=\omega_{J''}} \sim (1,1,g)^2(2g+1)W^2(JJ'11;gJ''). \tag{II.54}$$

More particularly, we have for $J' = J$:

$$\sum_{\sigma}|(f|c_{\sigma}^{0}|i)|^{2}:\sum_{\sigma}|(f|c_{\gamma}^{1}|i)|^{2}:\sum|(f\|c_{\sigma}^{2}\|i)|^{2} = \left.\begin{cases}1: & 2.9 \quad : \quad 0.69 \quad J''=J-1\\ 1: & 0.14 \quad : \quad 3.4 \quad J''=J\\ 1: & 2.8 \quad : \quad 0.12 \quad J''=J+1\end{cases}\right\} J=\frac{7}{2}.$$

$$= \left.\begin{cases}1: & 3 \quad : \quad 0 \quad J''=J\\ 1: & 0.75 \quad : \quad 0 \quad J''=J+1\end{cases}\right\} J=\frac{1}{2}.$$

Thus, the effects of vector polarizability can be large in this particular case.

The Tensor Polarizability of a Nucleus in the Hyperfine Structure of Atomic and Molecular Spectra

Townes, Geschwind, and Gunther-Mohr [55] suggested that it should be possible to determine the tensor polarizability of a nucleus with a radiospectroscopic technique in which the hyperfine structure of atomic and molecular spectra is measured.

The hyperfine structure of the spectrum of an atom or a molecule results from the interaction of the electron shell with the multiple moments of the nucleus. This interaction causes the energy of the system to depend upon the orientation of the electron shell relative to the spin of the nucleus.

The dependence can be described with a formula of [56]

$$E(JjF)=\sum_{q>0} B_g \frac{\sqrt{(2J+1)(2J+1)}}{(JgJ0|JJ)(jgj0|JJ)} W(JJjj;gF), \tag{II.55}$$

where $F = J + j$; J denotes the spin of the nucleus; and j denotes the total angular moment of the system without inclusion of the spin of the nucleus (j denotes the total angular moment of the electron shell in the case of atoms; in addition to this meaning, j includes the rotational moment of the nuclei in the case of molecules).

When we consider first-order perturbation theory and neglect the dimensions of the nucleus relative to the radius of the electron shell, the coefficients B_g are the energy of the electron interaction with the multipole moment of the nucleus, averaged over the state $|JM = J, j, m = j)$ (the interaction is of the order of g and we refer to electric interaction when g is even, and to magnetic, when g is odd).

The constant of quadrupole interaction is in first-order perturbation theory given by

$$B_2 = B_2^0 = \left(j, m=j\left|\sum_e \frac{3z_e^2 - r_e^2}{r_e^5}\right|j, \ m=j\right) Q_{JJ} \equiv q_{jj}\, Q_{JJ} \tag{II.56}$$

where the subscript e refers to electrons.

The correction which results from the static tensor polarizability of the nucleus [55] is one of the corrections to B_2, which are encountered in second-order perturbation theory:

$$\Delta B_2^c = \frac{1}{4}(JJ|c_{\mu}^{2}(\omega=0)|JJ)\left(jj\left|3\left(\sum_e \frac{z_e}{r_e^3}\right)^2 - \left(\sum_e \frac{r_e}{r_e^3}\right)^2\right|jj\right) = \frac{1}{4}c_{JJ}^{2}(\omega=0)p_{jj} \tag{II.57}$$

According to [55], this correction can be determined by measuring the ratio of the quadrupole constants B_2^{I}/B_2^{II} of two isotopes of the same element in two different states of a particular atomic species or in different molecules.

Since the energy of the interaction between electrons with quadrupole moments and the tensor polarizability of the nucleus depends upon the electron variables in various ways (the interaction energy is inversely proportional to the cube and the fourth power of the electron radius, respectively), the difference of the constants B_2^I/B_2^{II} measured for various electron configurations can indicate tensor-type polarizability.

As a matter of fact, we have

$$(B_2^0+\Delta B_2^c)^{\mathrm{I}}/(B_2^0+\Delta B_2^c)^{\mathrm{II}} = 1+\{[c_{JJ}^2(\omega=0)/Q_{JJ}]^{\mathrm{I}}-[c_{JJ}^2(\omega=0)/Q_{JJ}]^{\mathrm{II}}\}\,p_{jj}/(4q_{jj}), \tag{II.58}$$

i.e., changes in the ratio B_2^I/B_2^{II} upon changes in the electron configuration (which are expressed by changes in p/q) can be explained by the fact that $[c_{JJ}^2(\omega=0)/Q_{JJ}]^{\mathrm{I}} \neq [c_{JJ}^2(\omega=0)/Q_{JJ}]^{\mathrm{II}}$.

However, there exist other influences which result in a similar effect. Among them, there are the polarization of the electron shell by the multipole moments of the nucleus, the finiteness of the nuclear volume, and several other influences.

In order to estimate the ratio $c_{JJ}^2(\omega=0)/Q_{JJ}$, we use formulas of the spheroidal nuclear model. We can assume that the result provided by these formulas will be valid in all other cases, at least as far as the order of magnitude is concerned:

$$Q_{JJ}\approx P_2(J)\frac{4}{5}ZR_0^2\beta_0. \tag{II.59}$$

It follows from Eqs. (II.3) and (II.4) and the condition $\omega_b/\omega_a \approx 1+\beta_0$ (see Chapter 3):

$$c_{JJ}^2(\omega=0)\approx p_2(J)\frac{8}{3}c^0(\omega=0)\beta_0 f(A), \tag{II.60}$$

where $f(A)$ denotes an isotope-structure-dependent factor of the order of unity. In the case of a spheroidal nucleus, we have $f(A) = 1 + 0(\beta_0^2)$.

Migdahl's formula (2) is conveniently used in the case of static dipole polarizability c^0 $(\omega = 0)$.

The ratio $p/q \sim (j|r^{-4}|j)/(j|r^{-3}|j)$ depends upon the behavior of the wave function inside the shielding shells*

$$(j|r^{-4}|j)\,|\,(j/r^{-3}|j)\approx Z/a, \tag{II.61}$$

where $a = 0.53 \cdot 10^{-8}$ cm denotes the radius of the first Bohr orbit.

We obtain from Eqs. (2), (II.57), and (II.59)-(II.61):

$$\Delta B_2^c/B_2^0 \sim 2\,\frac{e^2/a}{920\ \mathrm{MeV}}A\sim 10^{-7}A, \tag{II.62}$$

which renders 10^{-5}-10^{-6} for A ≈ 30 and 10^{-4}-10^{-5} for A ≈ 200.

The accuracy (10^{-5}-10^{-7}) of contemporary radiospectroscopical measurements [56] allows, in principle, determinations of an effect of this magnitude.

Let us consider influences which may prevent the detection of tensor polarizability.

The correction which results from polarization of the electron shell by the magnetic or quadrupole moments of the nucleus and must be attached to the level energy JjF is

$$\Delta^2 E_{JjF} = -\sum_{j'}|(jJF\,|\,H_Q+H_\mu\,|\,j'JF)|^2/(E_{j'}-E_j), \tag{II.63}$$

* This follows, for example, from the "idea of penetrating orbits." Goldsmith's formula for $(j|r^{-3}|j)$ is obtained from this concept [57].

where H_Q and H_μ denote the potential of the interaction between the electrons and the electric quadrupole and magnetic dipole moments of the nucleus, respectively. The relative magnitude of the correction stated in Eq. (II.63) is

$$\Delta^2 E_{JjF} / B_2 \sim \mu_N / \mu_e = m / (NA) \approx 10^{-4} - 10^{-6}, \tag{II.64}$$

where μ_N and μ_e denote the Bohr magneton of the nucleus and the electron, respectively.

These corrections for atoms were calculated in detail by Schwartz [58]. Since several factors are hard to take into account (relativistic effects, polarization of the neutral core of the atoms, etc.), the error in the calculation of the correction $\Delta^2 E$ amounts to several per cent, according to Schwartz's estimates.

The correction for the finiteness of the core dimensions is the difference between the average value of the exact potential of the quadrupole interaction and the value of B_2^0:

$$\Delta B_2^V = \left(jjJJ \left| \frac{4\pi}{5} \sum_{e,p} e^2 Y_{20}\left(\frac{r_p}{r_p}\right) Y_{20}\left(\frac{r_e}{r_e}\right)\left(\frac{r_e^2}{r_p^3} - \frac{r_p^2}{r_e^3}\right) \right| jjJJ\right), \quad (r_e \leqslant r_p) \tag{II.65}$$

where the subscript p refers to the proton.

Since the wave function of an electron with orbital moment l behaves, at small r_e, like $(r_l Z/a)^l$, we can estimate the order of magnitude of the correction:

$$\Delta B_2^V \sim (R_0 Z/a)^{2l}, \tag{II.66}$$

and obtain 10^{-5}-10^{-6} for $A \approx 30$, and 10^{-3}-10^{-4}, for $A \approx 200$.

Obviously, the tensor polarizability is in the hyperfine structure of atoms "masked" by secondary effects which have the same order of magnitude and are known with an accuracy insufficient for the subtraction of these effects.

Since molecules have a more complicated structure than atoms, the number of "masking" influences is logically much greater for molecules than for atoms.

The dependence of the quadrupole constant upon the oscillations of a molecule is very important even in first-order perturbation theory. This dependence implies that the ratio of the quadrupole constants of two isotopes of the same element differs for various molecules and, in addition, depends upon the temperature [59]. This effect has a magnitude of the order of 10^{-4}-10^{-5}.

The theory of this effect agrees with the experimental results with high accuracy (within about 10%).

All these points make it very difficult, or even completely impossible, to separate the effect of tensor polarizability from the observed quadrupole constant of hyperfine splitting.

CHAPTER III

Parameters of a Giant Dipole Resonance of Strongly Deformed, Axially Symmetric Nuclei

The giant resonance of strongly deformed nuclei consists of two maxima and, hence, is characterized by twice as many parameters as the giant resonance of heavy spherical nuclei whose cross section of dipole photoabsorption exhibits a single peak.

TABLE 1

Isotope	σ_a, mbarn	$\hbar\omega_a$, MeV	$\hbar\Gamma_a$, MeV	σ_b, mbarn	$\hbar\omega_b$, MeV	$\hbar\Gamma_b$, MeV	$\frac{\sigma_a^{int}+\sigma_b^{int}}{0{,}06\frac{NZ}{A}\text{MeV}\cdot\text{barn}}$	$\frac{\sigma_a^{int}}{\sigma_b^{int}}$	$\frac{\omega_b}{\omega_a}$	$\frac{a}{b}$	$\frac{\Gamma_b}{\Gamma_a}$	Reference
$^{65}Tb^{159}$	258	12.5	2.4	310	16.3	4.0	1.27	2.00	1.30	1.30	1.67	[60]
	410	12.4±0,2	3.3	460	16.0±0.2	4,5	2.30	1.53	1.29	1.30	1.35	[63]
	267	12.5	3.4	317	16.4	3.4	1.35	1.19	1.31	1.30	1.0	[64]
	188±19	12.2±0.2	2.67±0.2	233±23	15.6±0.2	4.30±0.4	1.04±0.01	2.00	1.28	1.30	1.61	[65]
$^{67}Ho^{165}$	420	12.1±0.2	2.8	510	16.2±0.2	4.7	2.36	2.04	1.34	1.30	1.68	[63]
	318	12.2±0.2	2.33	328	16.0±0.5	4.5	1.47	2.00	1.31	1.30	1.94	[49]
	200	12.10	2.65	249	15.75	4.4	1.07	2.06	1.30	1.30	1.66	[48]
^{68}Er	318	12.2±0.2	2.33	328	16.0±0.5	4.5	1.47	2.00	1.31	1.31	1.94	[3]
$^{73}Ta^{181}$	308	12.45	2.3	348	15.45	4.4	1.35	2.16	1.24	1.22	1.91	[60]
	500	12.6	2.0	450	15.3	4	1.68	1.8	1.215	1.22	2.0	[61]
	317	12.5	2.3	444	15.5	3.6	1.40	2.18	1.24	1.22	1.57	[62]
	350	12.4	2.4	400	15.5	3.8	1.41	1.8	1.25	1.22	1.58	[64]
	198	12.75	3.00	224	15.5	5.0	1.03	1.89	1.22	1.22	1.67	[48]
$^{92}U^{235}$	400	10.85	2.45	505	14.10	4.00	1.4	2.2	1.30	1.25	1.63	[66]

* In all cases, except for uranium, the cross section of photoabsorption was defined as the sum of the cross sections of reactions (γ, n) and $(\gamma, 2n)$. In the case of uranium, photofission (γ, f) contributes to the absorption cross section: $\sigma = \sigma(\gamma, n) + \sigma(\gamma, 2n) + \sigma(\gamma, f)$. As can be inferred from the table, the frequencies and widths obtained by the various researchers are in rather good agreement. The most reliable experiments [48, 49] render values which are by approximately 30% smaller than the preceding values of the absolute cross section. The integral cross section agrees in this case with the Levinger–Bethe summation rule without exchange forces. This result is in good agreement with the latest scattering data [32].

In order to estimate the intensity, the energy, and, to some extent, the width of a dipole peak, summation rules and model calculations with averaging are useful, because it is then possible to describe the properties which are typical "on the average" for a large number of degrees of freedom.

When these estimates are applied to the parameters of the giant resonance of spheroidal nuclei, one can obtain understanding of the mechanism of dipole excitation without excessively detailed calculations. Comparisons with the mechanism obtained from research on the giant resonance in spherical nuclei are useful.

Experimental Data

The experimentally obtained parameters of the two maxima of the giant resonance of heavy deformed nuclei are listed in Table 1. The table includes data on nuclei of the rare-earth series and on the U^{235} nucleus.

The parameters σ_a, ω_a, Γ_a, σ_b, ω_b, and Γ_b, of the table are the parameters of the sum of two Lorentz curves:

$$\sigma = \frac{\sigma_a}{[(\omega_a^2 - \omega^2)/(\omega\Gamma_a)]^2 + 1} + \frac{\sigma_b}{[(\omega_b^2 - \omega^2)/(\omega\Gamma_b)]^2 + 1}. \tag{III.1}$$

The Lorentz sum describes adequately the experimental photoabsorption curve up to energies of 20-22 MeV.*

* See remark to Table 1.

The integral cross sections of the longitudinal and transverse maxima

$$\sigma_a^{\rm int} = \pi\sigma_a\Gamma_a/2, \quad \sigma_b^{\rm int} = \pi\sigma_b\Gamma_b/2 \tag{III.2}$$

satisfy the relation

$$\sigma_b^{\rm int}/\sigma_a^{\rm int} = 2:1, \tag{III.3}$$

and the total cross section area agrees with the Levinger–Bethe summation rule:

$$\sigma_a^{\rm int} + \sigma_b^{\rm int} = 0.06\,\frac{NZ}{A}(1+\varkappa)\ \text{MeV}\cdot\text{barn} \quad (0 \leqslant \varkappa \leqslant 0.4). \tag{III.4}$$

The resonance frequencies satisfy the relation

$$\omega_a/\omega_b = b/a, \tag{III.5}$$

where a and b denote the longitudinal and transverse axes of the nucleus, respectively.

The average energy of the total dipole maximum, $\hbar\omega_d = \hbar(\omega_a\omega_b^2)^{\frac{1}{3}}$, is 14.5 MeV in the case of the rare-earth nuclei and amounts to about 13 MeV in the case of uranium.

The widths $\hbar\Gamma_a$, and $\hbar\Gamma_b$ of the longitudinal and transverse maxima, respectively, amount to about 2.5 and 4-4.5 MeV. The transverse maximum is therefore by about 1.5-2 MeV broader than the longitudinal maximum.

We can conclude from the available experimental information on tensor polarizability of strongly deformed nuclei that the first maximum is produced by excitation of dipole oscillations parallel to a symmetry axis, whereas the second maximum (high-energy peak) corresponds to transverse dipole oscillations, i.e., we have

$$\sigma_a\{[(\omega_a^2-\omega^2)/(\omega\Gamma_a)]^2+1\}^{-1} = 8\pi 3^{-1}\hbar^{-1}c^{-1}\sum_r \omega_r\Gamma_r^{-1}|(0|d_{z'}|r)|^2\ \{[(\omega_r^2-\omega^2)/(\omega\Gamma_r)]^2+1\}^{-1}, \tag{III.6}$$

$$\sigma_b\{[(\omega_b^2-\omega^2)/(\omega\Gamma_b)]^2+1\}^{-1} = 8\pi 3^{-1}\hbar^{-1}c^{-1}\,\Sigma\omega_r\Gamma_r^{-1}[\,|(0|d_{x'}|r)|^2+|(0|d_{y'}|r)|^2]\,\{[(\omega_r^2-\omega^2)/(\omega\Gamma_r)]^2+1\}^{-1}. \tag{III.7}$$

Summation Rules

An estimate of the parameters of the longitudinal and transverse maxima can be obtained by calculating the sums

$$S_i^{(n)} \equiv \sum_r \hbar\omega_r|(0|d_i|r)|^2(\hbar\omega_r)^n, \tag{III.8}$$

which describe the various moments of the intensity distribution of longitudinal and transverse transitions, with $d_i = (\mathbf{d}e_i^{!})$.

Application of the closure theorem to the calculation of these quantities results in various summation rules.

In the case $-2 \leq n \leq 0$, the quantities $S_i^{(n)}$ have a simple experimental meaning, namely:

$$\int_0^\infty \sigma(\hbar\omega)^n\, d\hbar\omega = 4\pi^2(3c)^{-1}\sum_i S_i^{(n)} \tag{III.9}$$

and in the case of photoabsorption by oriented nuclei [22] [see Eq. (I.22) and Eq. (II.15)], we have

$$\int_0^\infty \sigma^J(\hbar\omega)^n\, d\hbar\omega = 4\pi^2(3c)^{-1}\Big\{\sum_i S_i^{(n)} - [3(\mathbf{Jk})^2 - J(J+1)][J(2J-1)]^{-1}P_2[S_{z'}^{(n)} - (S_{x'}^{(n)} + S_{y'}^{(n)}/2]\Big\}. \tag{III.10}$$

The $S_i^{(n)}$ values cannot be easily related to any experiment for n > 0, because the integrals $\int_0^\infty \sigma(\hbar\omega)^n d\hbar\omega$ $(n>0)$ (with n > 0) diverge at infinity (due to the Lorentz form of the partial cross sections). However, there may exist plausible models for which the series $S^{(n)}$ converge for n = 1, 2, etc. Calculation of these quantities can provide additional information on the energy distribution of the dipole transitions.

Of particular interest are sums $S^{(n)}$ for which the closure theorem renders results which, in their generality, exceed the usual model calculations.

As will be explained below, the quantity $S^{(2)}$, is, in addition to $S^{(0)}$ (for which the Levinger–Bethe summation rule is valid), one of the above-mentioned sums.

Calculations of the quantities $S^{(-2)(-1),(1)}$ require the same amount of initial information as the usual model calculations of dipole spectra.* The summation rules for these quantities are formal identities.

Let us consider some consequences which stem from the summation rules for $S^{(2)}$ [67]:

$$S_i^{(2)} = -\frac{1}{2}(0|[[Hd_i][H[Hd_i]]|0). \tag{III.11}$$

We assume for the sake of simplicity that paired forces do not depend upon the coordinates of the isospin of the nucleons and that N = Z. The Hamiltonian H of the nucleus is written in the form

$$H = -\frac{1}{2}\sum_\nu \frac{\hbar^2}{M}\nabla_\nu^2 + \frac{1}{2}\sum_{\nu,\nu'} v(s_\nu s_{\nu'}, |\mathbf{r}_\nu - \mathbf{r}_{\nu'}|). \tag{III.12}$$

From Eqs. (III.11) and (III.12) and from the summation rule for $S^{(0)}$

$$S_i^{(0)} = \hbar^2 A/(4M) \tag{III.13}$$

we find

$$\overline{E_{z'}^2} = S_3^{(2)}/S_3^{(0)} = \frac{1}{Z}\frac{\hbar^2}{M}\sum_p \frac{\partial^2}{\partial z_p'^2} 2\sum_n v_{np} = \frac{1}{N}\frac{\hbar^2}{M}\sum_n \frac{\partial^2}{\partial z_n'^2} 2\sum_p v_{pn}, \tag{III.14}$$

$$\overline{E_{x'}^2} = S_1^{(2)}/S_1^{(0)} = \frac{1}{Z}\frac{\hbar^2}{M}\sum_p \frac{\partial^2}{\partial x_p'^2} 2\sum_n v_{np} = \frac{1}{N}\frac{\hbar^2}{M}\sum_n \frac{\partial^2}{\partial x_n'^2} 2\sum_p v_{pn}. \tag{III.15}$$

When the sum $S^{(2)}$ includes only excited states which in their structure differ only slightly from the ground state, we can conclude from Eqs. (III.14) and (III.15) that the mean-square energy of the dipole oscillations is equal to the mean-square energy of the single-particle dipole transition of protons in a doubled neutron potential or of neutrons in a doubled proton potential:

$$U_p(n) \equiv \sum_n v_{pn}. \tag{III.16}$$

When we assume the latter potential to be local and when we make the logical assumption that the potential $U_p(n)$, as well as the density ρ, have the form of an ellipsoid

$$U_p(n) = U\left(\frac{z'^2}{a^2} + \frac{y'^2 + x'^2}{b^2}\right), \tag{III.17}$$

$$\rho = \rho\left(\frac{z'^2}{a^2} + \frac{y'^2 + x'^2}{b^2}\right) \tag{III.18}$$

* For the calculation of S^{-2}, S^{-1}, and $S^{(1)}$ see [1, 5, 11].

we obtain

$$(\overline{E^2_{z'}}/\overline{E^2_{x'}})^{1/2} = b/a \tag{III.19}$$

i.e., a relation which, to some extent, can be assumed to agree with experimental results.

Summation rules (III.14) and (III.15) allow also some conclusions on the absolute value of the energy of the dipole resonance.

This energy can be written in the form

$$\sqrt{\overline{E^2_d}} = \sqrt{\overline{\varepsilon^2_d}}\left[\left(0\left|\sum_p \frac{\partial^2}{\partial \mathbf{r}^2_p} 2\sum_n v_{np}\right|0\right)\right]^{1/2}\left[\left(0\left|\sum_p \frac{\partial^2}{\partial \mathbf{r}^2_p}\left(\sum_{p'} v_{p'p} + \sum_n v_{np}\right)\right|0\right)\right]^{-1/2}, \tag{III.20}$$

where $\sqrt{\overline{\varepsilon^2_d}}$ denotes the mean-square value of the single-particle dipole energies.

After rewriting the potential of paired forces in the form

$$v_{12} = F(\mathbf{s}_1\mathbf{s}_2) \text{ and } (|\mathbf{r}_1 - \mathbf{r}_2|), \tag{III.21}$$

we obtain from Eq. (III.20)

$$\sqrt{\overline{E^2_d}} = \sqrt{\overline{\varepsilon^2_d}}\,[2(3F^1 + F^0)/(3F^1 + 3F^0)]^{1/2}, \tag{III.22}$$

where $F^S \equiv (S|F(\mathbf{s}_1\mathbf{s}_2)|S)$ denotes the average value of the operator F in a state of two nucleons having the total spin S. (In deriving the last formula, we assumed that the probability of nucleon pairs existing with a certain total spin and isotope spin in the ground state of the nucleus is entirely determined by the Pauli principle.)

In the case of the forces which are frequently employed for shell-model calculations [10, 11]

$$v = (0.865 + 0.54\,\mathbf{s}_1\mathbf{s}_2)\,\delta(\mathbf{r}_1 - \mathbf{r}_2), \tag{III.23}$$

we obtain

$$\sqrt{\overline{E^2_d}} = 1.35\sqrt{\overline{\varepsilon^2_d}}.$$

Calculations with the "Particle – Hole" Model

General Relations

In the "particle–hole" approximation, excited states are considered as a superposition of states which differ from the ground state by the presence or absence of a particle–hole pair

$$|N) = \sum_{mi} x^N_{mi} a^+_m a_i |0) + \sum_{mi} y^N_{mi} a^+_i a_m |0). \tag{III.24}$$

where $|0)$ denotes the wave function of the ground state, and a^+_l and a_l (l = i, m) denote the generation and annihilation operators of the particle in state l.

The subscripts i (and, below, also the subscripts j and k) denote the lowest single-particle levels, which can be filled by nucleons of the nucleus in agreement with the Pauli principle

$$\varepsilon_i \leqslant \varepsilon^F, \tag{III.25}$$

where ε^F denotes the Fermi energy.

The subscripts m (and, below, also the subscript n) denote states with an energy greater than the Fermi energy

$$\varepsilon_m > \varepsilon^F. \tag{III.26}$$

For a state $|\Phi_0\rangle$ in which all levels i are filled and levels m are free, the relation

$$a_i^+ |\Phi_0) = a_m |\Phi_0) = 0. \tag{III.27}$$

holds. This state is the "vacuum" of the particles and "holes" (i.e., of the unfilled levels i). The operators a_i and a_i^+ are accordingly the generation and annihilation operators of the holes, respectively.

When we write the wave function in the form of Eq. (III.24), we indicate that the other interacting pairs which produce a particle–hole pair in the ground state were taken into account. The self-energy E_N and the coefficients x and y satisfy the equation (which was derived and discussed as approximation in [9, 13, 68]):

$$E_N x_{mi}^N = (\varepsilon_m - \varepsilon_i) x_{mi}^N + \sum_{nj} [(jm|v|ni) - (jm|v|in)] x_{nj}^N + \sum_{nj} [(mn|v|ij) - (mn|v|ji)] y_{nj}^N, \tag{III.28}$$

$$-E_N y_{mi}^N = (\varepsilon_m - \varepsilon_i) y_{mi}^N + \sum_{nj} [(ni|v|jm) - (in|v|jm)] y_{nj}^N + \sum_{nj} [(ij|v|mn) - (ji|v|mn)] x_{nj}^N, \tag{III.29}$$

where

$$(l_1 l_2 | v | l_3 l_4) = (\psi_{l_1}(1) \psi_{l_2}(2) | v(12) | \psi_{l_3}(1) \psi_{l_4}(2)). \tag{III.30}$$

The coefficients x and y satisfy the normalization conditions

$$\sum_{mi} (x_{mi}^{N*} x_{mi}^{N'} - y_{mi}^{N*} y_{mi}^{N'}) = \delta_{NN'}, \tag{III.31}$$

$$\sum_{N\,(E_N>0)} (x_{mi}^{N*} x_{nj}^N - y_{mi}^{*N} y_{nj}^N) = \delta_{mn}\delta_{ij} = - \sum_{N\,(E_N<0)} (x_{mi}^{N*} x_{nj}^N - y_{mi}^{N*} y_{nj}^N), \tag{III.32}$$

$$\sum_{N\,(E_N>0)} (x_{mi}^{N*} y_{nj}^N - y_{mi}^{N*} x_{nj}^N) = 0 = \sum_{N\,(E_N<0)} (x_{mi}^{N*} y_{nj}^N - y_{mi}^{N*} x_{nj}^N) \tag{III.33}$$

The matrix element of the single-particle operator Q for the transition from the ground state to excited states is

$$(N | Q | 0) = \sum_{mi} x_{mi}^N (m|q|i) + \sum_{mi} y_{mi}^N (m|q|i)^*, \tag{III.34}$$

where

$$Q = \sum_{ll'} (l'|q|l) a_{l'}^+ a_l. \tag{III.35}$$

This calculation procedure is called "random phase approximation."

In a simpler version of the "particle–hole" model, which is termed Tamm–Dankov approximation, the ground state is taken as the "vacuum" of particles and holes and all y are assumed to vanish.

The secular equation, the orthogonality relations, and the matrix elements have in this case the following forms:

$$E_N x_{mi}^N = (\varepsilon_m - \varepsilon_i) x_{mi}^N + \sum_{nj} [(jm|v|ni) - (jm|v|in)] x_{nj}^N, \tag{III.36}$$

$$\sum_{mi} x_{mi}^{N*} x_{mi}^{N'} = \delta_{NN'}, \tag{III.37}$$

$$\sum_N x_{mi}^{N*} x_{nj}^N = \delta_{mn}\delta_{ij}, \quad \text{(III.38)}$$

$$(N|q|0) = \sum_{mi} x_{mi}^N (m|q|i). \quad \text{(III.39)}$$

Our goal is to obtain qualitative conclusions which can be derived from a model when it is applied to deformed nuclei, i.e., when various averaged characteristics are calculated, as in the work of Balashov [11] and Fallieros [69] for spherical nuclei.

In order to simplify the calculations, we idealize the situation by assuming N = Z = A/2 and $s_3 = \sigma$, i.e., the projection of the spin upon a certain direction (the symmetry axis of the nucleus in our case) to be a good quantum number, i.e., we assume

$$\begin{aligned} |m) &\equiv |\tilde{m}, \sigma, \tau), \\ |i) &\equiv |\tilde{i}, \sigma, \tau), \end{aligned} \quad \text{(III.40)}$$

where the tilde denotes orbital quantum numbers and $\tau = t_3$.

The potential of the residual paired interactions is written in the form

$$v = F(\mathbf{s}_1\mathbf{s}_2, \mathbf{t}_1\mathbf{t}_2)\, u(|\mathbf{r}_1 - \mathbf{r}_2|). \quad \text{(III.41)}$$

The matrix elements of the transition operator are assumed to be real

$$(m|q|i) = (m|q|i)^*. \quad \text{(III.42)}$$

We introduce also operators q^- and q^+ such that

$$(m|q|i) = (m|q^+|i) = (i|q^-|m), \quad \text{(III.43)}$$

$$(i|q^+|m) = (m|q^-|i) = 0. \quad \text{(III.44)}$$

After multiplying the right sides and left sides of Eqs. (III.28) and (III.29) with (m|q|i) and summing over m and i, we obtain via several identity transformations

$$\begin{aligned} E_N \sum_{mi} x_{mi}^N (m|q|i) = \sum_{mi} (\varepsilon_m - \varepsilon_i) x_{mi}^N (m|q|i) + \sum_{mi} x_{mi}^N [(i|[Uq^-]|m) + \\ + (i|\Delta^-|m)] + \sum_{mi} y_{mi}^N [(m|[Uq^-]|i) + (m|\Delta^-|i)], \end{aligned} \quad \text{(III.45)}$$

$$\begin{aligned} - E_N \sum_{mi} y_{mi}^N (m|q|i) = \sum_{mi} (\varepsilon_m - \varepsilon_i) y_{mi}^N (m|q|i) + \\ + \sum_{mi} y_{mi}^N [(m|[q^+U]|i) - (m|\Delta^+|i)] + \sum_{mi} x_{mi}^N [(i|[q^+U]|m) - (i|\Delta^+|m)], \end{aligned} \quad \text{(III.46)}$$

$$U \equiv d^{S'T'} U_1 + e^{S'T'} U_2, \quad \text{(III.47)}$$

$$(\mathbf{r}_1|U_1|\mathbf{r}_2) = \delta(\mathbf{r}_1 - \mathbf{r}_2) \int d\mathbf{r}_2 \sum_{\tilde{j}} (r_2|\tilde{j})\, u(|\mathbf{r}_1 - \mathbf{r}_2|)(\tilde{j}|\mathbf{r}_1), \quad \text{(III.48)}$$

$$(\mathbf{r}_1|U_2|\mathbf{r}_2) = \sum_{\tilde{j}} (\mathbf{r}_1|\tilde{j})\, u(|\mathbf{r}_1 - \mathbf{r}_2|)\, (\tilde{j}|\mathbf{r}_2), \quad \text{(III.49)}$$

$$d^{S'T'} = \sum_{ST} \left[\frac{2S+1}{2}\delta_{S'0} + \frac{(-1)^{1+S}}{2}\delta_{S'1}\right]\left[\frac{2T+1}{2}\delta_{T'0} + \frac{(-1)^{1+T}}{2}\delta_{T'1}\right] F^{ST}, \quad \text{(III.50)}$$

$$e^{S'T'} = \sum_{ST}\left[\frac{2S+1}{2}\,\delta_{S'0} + \frac{(-1)^{1+S}}{2}\,\delta_{S'1}\right]\left[\frac{2T+1}{2}\,\delta_{T'0} + \frac{(-1)^{1+T}}{2}\,\delta_{T'1}\right](-1)^{1+S+T}F^{ST} \tag{III.51}$$

where

$$F^{ST} \equiv (ST\,|\,F(\mathbf{s}_1\mathbf{s}_2,\ \mathbf{t}_1\mathbf{t}_2)\,|\,ST) \tag{III.52}$$

denotes the average of operator F in a two-particle state with total spin S and isotope spin T; S' and T' denote the total spin and isotope spin of a particle–hole state, respectively.

$$(\mathbf{r}_1|\Delta^{\mp}|\mathbf{r}_2) = d^{S'T'}\delta(\mathbf{r}_1\mathbf{r}_2)\int d\mathbf{r}_2\sum_{\tilde{j}}(r_2|\tilde{j})\{[q_1^{\mp}u] + [q_2^{\mp}u]\}(\tilde{j}|\mathbf{r}_2) + e^{S'T'}\sum_{\tilde{j}}(\mathbf{r}_1|\tilde{j})\{[q_1^{\mp}u] + [q_2^{\mp}u\}(\tilde{j}|\mathbf{r}_2), \tag{III.53}$$

$$|m,\tilde{j}) \equiv (\mathbf{r}_1\mathbf{s}_1\,\mathbf{t}_1|m)(\mathbf{r}_2|\tilde{j}), \tag{III.54}$$

$$|\tilde{j},m) \equiv (\mathbf{r}_1|\tilde{j})(\mathbf{r}_2\mathbf{s}_1,t_1|m), \tag{III.55}$$

$$q_{1,2}^{\pm} = q^{\pm}\left(\mathbf{r}_{1,2},\ \frac{\partial}{\partial\mathbf{r}_{1,2}}\right). \tag{III.56}$$

The coefficients $d^{S'T'}$ and $e^{S'T'}$ for a state of electric dipole excitation (which, in accordance with our idealized assumptions, is characterized by S' = 0 and T' = 1) are

$$d^{01} \equiv d' = \frac{1}{4}\,[-F^{00} + F^{01} - 3F^{10} + 3F^{11}], \tag{III.57}$$

$$e^{01} \equiv e' = \frac{1}{4}\,[F^{00} + F^{01} - 3F^{10} - 3F^{11}]. \tag{III.58}$$

We obtain for a state with S' = T' = 0

$$d^{00} \equiv d = \frac{1}{4}\,[F^{00} + 3F^{01} + 3F^{10} + 9F^{11}], \tag{III.59}$$

$$e^{00} \equiv e = \frac{1}{4}\,[-F^{00} + 3F^{01} + 3F^{10} - 9F^{11}]. \tag{III.60}$$

The quantity

$$U^{HF} = dU_1 + eU_2 \tag{III.61}$$

is, obviously, the Hartree–Fock potential of the forces described by Eq. (III.41).

The quantity

$$U^{pn} = \frac{1}{2}\,(d-d')\,U_1 + \frac{1}{2}\,(e-e')\,U_2 \tag{III.62}$$

can be termed a "Hartree–Fock potential" which describes the effect of the protons of the nucleus upon a neutron or of the neutrons upon a proton.

Let us outline several general results which are obtained from Eqs. (III.45) and (III.46).

Summation Rule for S^0 in the Random Phase Approximation. Thouless [9] has shown that the relation

$$\sum_{E_N>0} E_N\,|(N|Q|0)|^2 = \frac{1}{2}\,(\Phi_0|\,[Q\,[HQ]]\,|\Phi_0), \tag{III.63}$$

holds for an Hermitian operator Q, where H denotes the Hamiltonian of the nucleus; E_N, $|N)$

denote the intrinsic energy and the eigenfunctions obtained in the random phase approximation, respectively; and $|\Phi_0)$ denotes the "vacuum" of the particles and holes.

Moreover, as will be shown below, if the initial single-particle potential is a Hartree–Fock potential, the calculation of S_0 in the random phase approximation leads to the well-known Levinger–Bethe formula of [1].

Based on Eqs. (III.45) and (III.46), a procedure, which is an analog to that of [9], can be used to rewrite the summation rule for an Hermitian operator in the form

$$\sum_{E_N>0} E_N |(N|Q|0)|^2 = \frac{1}{2} \sum_i (i|[q[H^{s.p} - U)q]]|i), \tag{III.64}$$

where $H^{s.p} = -\frac{\hbar^2}{2M}\nabla^2 + U^{s.p}$ denotes the single-particle Hamiltonian.

A sufficient condition for this reformulation of the summation rule is

$$[q_1 u] + [q_2 u] = 0 \tag{III.65}$$

and this relation is particularly valid for the dipole operator

$$\mathrm{q} = t_3 \mathbf{r}. \tag{III.66}$$

When we assume in this case a Hartree–Fock potential as single-particle potential and use Eqs. (III.48), (III.49), and (III.57)-(III.61), we obtain

$$\sum_{E_N>0} F_N |(N|d_z|0)|^2 = \frac{\hbar^2}{2N}\frac{A}{4} + \frac{1}{4}[-F^{00} + F^{01} + 3E^{10} - 3F^{11}] \sum_{\tilde{i}\tilde{j}} (\tilde{i}\,\tilde{j}|(z_1 - z_2)^2 u|\tilde{j}\,\tilde{i}). \tag{III.67}$$

Equation (III.67) is the Levinger–Bethe summation rule stated for the case in which $|\Phi_0)$ is the ground state.

The "Translational Invariance" Property of the Random Phase Approximation. When we assume $U^{s.p} = U^{HF}$ and $x^N_{mi} = -y^N_{mi}$ we obtain from Eqs. (III.45) and (III.46)

$$E^N \sum_{mi} x^N_{mi} \left(m\left|\frac{\partial}{\partial z}\right|i\right) = 0,$$

i.e., when the single-particle potential is self-consistent in the Hartree–Fock sense, the state describing the motion of the center of mass of the nucleus has eigenenergy zero.

Let us consider the case

$$[\bar{q_1} u] + [\bar{q_2} u] = 0. \tag{III.68}$$

This case occurs, for example, in dipole transitions when the single-particle wave functions are oscillator functions or when the paired forces' radius of action vanishes.

We obtain in this case in the Tamm–Dankov approximation

$$\sum_N E^N |(N|Q|0)|^2 = \sum_{mi} (i|[q^-(H^{s.p} - U)]|m)(m|q^+|i), \tag{III.69}$$

$$\sum_m (E^N)^2 |(N|Q|0)|^2 = \sum_{mi} |(i|[q^-(H^{s.p} - U)]|m)|^2. \tag{III.70}$$

The average energy and the mean-square energy of the set of states $|N)$, in which transitions Q take place, are

$$\overline{E_Q} = \sum_N E^N |(N|Q|0)|^2 \Big/ \sum_N |(N|Q|0)|^2 = \sum_{mi} (i|[q^-(H^{s.p} - U)]|m)(m|q^+|i) \Big/ \sum_{mi} (i|q^-|m)(m|q^+|i), \tag{III.71}$$

$$\overline{E_Q^2} = \sum_{mi} |(i|[q^-(H^{s.p} - U)]|m)|^2 \Big/ \sum_{mi} (i|q^-|m)(m|q^+|i) \tag{III.72}$$

and are, respectively, equal to the average energy and mean-square energy of single-particle transitions in the potential

$$U^{\text{eff}} = U^{s.p} - U. \tag{III.73}$$

Only matrix elements of U which are diagonal with respect to single-particle states are taken into account.

More particularly, it follows from Eq. (III.73) that the properties of a state describing the motion of the center of mass are (explicitly) independent of paired forces. We proved in this way the "translational invariance" property in the Tamm–Dankov approximation. This invariance had been previously postulated by S. Fallieros [69].

Of particular interest is a comparison of Eqs. (III.64) and (III.69), i.e., the "summation rules" for $S^{(0)}$ in the random phase approximation and in the Tamm–Dankov approximation.

q appears in the commutator of Eq. (III.64) and q^-, in the commutator of Eq. (III.69). This means that in calculations based on the Tamm–Dankov method, the summation rule for $S^{(0)}$ must be violated. However, the relative "inaccuracy" of the Tamm–Dankov approximation versus the random phase approximation is still within the limits between which the initial parameters of actual calculations vary (see [70]).

Calculations for Spheroidal Nuclei. Tamm–Dankov Method. Asymptotic approximation of [71]. The sinlge-particle states of the nucleons in spheroidal nuclei are assumed to be described with Nilsson wave function, i.e., with the eigenfunctions of the Hamiltonian

$$H^{s.p} = \frac{1}{2} M [\omega_\perp^2 (x^2 + y^2) + \omega_\parallel z^2] - \alpha\hbar\,\omega_0 \mathbf{ls} - \alpha\varkappa\,\hbar\omega_0 \mathbf{l}^2, \tag{III.74}$$

$$\omega_\perp | \omega_\parallel = a | b = 1 + \frac{3}{2} \sqrt{\frac{5}{4\pi}}\, \beta_0, \tag{III.75}$$

$$\omega_0 = (\omega_\perp^2 \omega_\parallel)^{1/3} = 41 A^{-1/3}\ \text{Mэв}, \ \alpha \approx 0{,}12, \ \alpha\varkappa \approx 0{,}25. \tag{III.76}$$

In the case of large deformations ($\beta_0 \gtrsim 0.3$), the effects of spin-orbit interaction and anharmonicity can be ignored in comparison to the effect of the deformed oscillator potential. (The corresponding small parameters are equal for $\beta_0 = 0.3$: $\alpha/(2\beta_0) = 0.2$ and $\alpha\varkappa/(2\beta_0) = 0.04$.)

The single-particle states can be described in this case with the aid of asymptotic quantum numbers $n_\parallel$, $n_\perp$, Λ, and $s = \sigma$. ($n_\parallel$, $n_\perp$ denote the principal quantum numbers which describe the motion parallel and perpendicular to the symmetry axis z of the nucleus; Λ and σ denote the projections of the orbital and spin moments upon the z axis, respectively.)

The ensuing calculations are made with the notation of [72]:

$$r \equiv (n_\perp + \Lambda)/2, \tag{III.77}$$

$$s \equiv (n_\perp - \Lambda)/2. \tag{III.78}$$

Intensive single-particle dipole transitions satisfy the asymptotic selection rules:

longitudinal

$$\Delta n_\parallel = 1, \quad \Delta r = \Delta s = \Delta\sigma = 0, \tag{III.79}$$

transverse

$$\Delta n_\parallel = 0, \quad \Delta r = 1, \quad \Delta s = \Delta\sigma = 0, \tag{III.80}$$

$$\Delta n_\parallel = 0, \quad \Delta r = 0, \quad \Delta s = 1, \ \Delta\sigma = 0. \tag{III.81}$$

The operators q^- and q^+ for the particular transitions can be immediately stated [72]. We obtain for the case defined by Eq. (III.79):

$$q_0^- = \frac{1}{\sqrt{2}}\left(\zeta + \frac{\partial}{\partial \zeta}\right), \tag{III.82}$$

$$q_0^+ = \frac{1}{\sqrt{2}}\left(\zeta - \frac{\partial}{\partial \zeta}\right), \tag{III.83}$$

where

$$\zeta = z\,(M\omega_{\parallel}/\hbar)^{1/2} \tag{III.84}$$

for the case defined by Eq. (III.80):

$$q_1^- = \frac{1}{2}\left[\left(\xi + \frac{\partial}{\partial \xi}\right) - i\left(\eta + \frac{\partial}{\partial \eta}\right)\right], \tag{III.85}$$

$$q_1^+ = \frac{1}{2}\left[\left(\xi - \frac{\partial}{\partial \xi}\right) + i\left(\eta - \frac{\partial}{\partial \eta}\right)\right] \tag{III.86}$$

and for the case defined by Eq. (III.81):

$$q_{-1}^- = \frac{1}{2}\left[\left(\xi + \frac{\partial}{\partial \xi}\right) + i\left(\eta + \frac{\partial}{\partial \eta}\right)\right], \tag{III.87}$$

$$q_{-1}^+ = \frac{1}{2}\left[\left(\xi - \frac{\partial}{\partial \xi}\right) - i\left(\eta - \frac{\partial}{\partial \eta}\right)\right], \tag{III.88}$$

where

$$\xi + i\eta = (x + iy)\,(M\omega_{\perp}/\hbar)^{1/2}. \tag{III.89}$$

(The subscript of q indicates projection of the total moment of the dipole state upon the z axis.)

The additional potential U which results from the other interactions is assumed to be local:

$$(\mathbf{r}_1|U|\mathbf{r}_2) = \delta(\mathbf{r}_1 - \mathbf{r}_2)\,U(\mathbf{r}_1). \tag{III.90}$$

The following expressions for the average energies of the longitudinal and transverse peaks of a giant resonance are obtained from Eqs. (III.71), (III.82)-(III.90), and (III.74):

$$\overline{E_{\parallel}} = \hbar\omega_{\parallel} + A^{-1}\int \rho(\mathbf{r}')\,d\mathbf{r}'\,\frac{\partial^2 U}{\partial \zeta^2}, \tag{III.91}$$

$$\overline{E_{\perp}} = \hbar\omega_{\perp} + A^{-1}\int \rho(\mathbf{r}')\,d\mathbf{r}'\,\frac{1}{2}\left(\frac{\partial^2 U}{\partial \xi^2} + \frac{\partial^2 U}{\partial \eta^2}\right), \tag{III.92}$$

where ρ denotes the density of the nucleus in the ground state:

$$\rho(\mathbf{r}) = 4\sum_{\tilde{j}} |(\mathbf{r}\,|\,\tilde{j})|^2, \tag{III.93}$$

with $\tilde{j}$ denoting the set of valid quantum numbers $n_{\parallel}$, r, and s.

It follows from Eqs. (III.91) and (III.92) that in the "self-consistent" independent particle model (which is self-consistent for describing the deformation of the nucleus), i.e., in a model in which the deformation of the potential U coincides with the deformation of the nuclear density, which, in turn, coincides with the nonspherical initial single-particle potential

$$U = U\left(\frac{x^2 + y^2}{b_1^2} + \frac{z^2}{a_1^2}\right), \tag{III.94}$$

$$\rho = \rho\left(\frac{x^2+y^2}{b_2^2}+\frac{z^2}{a_2^2}\right), \tag{III.95}$$

$$U^{sp} = U^{sp}\left(\frac{x^2+y^2}{b_3^2}+\frac{z^2}{a_3^2}\right), \tag{III.96}$$

where

$$a_1 = a_2 = a_3 = a; \quad b_1 = b_2 = b_3 = b, \tag{III.97}$$

the relation

$$E_{\parallel} / E_{\perp} = b / a \tag{III.98}$$

holds.

It is generally accepted that the self-consistent parameters of nonspherical nuclear shapes, which appear in the independent particle model [73] for strongly deformed nuclei (i.e., for rare-earth and transuranium nuclei), are in rather good agreement with experimental results. We can therefore conclude that the Tamm–Dankov approximation must lead to an energy which is consistent with the experimental results for the longitudinal and transverse peaks.

A simple analytical calculation for the residual δ forces in the Thomas–Fermi approximation can be used to illustrate this conclusion.

When we replace the quantity $(\mathbf{r}|\tilde{j})^2$ by the classical density distribution of the coordinate probability for the harmonic oscillator [74], we obtain for $\hbar\omega_{\parallel} + \hbar\omega_{\perp}(r+s) \leqslant \varepsilon^F$

$$\rho = 4\int dn_{\parallel}\, drds\,|(\mathbf{r}\,|\,\tilde{j})|^2 = \begin{cases} \dfrac{-4\sqrt{2}\,M^{3/2}}{3\pi^2\hbar^3}\,(\varepsilon^F - U^{osc})^{3/2} & \text{for } \varepsilon^F \geqslant U^{osc}, \\ 0 & \text{for } \varepsilon^F < U^{osc}, \end{cases} \tag{III.99}$$

where U^{osc} denotes the oscillator part of the Nilsson potential.

Equations (III.21), (III.48), (III.49), (III.57), (III.59), (III.91), and (III.92) lead to the following shifts of the longitudinal and transverse maxima, the shifts being induced by residual δ forces:

$$\Delta_{\parallel} = 2^{7/2}(35\pi^3)^{-1}V_0(3F^1 - F^0)(M\omega_0/\hbar)^{3/2}\sqrt{n^F}\,\omega_{\parallel}/\omega_0, \tag{III.100}$$

$$\Delta_{\perp} = 2^{7/2}(35\pi^3)^{-1}V_0(3F^1 - F^0)(M\omega_0/\hbar)^{3/2}\sqrt{n^F}\,\omega_{\perp}/\omega_0, \tag{III.101}$$

where

$$n^F \equiv \varepsilon^F / (h\,\omega_0).$$

We obtain for $n^F = 5$ and the δ forces of Eq. (III.23):

$$\Delta_{\parallel,\perp} = 7.5\,(\omega_{\parallel,\perp}/\omega_0)\ \text{MeV}.$$

The average energies of the longitudinal and transverse maxima for $A \approx 150$ and $\beta_0 = 0.3$ are therefore

$$\overline{E_{\parallel}} = \hbar\omega_{\parallel} + \Delta_{\parallel} \simeq 12\ \text{MeV},$$
$$\overline{E_{\perp}} = \hbar\omega_{\perp} + \Delta_{\perp} \simeq 16\ \text{MeV},$$

i.e., they are close to the above-mentioned experimental values.

We can obtain some idea of the energy spread of the longitudinal and transverse maxima and, hence, of the widths of these maxima when we calculate the dispersion of the energy distribution of the squares of matrix elements of dipole transitions:

$$D_{\parallel} = \overline{E_{\parallel}^2} - \overline{E}_{\parallel}^2, \tag{III.102}$$

$$D_{\perp} = \overline{E_{\perp}^2} - \overline{E}_{\perp}^2. \tag{III.103}$$

Nilsson and Mottelson [72] mentioned an interesting detail which attests to a lack of similarity in the shape of longitudinal and transverse peaks. They could show for the single-particle model, that in the case of asymptotic transitions,

$$D_{\perp} = 3D_{\parallel} + (\alpha\hbar\,\omega_0/2)^2. \tag{III.104}$$

holds.

We will show that this relation is conserved when residual interactions are taken into account in the Tamm–Dankov approximation.

As has been mentioned above, introduction of the other interactions in the calculation of $\overline{E}$ and $\overline{E}^2$ is equivalent to the addition of the potential U to the single-particle Hamiltonian, with the dipole matrix elements being conserved.

When the ratio of the dispersions is calculated, we ignore the "scale factor," i.e., we assume $\omega_{\parallel} = \omega_{\perp} = \omega_0$. The transitions are assumed to occur from a filled shell $n_{\parallel} + n_{\perp} = n^F$. In our ensuing formulas, we will assume $\omega_0 = M = \hbar = 1$.

Furthermore, we note that products of wave functions $|\,\tilde{l})(\tilde{m})$ can in this case be written in the form

$$re^{-r^2} \sum_{n=0}^{n^F} \Omega_n L_n^{3/2}(r^2), \tag{III.105}$$

where Ω_n denotes functions of the angles, and $L_n^{3/2}(r^2)$ denotes Chebyshev–Laguerre polynomials which are orthogonal and have the weight $r^4e^{-r^2}$.

This means that the general form of the anharmonicity of the potential U^{eff} (which anharmonicity causes the energy spread of the dipole peaks) is as follows:

$$\int_0^{r^2} d(r^2) \sum_n \Omega_n L_n^{3/2}(r^2) = \sum_{n=1}^{n^F+1} g_n L_n^{1/2}(r^2). \tag{III.106}$$

For $n^F = 4$ (which corresponds to rare-earth nuclei) we obtain from numerical calculations with any combination of coefficients g_n

$$D_{\perp} = kD_{\parallel} + (\alpha\hbar\omega_0/2)^2, \quad 2.7 \leqslant k \leqslant 3. \tag{III.107}$$

Eq. (III.107) indicates lack of similarity in the form of the longitudinal and transverse peaks. The experimentally observed difference in these peaks is therefore not surprising.

"Asymptotically Forbidden" Transitions. The relation between the dispersions of Eq. (III.107) is an important consequence of the fact that we considered only "asymptotically resolved" transitions and ignored asymptotically forbidden transitions.

The following arguments are, to some extent, a justification of the procedure.

Asymptotically forbidden are the following transitions:

longitudinal transitions

$$\Delta n_{\parallel} = 0,\ \Delta r - 0,\ \Delta s = 1,\ \Delta\sigma = 1, \tag{III.108}$$

$$\Delta n_{\parallel} = 0,\ \Delta r = 1,\ \Delta s = 0,\ \Delta\delta = -1; \tag{III.109}$$

transverse transitions

$$\Delta n_{\parallel} = 1,\ \Delta r = \Delta s = 0,\ \Delta\sigma = \pm 1, \tag{III.110}$$

and also the transverse transitions

$$\Delta n_{\parallel} = -1,\ \Delta r = \Delta s = 1,\ \Delta\sigma = 0; \tag{III.111}$$

and longitudinal transitions

$$\begin{gathered}\Delta n_{\parallel} = 2,\ \Delta r = -1,\ \Delta s = 0,\ \Delta\sigma = 0,\\ \Delta r = 0,\ \Delta s = -1,\ \Delta\sigma = 0.\end{gathered} \tag{III.112}$$

In the single-particle model, transitions (III.108)-(III.112) produce weak peaks (with a relative intensity $[\alpha/(2\beta_0)]^2 = 4\%$ and $[\alpha\varkappa/(2\beta_0)]^2 = 0{,}2\%$) which are located at distances $\hbar(\omega_{\perp} - \omega_{\parallel})$ and $2h(\omega_{\perp} - \omega_{\parallel})$, etc,, from the corresponding principal maximum, i.e., these peaks form an "uncorrelated background in space." It is advantageous to exclude these peaks from the considerations.

Central residual interactions do not mix the states resulting from transitions (III.108)-(III.110) with states formed by transitions having $\Delta\sigma = 0$.

A state with spin–isospin excitation is generated by transitions (III.108)-(III.110):

$$\sum_{\nu} \mathbf{d}_{\nu}(s_x + is_y)_{\nu}|(0). \tag{III.113}$$

The properties of this state were considered in [75], where it was shown that this state is characterized by a very large energy spread $D^{1/2}_{S=T=1} \simeq \hbar\omega_0$.

Consequently, even when the residual interactions are taken into account, the transitions (III.108)-(III.110) must form a weak uncorrelated three-dimensional background, and hence, these transitions must be excluded from our consideration.

States which result from asymptotically forbidden transitions (III.111) and (III.112) can be treated as if they resulted from transitions

$$\Delta n_{\perp} = -\Delta n_{\parallel} = 2 \tag{III.114}$$

from states produced by asymptotic transitions. Therefore, in order to correctly evaluate the importance of transitions (III.111) and (III.112), we must take into consideration "excitations" of the type defined by Eq. (III.114) even in the ground state, which goes beyond the model under consideration.*

Schematic Formalism in the Random Phase Approximation. Several interesting qualitative conclusions can be drawn by choosing the parameters of the "particle–hole" model so that the model provides a single unique state of collective excitation [8, 13].

When we consider dipole oscillations, the single-particle potential is assumed to be an oscillator potential and the residual forces are assumed to be dipole–dipole forces [11, 13]:

$$v_{12} \sim \mathbf{r}_1\mathbf{r}_2. \tag{III.115}$$

This idealization of the residual forces is not sensible in terms of physics when a giant resonance of spheroidal nuclei is described, because the idealization leads to inconsistencies with the experimental determination of the energy ratio of longitudinal and transverse maxima [11, 71]. It is in this case more convenient to introduce an additional potential U whose structure, as mentioned above, resembles that of a single-particle Hartree–Fock potential.

* We encounter now the problem of the coupling of dipole oscillations with the other degrees of freedom whose effect is essentially undistinguishable from the effect of the anharmonicity of the effective potential in microscopical considerations.

It is logical to assume that to the extent to which the single-particle potential can be replaced with the potential of an anisotropic oscillator

$$U^{HF} = dU_1 + eU_2 \to \frac{M\omega_\perp^2}{2}(x^2 + y^2) + \frac{M\omega_\parallel^2}{2} z^2, \qquad (\text{III}.116)$$

the potential U can be approximated by the anisotropic oscillator potential

$$U = d'U_1 + e'U_2 \to \frac{d'U_1 + e'U_2}{dU_1 + eU_2}\left[\frac{M\omega_\perp^2}{2}(x^2 + y^2) + \frac{M\omega_\parallel^2}{2} z^2\right]. \qquad (\text{III}.117)$$

By substituting Eqs. (III.116) and (III.117) into Eqs. (III.45) and (III.46), we obtain for a state of longitudinal dipole excitation

$$\frac{E_\parallel}{\hbar\omega_\parallel}\sum_{mi} x_{mi}(m|d_z|i) = f_1 \sum_{mi} x_{mi}(m|d_z|i) + f_2 \sum_{mi} y_{mi}(m|d_z|i), \qquad (\text{III}.118)$$

$$-\frac{E_\parallel}{\hbar\omega_\parallel}\sum_{mi} y_{mi}(m|d_z|i) = f_1 \sum_{mi} y_{mi}(m|d_z|i) + f_2 \sum_{mi} x_{mi}(m|d_z|i), \qquad (\text{III}.119)$$

where

$$f_1 = \frac{1}{2}\left[1 + \frac{(d-d')U_1 + (e-e')U_2}{dU_1 + eU_2}\right], \qquad (\text{III}.120)$$

$$f_2 = -\frac{1}{2}(d'U_1 + e'U_2)/(dU_1 + eU_2). \qquad (\text{III}.121)$$

We obtain immediately from Eqs. (III.118)-(III.121):

$$E_\parallel = \hbar\omega_\parallel \sqrt{f_1^2 - f_2^2} = \sqrt{\frac{(d-d')U_1 + (e-e')U_2}{dU_1 + eU_2}}. \qquad (\text{III}.122)$$

Thus, our conclusion agrees with the conclusion which was previously drawn from the summation rules [see Eq. (III.20)]. The energy of the dipole maximum is equal to the single-particle energy multiplied by the square root of the ratio of the doubled potential, which describes the effect of all protons upon a neutron (or of the neutrons upon a proton), to the single-particle potential.

We emphasize the following point in conclusion of the present chapter. Equation (III.5) can be obtained from very different models under the condition that the models correctly describe the deformation of the nucleus.

The widths of a giant resonance are a much more important characteristic of the nuclear structure than the energy.

It follows from our discussion that both the anharmonicity of the single-particle potential (including the spin-orbit part) and the residual interactions of the "particle–hole" type imply a larger energy spread of the transverse maximum than of the longitudinal maximum. However, the model used does not account for all influences contributing to the width of the giant resonance. One of these influences, namely, surface oscillations, will be considered in the following chapter.

CHAPTER IV

Coupling of Dipole Oscillations and Surface Oscillations in Atomic Nuclei

The Schematic Model [76-81]

Experiments prove that a correlation exists between the properties of the low-energy quadrupole spectrum of atomic nuclei and the dipole excitation of these nuclei.

The quadrupole spectrum of the nuclei of the rare-earth group, in which the giant resonance is split into two maxima and in which strong optical anisotropy is observed, is characterized by specific properties. The spectrum is identical with the spectrum of an axial rotator and is characterized by extremely low energies and high transition probabilities. On the other hand, the double magic nuclei whose quadrupole spectrum is characterized by high energies and small transition intensities exhibit a single, relatively narrow photoabsorption peak and are within the experimental error limits [38] free of any optical anisotropy.

Additional information on the coupling between the two collective excitations can be obtained from research on the giant resonance in nuclei whose collective quadrupole spectrum differs from the spectrum of the axial rotator of so-called soft and axially nonsymmetric nuclei.

In order to obtain a general idea of the effects which can be expected in this case, we greatly simplify the problem by selecting two collective degrees of freedom, a dipole and a quadrupole degree of freedom, from the entire set of possible degrees of freedom contributing to a giant resonance.

We represent the Hamiltonian of the system as a sum of the Hamiltonians of dipole and quadrupole excitations and the interaction potential of the dipole and quadrupole degrees of freedom

$$H = H_d + H_Q + V_{dQ}. \tag{IV.1}$$

We assume

$$H_d = \frac{M}{2}\omega_d^2 \mathbf{r}^2 - \frac{\hbar^2}{2M}\frac{\partial^2}{\partial \mathbf{r}^2}, \tag{IV.2}$$

where ω_d denotes the frequency of the giant resonance, M denotes the mass of the nucleon, and $er = d[A/(NZ)]^{1/2}$. This notation reflects the experimental result that a dipole resonance of heavy magic nuclei manifests itself by a single, relatively narrow maximum which accounts for the entire dipole sum.

We assume that influences resulting in the energy spread of the giant resonance of double magic nuclei can be accounted for by ascribing widths $\hbar\Gamma$ to the dipole levels of the Hamiltonian stated in Eq. (IV.1).

It is logical to identify the Hamiltonian H_Q with the model Hamiltonian which describes the low-energy quadrupole spectrum of the nucleus.

The quadrupole coordinates are defined as parameters which appear in the equation of the nuclear surface [44]

$$R(\varphi\theta) = R_0\Big(1 + \sum_{\mu=-2}^{2} \alpha_\mu (-1)^\mu Y_{2,-\mu}(\varphi\theta)\Big). \tag{IV.3}$$

The form of the potential V_{dQ} is in the lowest order of small amplitudes of r and α_μ uniquely defined by the condition that the potential must be a scalar:

$$V_{dQ} = C\sum_{\mu\lambda}(11\mu\nu\,|\,2\lambda)\, r_\mu r_\nu (-1)^\lambda \alpha_{-\lambda}. \tag{IV.4}$$

Coefficient C is chosen so that the potential V_{dQ} in the sum with the Hamiltonian H_d describes dipole oscillations parallel to the principal deformation axis. These dipole oscillations have frequencies which are inversely proportional to the lengths of the axes. We obtain from this condition and from Eq. (IV.3):

$$C = -[15/(8\pi)]^{1/2} M\omega_d^2. \tag{IV.5}$$

This particular choice of coefficient C assumes that the change in the quadrupole coordinates is adiabatically slow compared to the change in the dipole coordinates so that the form of the Hamiltonian $H_d + V_{dQ}$ can be established under the assumption that the shape of the nucleus is stationary. Since the frequency of the quadrupole oscillations does usually not exceed 1 MeV/$\hbar$ and since $\hbar\omega_d \approx 15$ MeV, the condition that the change be adiabatic is rather well satisfied.

In low-energy states of the nucleus, in which the number n_d of dipole quanta vanishes, the dipole oscillations are uncoupled from the quadrupole oscillations:

$$(n_d = 0 | V_{dQ} | n_d = 0) = 0.$$

Coupling occurs in the state of dipole excitation with $n_d = 1$. Consequently, when a dipole photon is absorbed, nuclear surface oscillations are excited along with the "intrinsic dipole oscillations" of the nucleus.

This effect* and its importance for the formation of a giant dipole resonance were for the first time mentioned by Balashov and Chernov [82].

The model refers to even–even nuclei. The model can be generalized to odd nuclei by including the coordinates of the external nucleon as another degree of freedom in the considerations.

Notation and Terminology

Before we consider the main consequences of the above model, it is convenient to establish order in the notation, symbols, and terminology which we have to use. Depending upon the details of the problem, one of the following coordinate system is conveniently used.

1. Laboratory System. The dipole and quadrupole coordinates are in this system $\mathbf{r}(x, y, z)$ and α_μ, respectively.

2. System of the Principal Deformation Axes. The Euler angles, which determine the position of the system relative to the laboratory system, are denoted by φ, θ, and ψ. We use primed symbols for denoting the coordinates in the systems is conveniently used. mation axes:

$$\alpha'_1 = \alpha'_{-1} = 0. \tag{IV.6}$$

The parameters β and γ, which characterize the shape of the nucleus, are defined by the relations

$$\alpha'_0 = \beta \cos\gamma, \tag{IV.7}$$

$$\alpha'_2 = -\alpha'_{-2} = 2^{-1/2}\beta \sin\gamma. \tag{IV.8}$$

The lengths of the principal axes are

$$R_i = R_0\left[1 + \sqrt{\frac{5}{4\pi}}\,\beta\cos(\gamma - 2\pi i/3)\right], \quad i = 1, 2, 3.$$

---

* Including this effect in a microscopic theory is equivalent to including into the number of basis vectors, states of the type dipole pair of "particle–hole" ± several quadrupole particle–hole pairs, formed by transitions between unfilled shell levels; these states are included in addition to states which differ from the ground state by the presence or absence of a dipole "particle–hole" pair [18].

The components T_{lm} of a tensor of rank l are transformed to the laboratory system with the formula

$$T_{lm} = \sum_{m'} D^{l}_{mm'}(\varphi\theta\psi)\, T'_{ml'}. \tag{IV.9}$$

It is generally accepted that the selection of the system of principal deformation axes is not unique. There exist 24 transformations of the Euler angles and of the primed coordinates at which the coordinates expressed by these parameters in the laboratory system remain unchanged [44, 45].

3. System of the Spheroidal Core. In the case of weak surface oscillations around a strongly deformed axially symmetric core, Eqs. (IV.7) and (IV.8) can be linearized:

$$\alpha'_0 = \beta_0 + \beta', \tag{IV.10}$$

$$\alpha'_{2-} = \alpha'_2 = \beta_0\gamma/\sqrt{2}. \tag{IV.11}$$

The surface oscillations which are described by changes in the parameter α'_0 are termed β oscillations.

A change in the parameter γ, along with a rotation around the axis $\mathbf{e}'_3$ (described by a change in the parameter ψ), will be termed γ oscillations. Thus, γ oscillations are two-dimensional oscillations.

Rotations of spheroidal nuclei around an axis perpendicular to the symmetry axis $\mathbf{e}'_3$ are much slower than β and γ oscillations:

$$\frac{\hbar^2}{I_{xx}} \equiv \frac{\hbar^2}{I} \ll \hbar\Omega_\beta, \hbar\Omega_\gamma. \tag{IV.12}$$

In a first approximation, the β and γ oscillations can be assumed to occur independently of the rotations. The projection $K_{\beta,\gamma}$ of the moment of the surface oscillations upon the axis $\mathbf{e}'_3$ is a good quantum number. The photons of the β and γ oscillations are characterized by $K_\beta = 0$ and $K_\gamma = 2$, respectively.

When the coupling between dipole oscillations and surface oscillations in a spheroidal nucleus is considered, it is convenient to use the "coordinate system of the spheroidal core." The position of the unit vectors $\tilde{\mathbf{e}}_1$, $\tilde{\mathbf{e}}_2$, and $\tilde{\mathbf{e}}_3$ of this system is defined by the Euler angles φ, θ, and $\psi = 0$. The system of principal axes is obtained by rotating the system an angle ψ around the $\tilde{\mathbf{e}}_3$ axis. The variables in the system of the spheroidal core are denoted by a tilde. It is easy to verify that

$$T_{lm} = \sum_{m'} D^{l}_{mm'}(\varphi, \theta, 0)\, \tilde{T}_{lm'}, \tag{IV.13}$$

$$\tilde{T}_{lm} = e^{im\psi} T'_{lm}. \tag{IV.14}$$

More particularly,

$$(\tilde{x} \pm i\tilde{y})\, e^{\pm i\psi}(x' \pm iy'), \quad \tilde{z} = \tilde{z}', \ \tilde{\alpha}_0 = \alpha'_0. \tag{IV.15}$$

$$\tilde{\alpha}_{\pm 2} = \pm e^{2i\psi}\beta_0\gamma/\sqrt{2}. \tag{IV.16}$$

Equations (IV.13)-(IV.16) are invariants with respect to the following transformation [45]:

$$\psi \to \psi + \pi/2, \quad x' \to y', \quad y' \to -x', \quad \gamma \to -\gamma, \tag{IV.17}$$

which refers to the set of aforementioned symmetry transformations.

The Role of Surface Oscillations in the Development of the Giant Dipole Resonance of Spheroidal Nuclei [78, 79]

This problem is of interest in relation to the reasons for the broadening of the transverse maximum relative to the broadening of the longitudinal maximum.

We use an approximation which is compatible with condition (IV.12).

The dipole eigenfunctions are in this case

$$[3/(8\pi^2)]^{1/2} D^1_{M0}(\varphi\theta;\,0)\tilde{f}_{K=0},$$

$$[3/(16\pi^2)]^{1/2}[D^1_{M1}(\varphi\theta;\,0)\tilde{f}_{K=1} + D^1_{M,-1}(\varphi\theta;\,0)\tilde{f}_{K=-1}],$$

where $\tilde{f}_{K=0}$ and $\tilde{f}_{K=\pm 1}$ denote the eigenfunctions of the states of the longitudinal and transverse dipole excitation in the system of the spheroidal core, respectively.

The Hamiltonian of the dipole and quadrupole oscillations in this system is

$$H = H(\tilde{z}) + H(\tilde{x}, \tilde{y}) + H(\beta') + H(\gamma, \psi) + V(\tilde{x}, \tilde{y}, \tilde{z}, \beta, \gamma, \psi), \tag{IV.18}$$

$$H(\tilde{z}) = \frac{M\omega_a^2 \tilde{z}^2}{2} - \frac{\hbar^2}{2M}\frac{\partial^2}{\partial \tilde{z}^2}, \tag{IV.19}$$

$$H(\tilde{x}, \tilde{y}) = \frac{M\omega_b^2(\tilde{x}^2 + \tilde{y}^2)}{2} - \frac{\hbar^2}{2M}\left(\frac{\partial^2}{\partial \tilde{x}^2} + \frac{\partial^2}{\partial \tilde{y}^2}\right), \tag{IV.20}$$

$$H(\beta') = \frac{B'\Omega'^2\beta'^2}{2} - \frac{\hbar^2}{2B'}\frac{\partial^2}{\partial \beta'^2}, \tag{IV.21}$$

$$H(\gamma, \psi) = -\frac{\hbar^2}{8B\beta_0^2\gamma^2}\frac{\partial^2}{\partial \psi^2} - \frac{\hbar}{2B\beta_0^2}\frac{1}{\gamma}\frac{\partial}{\partial \gamma}\gamma\frac{\partial}{\partial \gamma} + B\beta_0^2\Omega^2\gamma^2/2 \ [83], \tag{IV.22}$$

$$V = -M\omega_d^2[5/(4\pi)]^{1/2}[\tilde{z}^2 - (\tilde{x}^2 + \tilde{y}^2)/2]\beta' - M\omega_d^2[15/(8\pi)]^{1/2}[(i\tilde{y} + \tilde{x})^2\tilde{\alpha}_{-2}/2 + (i\tilde{y} - \tilde{x})^2\tilde{\alpha}_2/2] = -M\omega_d^2[5/(4\pi)]^{1/2}[z'^2 - (x'^2 + y'^2)/2]\beta' - M\omega_d^2[15/(16\pi)]^{1/2}(x'^2 - y'^2)\beta_0\gamma, \tag{IV.23}$$

where

$$\omega_a = \omega_d\left(1 - \sqrt{\frac{5}{4\pi}}\beta_0\right); \quad \omega_b = \omega_d\left(1 + \frac{1}{2}\sqrt{\frac{5}{4\pi}}\beta_0\right).$$

The Hamiltonian of the surface oscillations of the spheroidal nucleus comprises four parameters Ω', B', Ω, and B. In place of B' and B, we can introduce the quantities

$$\bar{\beta}' = [\hbar/(2B'\Omega')]^{1/2} = (0|\beta'^2|0)^{1/2}, \tag{IV.24}$$

$$\bar{\gamma} = [\hbar/(B\beta_0^2\Omega)]^{1/2} = (0|\gamma^2|0)^{1/2}, \tag{IV.25}$$

which are the mean-square amplitude of the zeroth β and γ oscillations, respectively.

Since the frequency of the β oscillations is high and their amplitude is small ($\hbar\Omega' \approx 1.5$ MeV and $\bar{\beta}' \approx 0.05$–$0.1$), the influence of β oscillations upon a giant resonance can be ignored.

Of greatest interest is the importance of γ oscillations in terms of physics.

We infer from Eq. (IV.23) that the γ oscillations affect only the state of a transverse dipole excitation.

The part of the Hamiltonian stated in Eq. (IV.18) which describes transverse surface and dipole oscillations is

$$H_{\perp} = \frac{M\omega_b^2(x'^2+y'^2)}{2} - \frac{\hbar^2}{2M}\left(\frac{\partial^2}{\partial x'^2}+\frac{\partial^2}{\partial y'^2}\right) + \frac{B\beta_0^2\Omega^2\gamma^2}{2} -$$

$$-\frac{\hbar^2}{2B\beta_0^2}\frac{1}{\gamma}\frac{\partial}{\partial\gamma}\gamma\frac{\partial}{\partial\gamma} - \frac{\hbar^2}{8B\beta_0^2\gamma^2}\frac{\partial^2}{\partial\psi^2} - \sqrt{\frac{15}{16\pi}}\,M\omega_d^2(x'^2-y'^2)\beta_0\gamma. \qquad \text{(IV.26)}$$

The dipole eigenfunctions of the Hamiltonian stated in Eq. (IV.26) are related to the following integrals of motion: K = ±1 denote the projection of the total moment upon the e_3' axis; $n_{\perp} = 1$ denotes the quantum number of the transverse dipole excitation; and $\hbar\omega_b + \varepsilon_\lambda$ –energy = $\hbar\omega_b + \lambda\ \hbar\Omega$. The dipole eigenfunctions can be written in the form:

$$\tilde{f}_{K=\pm 1} \equiv (\psi\gamma x'y'|K,\lambda,\ n_{\perp}=1) =$$

$$= (2\pi)^{-1/2}\exp(iK\psi)[i(x'y'|n_{x'}=0,\ n_{y'}=1)\,\xi_\lambda(\gamma) + (x'y'|n_{x'}=1,\ n_{y'}=0)\,\xi_\lambda(-\gamma)], \qquad \text{(IV.27)}$$

where $n_{x'(y')}$ denotes the number of dipole quanta parallel to the principal deformation axis $e_1'(e_2')$.

The wave function stated in Eq. (IV.27) is invariant relative to the operation defined by Eq. (IV.17), as can be easily verified.*

After substituting Eqs. (IV.26) and (IV.27) into the Schrödinger equation, we obtain a system of second-order differential equations for $\xi_\lambda(\pm\gamma)$:

$$\left(-\frac{\hbar^2}{2B\beta_0^2}\frac{1}{\gamma}\frac{\partial}{\partial\gamma}\gamma\frac{\partial}{\partial\gamma} + \frac{B\beta_0^2\Omega^2\gamma^2}{2}\right)\xi_\lambda(\pm\gamma) + \frac{\hbar^2}{4B\beta_0^2\gamma^2}[\xi_\lambda(\pm\gamma) - \xi_\lambda(\mp\gamma)]\mp$$

$$\mp\hbar\omega_d[15/(16\pi)]^{1/2}\beta_0\gamma\xi_\lambda(\pm\gamma) = \hbar\Omega(\lambda+1)\,\xi_\lambda(\pm\gamma). \qquad \text{(IV.28)}$$

With the notation

$$\varphi_\lambda^0 \equiv [\xi_\lambda(+\gamma) + \xi_\lambda(-\gamma)]/2, \qquad \text{(IV.29)}$$

$$\varphi_\lambda^1 \equiv [\xi_\lambda(+\gamma) - \xi_\lambda(-\gamma)]/2, \qquad \text{(IV.30)}$$

we can transform the system of Eq. (IV.28) into the following form:

$$\left(-\frac{\hbar^2}{2B\beta_0^2}\frac{1}{\gamma}\frac{\partial}{\partial\gamma}\gamma\frac{\partial}{\partial\gamma} + \frac{B\beta_0^2\Omega^2\gamma^2}{2}\right)\varphi_\lambda^0 - \hbar\omega_d[15/(16\pi)]^{1/2}\beta_0\gamma\varphi_\lambda^1 = \hbar\Omega(\lambda+1)\,\varphi_\lambda^0, \qquad \text{(IV.31)}$$

$$\left(-\frac{\hbar^2}{2B\beta_0^2}\frac{1}{\gamma}\frac{\partial}{\partial\gamma}\gamma\frac{\partial}{\partial\gamma} + \frac{B\beta_0^2\Omega^2\gamma^2}{2} + \frac{\hbar^2}{2B\beta_0^2\gamma^2}\right)\varphi_\lambda^1 - \hbar\omega_d[15/(16\pi)]^{1/2}\beta_0\gamma\varphi_\lambda^0 = \hbar\Omega(\lambda+1)\,\varphi_\lambda^1. \qquad \text{(IV.31)}$$

The last term on the left side of Eqs. (IV.31) and (IV.32) describes the coupling between the dipole oscillations and γ oscillations. Without coupling, Eqs. (IV.31) and (IV.32) can be separated. The eigenfunctions of the γ oscillation Hamiltonian are the solutions. These eigen-

* The wave functions of coupled dipole oscillations and γ oscillations of [77] are not invariant with respect to the transformation defined by Eq. (IV.17) and, hence, the equations for the vibration functions of the state of transverse dipole excitation of [77] are incorrect.

TABLE 2

	β_0	$\hbar\Omega$, MeV	$\bar{\gamma}$	S		β_0	$\hbar\Omega$, MeV	$\bar{\gamma}$	S
$_{62}Sm^{152}$	0.28	1.1	0.18	0.39	$^{68}Er^{170}$	0,30	0.95	0.22^T	0.55
$_{62}Sm^{154}$	0.30	1.4	0.18^T	0.30	$_{70}Yb^{170}$	0,29	1.2	0.21^T	0.51
$_{64}Gd^{156}$	0.30	1.2	0,20	0.40	$_{70}Yb^{172}$	0,30	1.5	0.19^T	0.31
$_{64}Gd^{158}$	0.31	1.2	0.18	0.39	$_{70}Yb^{174}$	0,30	1.1^T	0.20^T	0.45
$_{64}Gd^{160}$	0.33	1.0	0.16	0.43	$_{70}Yb^{176}$	0,27	1.3	0.32^T	0.54
$_{66}Dy^{160}$	0.29	0.97	0.33^T	0.81	$_{72}Hf^{176}$	0,29	1.9^T	0.31^T	0,39
$_{66}Dy^{162}$	0.31	0.88	0.27^T	0.77	$_{72}Hf^{178}$	0,24	1.4	0.14^T	0.19
$_{66}Dy^{164}$	0.32	0.76	0.22^T	0.75	$_{74}W^{182}$	0,23	1.2	0.24	0.38
$_{68}Er^{164}$	0.29	0.84	0.30^T	0.85	$_{74}W^{184}$	0,21	0.90	0.23	0.44
$_{68}Er^{166}$	0.31	0.79	0.18	0.60	$_{74}W^{186}$	0,20	0,73	0.23	0,52
$_{68}Er^{168}$	0.31	0.82	0.19	0.59					

Note. The parameters of the γ oscillations and the parameter S of the coupling between dipole and γ oscillations for nuclei with N = 90-110, γ, and $\hbar\Omega$ are experimental values of [84]. The superscript T indicates that a theoretical value calculated in [84] with the superfluidity model is cited.

functions are characterized by the values $K_\gamma = 0$ and 2, respectively:

$$(\gamma|K_\gamma, N) \equiv (\gamma|2p + |K_\gamma|/2) = 2^{1/2}\bar{\gamma}^{-1}(-1)^{p+|K_\gamma|/2}\,[p!/(p+|K_\gamma|/2)!^3]^{1/2}\exp[-\gamma/\bar{\gamma}]^2]\,L_p^{|K_\gamma|/2}[(\gamma/\bar{\gamma})^2], \quad \text{(IV.33)}$$

where $N = 2p + |K_\gamma|/2$, $p = 0, 1, 2, \ldots$, and $L_p^{|K_\gamma|/2}$ denotes Chebyshev–Laguerre polynomials.*

In order to solve system (IV.31) and (IV.32), we use the conventional diagonalization procedure.

We represent φ_λ^0 and φ_λ^1 as expansions

$$\varphi_\lambda^0 = \sum_N (\gamma|N)(N|\lambda), \qquad N = 2p, \quad \text{(IV.34)}$$

$$\varphi_\lambda^1 = \sum_N (\gamma|N)(N|\lambda), \qquad N = 2p + 1, \quad \text{(IV.35)}$$

which corresponds to rewriting the eigenfunction of the Hamiltonian stated in Eq. (IV.26) in the form

$$(\psi\gamma\bar{x}\bar{y}|K, n_1 = 1, \lambda) = \sum_N (\psi\gamma|NK_\gamma)(\bar{x}\bar{y}|n_1 = 1, K - K_\gamma)(N|\lambda) \quad \text{(IV.36)}$$

After substituting Eqs. (IV.34) and (IV.35) into the system (IV.31) and (IV.32) [or the expansion (IV.36) and the Hamiltonian of (IV.26) into the Schrödinger equation], we obtain the following infinite equation system:

$$\left.\begin{aligned} S\sqrt{p}(2p-1|\lambda) + (\lambda - 2p)(2p|\lambda) + S\sqrt{p+1}(2p+1|\lambda) = 0, \\ S\sqrt{p+1}(2p|\lambda) + (\lambda - 2p - 1)(2p+1|\lambda) + S\sqrt{p+1}(2p+2|\lambda) = 0. \end{aligned}\right\} \quad \text{(IV.37)}$$

* The eigenfunctions of the Hamiltonian (IV.22) can be obtained from the eigenfunctions $(\rho\,\varphi\,|\,n_\perp\Lambda)$ of the two-dimensional dipole oscillator, after replacing in the latter eigenfunctions the quantity ρ (radius vector) by γ; φ (azimuthal angle) by 2ψ; $n_\perp$ by N; and Λ (projection of the moment upon the e_3' axis) by $K_\gamma/2$. Nilsson and Mottelson [72] developed a simple method of constructing the functions $(\rho\varphi/n_\perp\,\Lambda)$. The phases which we have adopted in Eq. (IV.33) agree with the phases of the functions of [72].

The solution of the problem depends upon the coupling parameter

$$S = (n_{\perp} = 1,\ N = 1|V|n_{\perp} = 1,\ N = 0)/(\hbar\Omega) = [15/(16\pi)]^{1/2}\omega_d\beta_0\bar{\gamma}/\Omega. \quad \text{(IV.38)}$$

The quantity S is the ratio of the matrix element expressing the transition between a "pure" dipole state and a state comprising a single dipole and quadrupole photon to the energy difference between these states.

Depending upon S, the coupling between the two excitations can be classifed as weak coupling (when $S \ll 1$ so that perturbation theory can be used), intermediate coupling ($S \simeq 1$), and strong coupling ($S \gg 1$).

Table 2 lists the β_0, $\bar{\gamma}$, $\hbar\Omega$ values which were obtained from experimental low-energy spectra and calculations with the superfluidity model of [84, 85]. Table 2 includes the S value (for $\hbar\omega_d = 15$ MeV) of the rare-earth nuclei.

It follows from the table that $0.4 \leq S \leq 0.8$, i.e., weak or intermediate coupling between dipole and γ oscillations exists.

The secular equation (IV.37) was solved for the following values of the parameter S: $1/\sqrt{2}$, 1, and $\sqrt{2}$. (In the latter case, the value was greater than the experimentally obtained value.) The solution was made with the Tamm–Dankov method, i.e. by cutting off an infinite system.

The order of magnitude of the cut-off system was increased until the results of the successive interactions for the first five eigenvalues and the corresponding expansion coefficients $(0|\lambda)$, $(1|\lambda)$, and $(2|\lambda)$ differed by less than 1%. One-percent agreement begins at the 10th (for $S = 1/\sqrt{2}$), the 14th (for $S = 1$), and the 20th (for $S = \sqrt{2}$) order of the system. The results of the solution are listed in Table 3. Figure 2a,b,c shows the distribution of the relative intensities of the transverse dipole transitions, $I_\lambda = (0|\lambda)^2$ for $S = S/\sqrt{2}$, 1, and $\sqrt{2}$. The cross section of dipole photoabsorption assumes the following form:

$$\sigma = \frac{\sigma_a}{[(\omega_a^2 - \omega^2)/(\omega\Gamma_a)]^2 + 1} + \sigma_b' \sum_\lambda \frac{(0|\lambda)^2}{[(\omega_\lambda^2 - \omega^2)/(\omega\Gamma_b')]^2 + 1} \quad \text{(IV.39)}$$

where $\omega_\lambda = \omega_b + \lambda\Omega$, $\sigma_a = 8\pi(9\hbar c)^{-1}(0|d^2|0)\,\omega_d/\Gamma_a$, and $\sigma_b' = 16\pi(9\hbar c)^{-1}(0|d^2|0)\,\omega_d/\Gamma_b'$.

TABLE 3

λ	$(0\|\lambda)$	$(1\|\lambda)$	$(2\|\lambda)$	$(0\|\lambda)^2$
		$S = \frac{1}{\sqrt{2}}$		
−0.423	0.843	0.504	0.168	0.715
0.820	0.466	−0.54	−0.604	0.218
1.73	0.250	−0.62	0.382	0.063
2.74	0.064	−0.25	0.590	0.004
3.78	0.019	−0.102	0.376	0.000
		$S = 1$		
−0.767	0.757	0.581	0.300	0.574
0.485	0.490	−0.244	−0.617	0.242
1.57	0.382	−0.600	−0.041	0.146
2.43	0.182	−0.442	0.451	0.033
3.54	0.055	−0.199	0.448	0.003
		$S = \sqrt{2}$		
−1.37	0.634	0.612	0.393	0.402
−0.181	0.503	0.065	−0.449	0.253
1.03	0.398	−0.292	−0.390	0.158
2.05	0.370	−0.540	0.0109	0.136
2.93	1.200	−0.410	0.361	0.040
4.03	0.069	−0.218	0.399	0.005

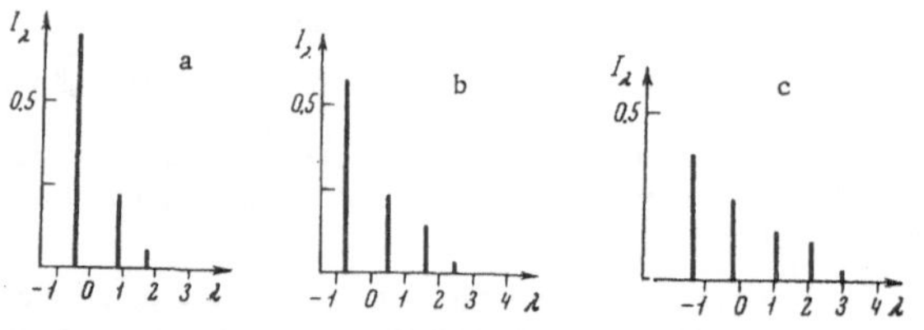

Fig. 2. Distribution of the relative intensities $I_\lambda = |(\lambda|d_1|0)|^2| / (0||d_1|^2|0)$ of transverse dipole transitions in spheroidal nuclei. The distribution is obtained from a schematic model for the following parameters of the coupling between the dipole oscillations and γ oscillations: $S = 1/\sqrt{2}$ (a); $S = 1$ (b); and $S = \sqrt{2}$ (c). $\lambda = (\varepsilon_\lambda - \hbar\omega_b)(\hbar\Omega)$, where ε_λ denotes the eigenenergy of dipole state λ.

Figure 3 displays the experimental data of [48] for the photoabsorption cross section of the Ho^{165} nucleus. The solid line denotes the sum of the two Lorentz curves which approximate the cross section [see Eq. (III.1)], with the resonance parameters $\sigma_a = 200$ mbarn, $\sigma_b = 249$ mbarn, $\hbar\omega_a = 12.1$ MeV, $\hbar\omega_b = 15.75$ MeV, $\hbar\Gamma_a = 2.65$ MeV, and $\hbar\Gamma_b = 4.4$ MeV.

Calculation of the cross section with Eq. (IV.39) and with parameters characterizing the transverse maximum results in:

calculation I

$$\bar{\gamma} = 0.2;\ \hbar\Omega = 0.75\ \text{MeV}\ (S = 1/\sqrt{2}),\ \Gamma_b' = 3.9\ \text{MeV},\ \sigma_b' = 270\ \text{mbarn};$$

calculation II

$$\bar{\gamma} = 0.3;\ \hbar\Omega = 0.75\ \text{MeV}\ (S = 1),\ \Gamma_b' = 3.7\ \text{MeV},\ \sigma_b' = 270\ \text{mbarn}.$$

The results differ by less than 3% from the values represented by the solid line.

The broadening which the γ oscillations induce in the transverse maximum is $\Delta\Gamma = 0.5$–0.7 MeV for the greater $\bar{\gamma}$ value ($\bar{\gamma} = 0.2$–0.3) and smallest $\hbar\Omega$ value ($\hbar\Omega = 0.75$ MeV).

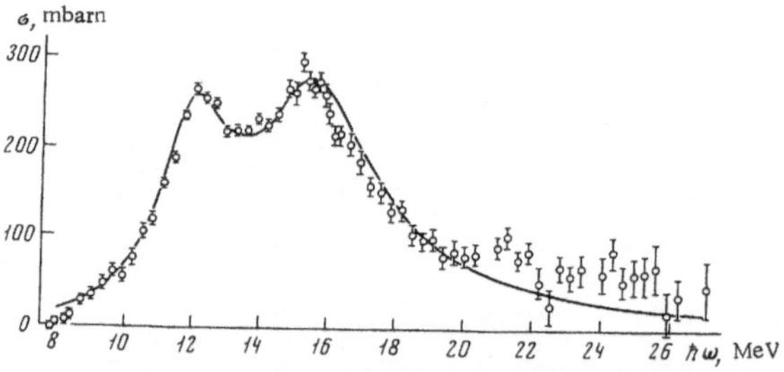

Fig. 3. Photoabsorption cross section $\sigma = \sigma(\gamma, n) + \sigma(\gamma, 2n)$ for Ho^{165}; measurements of [48]. The experimental data are in good agreement with the sum of two Lorentz-type resonance curves calculated in [48] (solid curve). The scale of the y axis was corrected relative to that of [48].

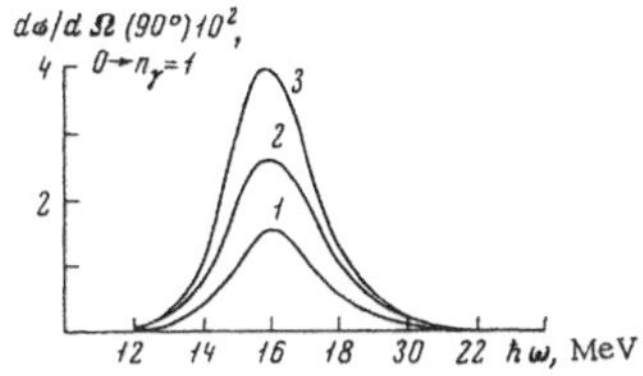

Fig. 4. Cross section of photon scattering with excitation of the first γ oscillation level (expressed in mbarn/sr). The cross section was calculated for the parameter sets I, II, and III (curves 1, 2, and 3, respectively).

However, this conclusion cannot be checked by comparing theoretical calculations and experimental data on photoabsorption, because the relative value $\Delta\Gamma/\Gamma_b$ of the broadening is small.

By varying the parameters $\Gamma_b' = \Gamma_b - \Delta\Gamma$, we can obtain agreement with the photoabsorption experiment for very dissimilar γ oscillation parameters.

For example, when the following parameter values are chosen for the case $\hbar\omega$ = 14-18 MeV, the results are almost the same as in the preceding calculations:

calculation III

$$\gamma = 0.3;\ \hbar\Omega = 0.53\ \text{MeV}\ (S = \sqrt{2}),\ \Gamma_b' = 3.1\ \text{MeV},\ \sigma_b' = 270\ \text{mbarn}.$$

This ambiguity can be eliminated by investigations of Raman scattering in which γ oscillation levels are excited.

The cross section of scattering with excitation of the first γ oscillation level is

$$\frac{d\sigma}{d\Omega}_{0\to N=1} = \frac{(\omega-\Omega)^3\omega}{40c^4}\,|(N=1\,|\,c_2^2\,|\,0)|^2\,(13+\cos^2\theta^S), \quad \text{(IV.40)}$$

where

$$(N=1\,|\,c_2^2\,|\,N=0) = 4\hbar^{-1}3^{-1/2}(0\,|\,d^2\,|\,0)\,\omega_d \sum_\lambda (0|\lambda)\,(1|\lambda)\,(\omega_\lambda^2-\omega^2-i\omega\Gamma_b')^{-1} \quad \text{(IV.41)}$$

denotes the corresponding matrix element of tensor polarizability. Figure 4 shows the cross section curves of photon scattering with excitation of the first γ oscillation level.* The curves were calculated with the parameter sets of calculations I, II, and III.

According to the figure, the magnitude of the effect differs in all three cases. The overall effect is roughly proportional to $\Delta\Gamma/\Gamma_b'$.

It is interesting to compare the results of the γ oscillation model with calculations based on the model of a rotator with a small static asymmetry in axial direction [47]. Expansion (IV.36) consists in this case only of two terms: the wave function describing the rotational state with projection of moment 2 upon the axis of approximate symmetry and with the energy $\hbar\Omega$ is denoted by $|1)$.

The secular equation is identical with equation system (IV.37), the equations being cut off at the second-order terms. The secular equation has the following solution [79]:

$$\lambda_{0,1} = [1 \mp (4S^2+1)^{1/2}]/2, \quad \text{(IV.42)}$$

$$(0|\lambda_0) = -(1|\lambda_1) = [1+(4S^2+1)^{-1/2}]^{1/2}2^{-1/2}, \quad \text{(IV.43)}$$

$$(1|\lambda_0) = (0|\lambda_1) = [1-(4S^2+1)^{-1/2}]^{1/2}2^{-1/2}, \quad \text{(IV.44)}$$

* Calculations of the cross section of photon scattering with excitation of the second γ oscillation level indicate that for the parameter set of our calculations, the cross section amounts to 10% of the scattering cross section with excitation of the first γ oscillation level.

TABLE 4

S	$1/\sqrt{2}$		1		$\sqrt{2}$	
λ	−0.366	1.366	−0.618	1.618	−1	2
$(0 \mid \lambda)$	0.895	0.46	0.85	0.524	0.815	0.577
$(1 \mid \lambda)$	0.46	−0.895	0.524	−0.85	0.577	−0.815
$(0 \mid \lambda)^2$	0.79	0.21	0.725	0.275	0.666	0.333

where

$$S \equiv [15/(16\pi)]^{1/2}\,\omega_d\beta_0\gamma_0/\Omega. \tag{IV.45}$$

It follows from Eqs. (IV.42)-(IV.45) that the case considered by Iponin [30], i.e., splitting of the transverse maximum into two identical peaks with energies $h(\omega_b = \mp\sqrt{15/(16\pi)}\omega_d\beta_0\gamma_0)$ occurs for S ≫ 1, which does not correspond to real conditions because we have in reality S ≲ 1.

The results obtained from the solution of the secular equation for the S values used in the oscillator model are listed in Table 4.

The results listed in the table lead us to the conclusion that in the rotation model, which differs in details of the "fine structure" from the oscillator model, agreement is observed with the oscillator model, as far as the total broadening is concerned.

The cross section calculations for Raman scattering with excitation of the second 2^+ level differ also slightly (by less than 20%) from the calculations which were made with the same sets of parameters and the γ oscillator model.

The large widths $\hbar\Gamma$ (about 4 MeV) of the λ levels smear out the details and make oscillator and rotator models indistinguishable, at least with the present accuracy of experiments.

Effects of the Optical Anisotropy of Spherical Nuclei

"Soft" nuclei, i.e., spherical or almost spherical nuclei with a large amplitude of the zeroth surface oscillations are of great interest for experimental studies of the coupling between dipole and quadrupole oscillations.

The numerical solution to the problem of coupling between dipole and quadrupole oscillations in spherical nuclei was obtained by Le Tourneux [80]. The solution method is similar to that described on p. 182.

The intrinsic wave function is stated as an expansion in terms of products of eigenfunctions of the Hamiltonians H_d and H_Q, the eigenfunctions being combined by vector products:

$$(\mathbf{r}, \alpha_\mu | J = 1, M, n_d = 1, \varepsilon) = \sum_{jmNv} (1jM - m, m|1M)(\mathbf{r}\,|\,n_d = 1,$$

$$J_d = 1, M - m)(\alpha_\mu|Nvjm)(jNv|\varepsilon), \tag{IV.46}$$

where j, N, and v denote the total moment of momentum, the oscillator number, and seniority, respectively, which quantities characterize the eigenfunction of the Hamiltonian of quadrupole oscillations in a spherical nucleus:

$$H_Q = -\frac{\hbar^2}{2B}\sum_\mu \left|\frac{\partial}{\partial\alpha_\mu}\right|^2 + \frac{B\Omega^2}{2}\sum_\mu |\alpha_\mu|^2.$$

The coupling coefficient S is in this case

$$S = \sum_m (12M - m, m|1M)(n_d = 1, M - m, N = 1, j = 2, m|V_{dQ}|n_d = 1, N = 0, M)/(\hbar\Omega) = (5/8\pi)^{1/2}\omega_d\bar{\beta}/\Omega, \tag{IV.47}$$

where

$$\bar{\beta} = \left(0 \left| \sum_\mu |\alpha_\mu|^2 \right| 0\right)^{1/2}.$$

We have for the case of Se^{76}, which was considered by Le Tourneux: $\hbar\omega_d = 18$ MeV, $\hbar\Omega = 0.56$ MeV, $\bar{\beta} = 0.3$, and $S = 3$, i.e., we have an almost strong coupling.

For the elastic scattering cross section and for the cross section of scattering with excitation of the lowest levels of the quadrupole spectrum, we have JN = 21, 22, 02. We obtain for $\hbar\omega = 20$ MeV, $\theta^S = \pi/2$, and $\hbar\Gamma = 3$ MeV from Le Tourneux's calculations [81]:

$$\frac{d\sigma}{d\Omega}_{21\leftarrow 0} \Big/ \frac{d\sigma}{d\Omega}_{0\leftarrow 0} = 0.35, \quad \frac{d\sigma}{d\Omega}_{22\leftarrow 0} \Big/ \frac{d\sigma}{d\Omega}_{0\leftarrow 0} = 0.03, \quad \frac{d\sigma}{d\Omega}_{02\leftarrow 0} \Big/ \frac{d\sigma}{d\Omega}_{0\leftarrow 0} = 0.10.$$

Approximation of Strong Coupling [76]

In the limit $S \to \infty$ we can obtain* a simple analytic solution of the coupling between dipole and quadrupole oscillations on the basis of the following considerations.

As has been shown above, the coupling between the dipole oscillations and the surface oscillations implies that the dipole spectrum comprises an infinite number of levels spaced by intervals of the order $\hbar\Omega$. The condition $S \gg 1$ means that in an energy interval $\hbar\omega_d\alpha_\mu$ a very large number of such levels can be accommodated. In order to describe the effect averaged over an interval $n\hbar\Omega$ $(1 \ll n < S)$, the discrete spectrum can be approximated by a continuous spectrum under the assumption $\hbar\Omega = d\varepsilon \to 0$.

In accordance with the substitution of the dipole state in the equation of the wave functions, we assume $H_Q = 0$ (because $H_Q/V_{dQ} \sim 1/S$). The variables β and γ, as well as the directions of the dipole oscillations parallel to the principal deformation axes, are therefore integrals of the motion of the dipole state. The form of the unknown wave functions is therefore uniquely determined.

Let us consider concrete examples.

1. Coupling between Dipole Oscillations and β Oscillations in Spheroidal Nuclei

This case is uniquely determined when it is possible to compare the results obtained in the approximation of strong coupling with the accurate analytic results. By taking the limit from the analytic results to the approximation of strong coupling, the meaning of the approximation of strong coupling for oscillator models can be established in a concrete example.

The wave function of coupled $\tilde{z}$ and β oscillations has the following form for $S \to \infty$:

$$(\tilde{z}\beta'|n_{\tilde{z}} = 1, \varepsilon) = [5/(4\pi)]^{+1/4}(\hbar\omega_d)^{1/2}(\tilde{z}|n_{\tilde{z}} = 1)\,\delta(\varepsilon + \sqrt{5/(4\pi)}\hbar\omega_d\beta'). \tag{IV.48}$$

* The actual form of the coupling parameter S differs in the various models [see Eqs. (IV.38), (IV.45), (IV.47), and also (IV.54)].

The function of Eq. (IV.48) is normalized so that

$$\int d\tilde{z}\, d\beta' (n_{\tilde{z}} = 1,\ \varepsilon|\tilde{z},\ \beta')(\tilde{z},\ \beta'|n_{\tilde{z}} = 1,\ \varepsilon') = \delta(\varepsilon - \varepsilon'). \quad \text{(IV.49)}$$

The expansion coefficients of the function (IV.48) in terms of the wave functions of uncoupled $\tilde{z}$ and β oscillations are

$$(n_{\tilde{z}} = 1,\ N|n_{\tilde{z}} = 1,\ \varepsilon) = (\sqrt{5/(4\pi)}\,\hbar\omega_d\bar{\beta}')^{-1/2}(-1)^N\ (\sqrt{2\pi}2^N N!)^{-1/2}\exp(-\chi^2/4)\,H_N(\chi/\sqrt{2}), \quad \text{(IV.50)}$$

where

$$\chi = \varepsilon/(\sqrt{5/(4\pi)}\,\hbar\omega_d\bar{\beta}'); \quad \text{(IV.51)}$$

and H_N denotes a Hermitian polynomial (N denotes the number of β oscillation photons).

Consequently, the square of the dipole matrix element obeys a Gauss distribution:

$$(\varepsilon|d_{\tilde{z}}|0)^2 = (0|d^2_{\tilde{z}}|0)(\sqrt{5/(4\pi)}\,\hbar\omega_d\bar{\beta}')^{-1}(2\pi)^{-1/2}\exp(-\chi^2/2). \quad \text{(IV.52)}$$

The accurate wave function of coupled $\tilde{z}$ and β oscillations has the form of [77]:

$$(\tilde{z},\ \beta'|n_{\tilde{z}} = 1,\ \nu) = (\tilde{z}|n_{\tilde{z}} = 1)(\beta'|\nu) = (\tilde{z}|n_{\tilde{z}} = 1)(\sqrt{2\pi}\,2^\nu \nu!\,\bar{\beta}')^{-1/2} \times$$
$$\times \exp\{-[(\beta' - 2S\bar{\beta}')/\bar{\beta}']^2\}\,H_\nu[(\beta' - 2S\bar{\beta}')/(\bar{\beta}'\sqrt{2})], \quad \text{(IV.53)}$$

where

$$S(n_{\tilde{z}} = 1,\ N = 1|V_{dQ}|n_{\tilde{z}} = 1,\ N = 0) = \sqrt{5/(4\pi)}\,\omega_d\bar{\beta}'/\Omega', \quad \text{(IV.54)}$$

H_ν denotes a Hermitian polynomial ($\nu = 0, 1, 2, \ldots$); and

$$\varepsilon_\nu = \left(-S^2 + \nu + \frac{1}{2}\right)\hbar\Omega' \quad \text{(IV.55)}$$

denotes the eigenenergy.

We have

$$(n_{\tilde{z}} = 1, N = 0|n_{\tilde{z}} = 1, \nu) = S^\nu \exp(-S^2/2)(\nu!)^{-1/2}, \quad \text{(IV.56)}$$

where $|n_{\tilde{z}} = 1,\ N)$ denotes the wave function of uncoupled dipole oscillations and β oscillations.

The distribution of the square of the dipole matrix element for the transition between the ground state of the nucleus and the state which is described by the function of Eq. (IV.53) has the following form:

$$|(n_{\tilde{z}} = 1,\ \nu\,|\,d_{\tilde{z}}\,|\,0)|^2 = (0\,|\,d^2_{\tilde{z}}\,|\,0)\,S^{2\nu}\exp(-S^2)/(\nu!). \quad \text{(IV.57)}$$

We perform now the transition to the limit, i.e., the transition from Eq. (IV.57) to Eq. (IV.52).

After we have set $\nu = \chi S + S^2$ in accordance with Eqs. (IV.55) and (IV.51), we take the limit $S \to \infty$ in Eq. (IV.57):

$$\lim_{S\to\infty} S^{2(\chi S+S^2)}\exp(-S^2)/[(\chi S + S^2)!] = S^{-1}\exp(-\chi^2/2)/\sqrt{2\pi} = (\sqrt{5/(4\pi)}\ \hbar\omega_d\,\bar{\beta}')^{-1}(2\pi)^{-1/2}d\varepsilon\exp(-\chi^2/2). \quad \text{(IV.58)}$$

Eq. (IV.58) results in

$$(N = 0|\nu) = (\sqrt{5/(4\pi)}\,\hbar\omega_d\bar{\beta}')^{-1/2}(2\pi)^{-1/4}\sqrt{d\varepsilon}\exp(-\chi^2/2). \quad \text{(IV.59)}$$

In order to generalize Eq. (IV.59) to the other N, we use the secular equation

$$\sqrt{N}\,(N-1|\varepsilon)+\left[\chi-\frac{N+{}^1\!/\!_2}{S}\right](N/\varepsilon)+\sqrt{N+1}\,(N+1/\varepsilon)=0, \tag{IV.60}$$

which for S → ∞ becomes the recurrence relation for normalized Hermitian polynomials. We obtain from Eqs. (IV.59) and (IV.60):

$$(N|\nu)=(\sqrt{5/(4\pi)}\,\hbar\omega_d\bar{\beta}')^{-1/2}(2\pi)^{-1/4}\sqrt{d\varepsilon}\,(2^N N!)^{-1/2}\exp\left[-\chi^2/4\right]H_N(\chi/\sqrt{2}). \tag{IV.61}$$

The function of Eq. (IV.61), divided by $\sqrt{d\varepsilon}$ (i.e., normalized to the δ function of the energy), is identical with the functions stated in Eq. (IV.50).

Noteworthy enough, the approximation which is linear in the deformation parameters (and which determines the form of potential V_{dQ}) and the approximation of strong coupling are compatible when the following conditions are satisfied: $\sqrt{\omega_d/\Omega}>S\gg 1$.

2. Coupling of Dipole Oscillations and γ Oscillations in a Spheroidal Nucleus

The dipole eigenfunction of coupled transverse dipole oscillations and γ oscillations in a spheroidal core assumes the following form for S → ∞:

$$\begin{aligned}(x', y', \psi, \gamma\,|\,K=\pm 1,\ n_\perp=1,\ \varepsilon) &= (\sqrt{15/(16\pi)}\,\hbar\omega_d\,\beta_0/\gamma)^{1/2}(2\pi)^{-1/2}\times\\ &\times\exp(iK\psi)\,[i\,(x'y'|n_{x'}=0,\ n_{y'}=1)\,\delta(\varepsilon+\sqrt{15/(16\pi)}\,\hbar\omega_d\,\beta_0\gamma)+\\ &+K\,(x'y'|n_{x'}=1,\ n_{y'}=0)\,\delta(\varepsilon-\sqrt{15/(16\pi)}\,\hbar\,\omega_d\beta_0\gamma)].\end{aligned} \tag{IV.62}$$

The wave function of Eq. (IV.62) is invariant with respect to the transformation stated in Eq. (IV.17) and satisfies the normalization condition*

$$\int (K,\ n_\perp=1,\varepsilon|x'y'\ \psi\gamma)\,dx'dy'\,d\psi\gamma\,d\gamma\,(x'y'\psi\gamma|K',\ n_\perp=1,\ \varepsilon')=\delta_{K'K}\delta(\varepsilon-\varepsilon').$$

We use Eqs. (I.15) and (IV.62) to calculate the matrix elements of the polarizability. Since the spectrum of resonance states is assumed to be continuous, the resonance states are conveniently written in the form

$$(f\,|\,c\,(\omega,\ \Gamma)\,|\,i)=\frac{1}{\pi}\int_0^\infty \operatorname{Im}(f\,|\,c\,(\omega',\Gamma=0)\,|\,i)\ \frac{2\omega'\,d\omega'}{\omega'^2-\omega^2-i\omega\Gamma}, \tag{IV.63}$$

where $\Gamma\equiv\Gamma(\omega')$ denotes the width of the dipole level $\hbar\omega'$.

For the scalar polarizability part which describes the transverse oscillations, and for the tensor polarizability part which describes photon scattering with excitation of the first γ oscillation level, we obtain

$$\operatorname{Im} c^0_\perp(\omega,\ 0)=(0\,|\,\mathbf{d}^2\,|\,0)\,2\pi\,(9\sqrt{15/(16\pi)}\,\hbar\omega_a\beta_0\bar{\gamma})^{-1}\,|\,\chi\,|\,e^{-\chi^2}, \tag{IV.64}$$

$$\operatorname{Im}(J=1,\ M,\ K=2,\ N=1\,|\,c^2_\sigma(\omega,\ 0)|0)=-\,\delta_{\sigma M}\,(0\,|\,\mathbf{d}^2\,|\,0)\,2\pi\ [45\,(16\pi)^{-1/2}\hbar\,\omega_d\,\beta_0\bar{\gamma}]^{-1}\chi\,|\,\chi\,|\,e^{-\chi^2}, \tag{IV.65}$$

respectively.

---

* The two terms in the wave function of Eq. (IV.62) result from the condition that the wave function be invariant with respect to the transformation stated in Eq. (IV.17). But it is easy to verify that these terms correspond to different ranges of energy variation. This means that the use of unsymmetrized wave functions does not cause errors in the results obtained in this case. This is a specific property of the strong coupling approximation in which the direction of dipole oscillations parallel to the principal deformation axes is conserved. Unsymmetrized wave functions do usually not satisfy the Schrödinger equation [see remark to Eq. (IV.27)].

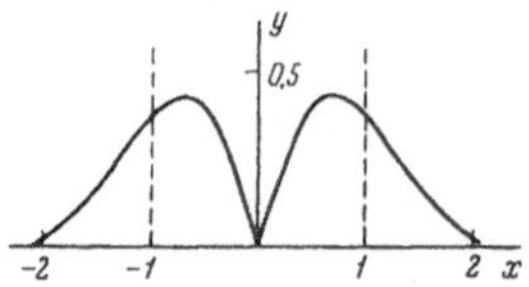

Fig. 5. Distribution of the relative intensities $y = |(\varepsilon|d_1|0)|^2 \square / [(0 \| d_1|^2|0) dx]$ of transverse dipole transitions in a spheroidal nucleus; schematic model in the limit of strong coupling $(S \to \infty)$. $x = (\varepsilon - \hbar\omega_b)/[\sqrt{15/(16\pi)}\, \hbar\omega_d \beta_0 \bar{\gamma}]$, where ε denotes the energy of the dipole state. The dashed line denotes the position of the resonance energies (for $S \to \infty$) in the model of a rigid nonaxial core with $\gamma_0 = \bar{\gamma}$.

We recognize that for $S \to \infty$, the function describing the energy distribution of the square of the transverse dipole matrix element is split into two identical, smeared maxima which are separated by the interval $\sqrt{15/(8\pi)}\, \hbar\omega_d \beta_0 \bar{\gamma}$ (see also Fig. 7).

3. Coupling of Dipole Oscillations and Quadrupole Oscillations in a Spherical Nucleus

The approximation of strong coupling can be of definite interest for this case, because there exist spherical nuclei with a rather large amplitude but low frequency of surface oscillations ("soft" nuclei) for which we can assume $S \gg 1$ (e.g., in the case of Sm^{150}, we have $S \approx 5$).

The dipole eigenfunctions of coupled dipole oscillations and quadrupole oscillations of a spherical nucleus assume the form

$$(\varphi\theta\psi\, \beta\gamma\, x'y'z'|J = 1, M, n_d = 1, \eta, \varepsilon) = \sum_{K,i} [3/(8\pi^2)]^{1/2} D^1_{MK}(n_i = 1|K) \times$$
$$\times (x'y'z'|n_i = 1)\, \delta\, [\varepsilon + \sqrt{5/(4\pi)}\, \hbar\omega_d \beta \cos(\gamma - 2\pi i/3)] \times$$
$$\times \delta(\eta - \beta)(\sqrt{5/(4\pi)}\, \hbar\omega_d)^{3/2} \eta^{-1/2} |[4\varepsilon^2 - 5(\hbar\omega_d \beta)^2 | (4\pi)]|^{-1/4}, \qquad (IV.66)$$

in the limit $S \to \infty$, where K denotes the projection of the angular moment of the dipole oscillations upon the principal deformation axis $i = 3$, $n_{i=1,2,3} \equiv n_{x',y',z'}$ $(\sum_i n_i = 1)$. The symbols $n_{x',y',z'}$ have the same meaning as in Eqs. (IV.27) and (IV.48).

The wave function stated in Eq. (IV.66) is invariant with respect to all transformations of φ, θ, ψ, γ, x', y', and z' which do not affect the coordinates in the laboratory system. These wave functions satisfy the normalization condition

$$\int (1M'n_d = 1, \eta', \varepsilon' | x'y'z'\, \varphi\theta\psi\gamma)\, dx'dy'dz'\, d\varphi\, d\psi \sin 3\gamma\, d\gamma \beta^4 d\beta \times$$
$$\times (x'y'z'\varphi\theta\,\psi\gamma | 1Mn_d = 1, \eta, \varepsilon) = \delta_{MM'} \delta(\eta - \eta')\, \delta(\varepsilon - \varepsilon').$$

Let us state the expressions for the matrix elements of polarizability which describe elastic scattering and scattering with excitation of the first and second 2+ levels:

$$\mathrm{Im}\,(0|c^0(\omega, 0)|0) = (0|\mathbf{d}^2|0)\, \pi \sqrt{8}\, (9\hbar\omega_d \bar{\beta})^{-1} [(3\chi^2 - 1) e^{-\chi^2} + 2e^{-4\chi^2}], \qquad (IV.67)$$

$$\mathrm{Im}\,(N = 1, J = 2, M|c^2_\sigma(\omega, 0)|0) = \delta_{\sigma M} (0|\mathbf{d}^2|0)\, 8\pi (45\hbar\omega_d \bar{\beta})^{-1} \chi [(3\chi^2 - 1) e^{-\chi^2} + 2e^{-4\chi^2}], \qquad (IV.68)$$

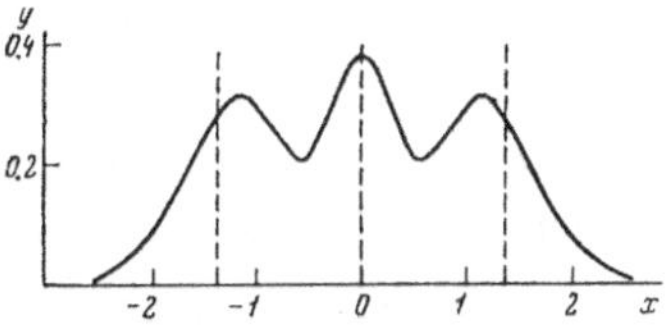

Fig. 6. Distribution of the relative intensities of dipole transitions of a spherical nucleus, $y = |(\varepsilon|d|0)|^2 / [(0|d^2|0)dx]$ in the schematic model; limit of strong coupling ($S \to \infty$); $x = (\varepsilon - \hbar\omega_d)(2\pi)^{1/2}/(\hbar\omega_d\bar{\beta})$. The dashed lines denote the position of the resonance energies in the limit of strong coupling for a rigid nonaxial core with $\beta_0 = \bar{\beta}$; $\gamma_0 = \pi/6$.

$$\mathrm{Im}\,(N=2, J=2, M\,|\,c_\sigma^2(\omega, 0)\,|\,0) = \delta_{\sigma M}(0\,|\,\mathbf{d}^2\,|\,0)\,8\pi\sqrt{2}\,(45\sqrt{7}\,\hbar\,\omega_d\bar{\beta})^{-1}\,[(3\chi^4 - 2\chi^2 + 2)\,e^{-\chi^2} - 4(\chi^2+1)\,e^{-4\chi^2}]. \quad \text{(IV.69)}$$

where N denotes the number of quadrupole quanta; and

$$\bar{\beta} \equiv (0\,|\,\beta^2\,|\,0)^{1/2}; \qquad \chi = (\omega - \omega_d)\sqrt{2\pi}/(\omega_d\bar{\beta}).$$

According to Eq. (IV.69) (see also Fig. 6), the distribution of the dipole matrix element $|(\varepsilon|d|0)|^2$ is for $S \to \infty$ a curve with three broadened maxima.

The approximation of strong coupling was also considered by Kerman and Quang [86] and by Le Tourneux in the case of spherical nuclei [80].

4. Comparison of the Results of the Approximation for the Oscillator Model with the Calculations of the Nonaxial Rotator Model

Let us compare the basic results of the above approximation for the oscillator model with the calculations based on the nonaxial rotator model for $S \to \infty$.

The formal wave function of the dipole state remains as in Eq. (IV.66) [or Eq. (IV.62)]. A difference to the preceding form is found in the properties of the wave functions of the low-energy quadrupole spectrum, because the parameters of these wave functions are constants. The dipole spectrum consists in the latter case of a total of three correlated three-dimensional levels.

The scalar polarizability and the matrix elements of tensor polarizability, which describe the transition from the ground state to the two rotational 2^+ levels are in this case*

$$(0\,|c^0|0) = (0\,|\,\mathbf{d}^2\,|\,0)\,2\omega_d\,(9\hbar)^{-1}\,[(\omega_3^2 - \omega^2 - i\omega\Gamma_3]^{-1} + (\omega_1^2 - \omega^2 - i\omega\Gamma_1)^{-1} + (\omega_2^2 - \omega^2 - i\omega\Gamma_2)^{-1}], \quad \text{(IV.70)}$$

$$(J^{(n)} = 2, M\,|\,c_\sigma^2\,|\,0) = \delta_{\sigma M}(0\,|\,\mathbf{d}^2\,|\,0)\,2\omega_d\,(9\sqrt{5}\,\hbar)^{-1}\,[2a^{(n)}(\omega_3^2 - \omega^2 - i\omega\Gamma_3)^{-1} - (a^{(n)} - b^{(n)}/\sqrt{3})(\omega_1^2 - \omega^2 - i\omega\Gamma_1)^{-1} - (a^{(n)} + b^{(n)}/\sqrt{3})(\omega_2^2 - \omega^2 - i\omega\Gamma_2)^{-1}]. \quad \text{(IV.71)}$$

respectively. Here ω_i ($i = 1, 2, 3$) denotes the frequency of the dipole oscillations parallel to the principal deformation axis $\mathbf{e}_i'$; and $a^{(n)}$ and $b^{(n)}$ (with $n = 1, 2$) denote the coefficients in the expressions for the wave functions of the first and second 2^+ levels of the asymmetric rotator:

$$\Psi_{2M}^{(n)} = \sqrt{\frac{5}{8\pi^2}}\,\varphi(\beta_0\gamma_0)\,[a^{(n)}D_{M0}^2 + \frac{b^{(n)}}{\sqrt{2}}(D_{M2}^2 + D_{M,-2}^2]. \quad \text{(IV.72)}$$

* These formulas were derived by Tulupov [42].

In the case of a slightly nonaxial shape ($\gamma_0 << 1$) we have

$$a^{(1)} = 1, \quad b^{(1)} = 0;$$
$$a^{(2)} = 0, \quad b^{(2)} = 1.$$

For $\gamma_0 = \pi/6$ (maximum deviation from axial shape), we have

$$a^{(1)} = -b^{(2)} = \sqrt{3}/2,$$
$$a^{(2)} = b^{(1)} = 1/2.$$

For a comparison with the oscillator model, we bear in mind that, for $\Gamma_i = 0$, the imaginary parts of Eqs. (IV.70) and (IV.71) transform into the following combinations of δ functions:

$$\mathrm{Im}\,(0\,|\,c^{(0)}(\omega, 0)\,|\,0) = \pi\,(0\,|\,\mathbf{d}^2\,|\,0)\,\sum_i \delta(\hbar\omega - \hbar\omega_i),$$

$$\mathrm{Im}\,(2^{(n)}M\,|\,c_\sigma^2\,|\,0) = \pi\,(0\,|\,\mathbf{d}^2\,|\,0)\,\delta_{\sigma M}\,(9\sqrt{5})^{-1}\,[2a^{(n)}\delta(\hbar\omega - \hbar\omega_3) -$$
$$- (a^{(n)} - b^{(n)}\sqrt{3})\,\delta(\hbar\omega - \hbar\omega_1) - (a^{(n)} + b^{(n)}\sqrt{3})\,\delta(\hbar\omega - \hbar\omega_2)].$$

We can conclude from these formulas and from Figs. 7 and 8 that (in the limit $S \to \infty$) the oscillator model, in distinction to the rotator model, predicts not only splitting of the giant resonance into correlated three-dimensional peaks but also a broadening of these peaks due to the zero oscillations of the deformation parameters. The "overall structure" of the curves corresponding to the oscillator model can be obtained with the formulas of the nonaxial rotator model when the static deformation parameters are assumed to be approximately equal to the mean-square amplitudes of the oscillator model (for approximately spherical nuclei $\gamma_0 = \pi/6$). Small additional widths $\Delta\Gamma \approx \beta_0\,\hbar\omega_d/\sqrt{8\pi}$ are ascribed to the three-dimensional correlated maxima.

Some Possible Improvements of the Theory

Effects of the Higher Orders of α_μ

We try to establish how the results of the schematic model may be affected when higher terms in α_μ are taken into account in the energies of the dipole states.

To do this, we must first of all clarify the meaning of the deformation parameters proper.

When we use the definition of Eq. (IV.3), the nucleus becomes compressible even in the second order in α_μ:

$$v = \frac{4\pi}{3} R_0^3 \left[1 + \frac{3}{4\pi} \sum_\mu |\alpha_\mu|^2 + \ldots\right].$$

The assumption that the nucleus is incompressible therefore implies that the definition of Eq. (IV.3) must be replaced with

$$R = R_0 \left(1 + \sum_\mu \alpha_\mu (-1)^\mu Y_{2-\mu}(\varphi\theta) - \frac{1}{4\pi} \sum_\mu |\alpha_\mu|^2\right). \qquad \text{(IV.73)}$$

Below we state and compare the formulas for the two definitions stated in Eqs. (IV.3) and (IV.73).

We use the adiabatic approximation and consider dipole excitations at fixed deformation parameters. The dependence of the energies of the dipole oscillations upon the deformation parameters can be conveniently expressed in the following form:

$$
\begin{aligned}
(i|H_d + V_{dQ}|j) - (0|H_d + v_{dQ}|0) &= \varepsilon_i \delta_{ij} = \varepsilon_i (\alpha'_\mu = 0)\, \delta_{ij} + \\
&+ G_1 \sum_\mu (11\mu\nu|2\sigma)\, \alpha'_{-\sigma} (-1)^\sigma e_{i\mu} e_{j\nu} + G_{20} \sum_{\mu\rho} (11\mu\nu|00)\, e_{i\mu} e_{j\nu} (22\rho\sigma|00) \times \\
&\times \alpha'_\rho \alpha'_\sigma + G_{22} \sum_{\mu\rho\varkappa} (11\mu\nu|2\varkappa)\, e_{i\mu} e_{j\nu} (22\rho\sigma|2-\varkappa)(-1)^\varkappa \alpha'_\rho \alpha'_\sigma,
\end{aligned}
\tag{IV.74}
$$

where i denotes the number of the principal deformation axis; $|i) = |n_i = 1)$; and $e_{i\mu} = (i\,|\,K = \mu)$.

Obviously, Eq. (IV.74) is invariant with respect to rotations. The transition to the laboratory system is made by the following substitutions: $\alpha' \to \alpha$ and $e_{i\mu} \to r_\mu/|r|$, where **r** denotes the dipole coordinate.

It suffices to consider a deformation with axial symmetry in order to determine the coefficients G_1, G_{20}, and G_{22}.

The following are numerical calculations for two real examples.

Oscillatory Dipole Potential. In the first approximation in α_μ, the potential $V_d + V_{dQ}$ of the above model is an oscillator potential whose form resembles the shape of the nucleus. We assume that this approximation is valid and rewrite $V_d + V_{dQ}$ for higher orders in α_μ as follows:

$$V_d + V_{dQ} = M\,\omega_d^2 \mathbf{r}^2/2 + V_{dQ} = M\omega_d^2 \mathbf{r}^2/[2\,(R/R_0)^2], \tag{IV.75}$$

where R/R_0 is given by Eq. (IV.3) or Eq. (IV.73).

The following formulas for the energy of the longitudinal and transverse oscillations of a nucleus with axial symmetry result from Eqs. (IV.75) and (IV.3):

$$\omega_\parallel = \omega_d \left(1 - \sqrt{\frac{5}{4\pi}}\,\beta + 0.091\beta^2\right), \tag{IV.76}$$

$$\omega_\perp = \omega_d \left(1 + \frac{1}{2}\sqrt{\frac{5}{4\pi}}\,\beta - 0.016\beta^2\right), \tag{IV.77}$$

which results in

$$G_1 = -\sqrt{15/(8\pi)}\,\hbar\omega_d, \tag{IV.78}$$

$$G_{20} = -0.08\hbar\omega_d, \tag{IV.79}$$

$$G_{22} = -0.17\hbar\omega_d. \tag{IV.80}$$

When the deformation parameters are defined by Eq. (IV.73), we have

$$\omega_\parallel = \omega_d \left(1 - \sqrt{\frac{5}{4\pi}}\,\beta + 0.17\beta^2\right), \tag{IV.81}$$

$$\omega_\perp = \omega_d \left(1 + \frac{1}{2}\sqrt{\frac{5}{4\pi}}\,\beta + 0.08\beta^2\right). \tag{IV.82}$$

Consequently,

$$G_{20} = -0.385\hbar\omega_d. \tag{IV.83}$$

The other parameters are unchanged.

Hydrodynamic Model. According to the hydrodynamic model of [7], the difference between the proton and neutron densities

$$\rho_p - \rho_n = \chi(\mathbf{r})\, e^{-i\omega t} \tag{IV.84}$$

satisfies the equation

$$\Delta\chi + \varkappa^2 \chi = 0 \tag{IV.85}$$

with the boundary condition

$$(\mathbf{n}\nabla)\chi = 0 \tag{IV.86}$$

on the surface of the nucleus.

We have $\omega = \varkappa\sqrt{2k/M}$, with k denoting the coefficient of the symmetry energy in the Weizsäcker formula.

The condition stated in Eq. (IV.86) means that the shift of the protons relative to the neutrons is maximal on the surface of the nucleus.

The condition stated in Eq. (IV.86) assumes the following form on the surface $R(\theta, \varphi)$:

$$(\mathbf{n}\nabla)_R\chi = \left\{\frac{\partial\chi}{\partial r} - \frac{1}{R^2\sin^2\theta}\cdot\frac{\partial R}{\partial\varphi}\cdot\frac{\partial\chi}{\partial\psi} - \frac{1}{R^2}\frac{\partial R}{\partial\theta}\frac{\partial\chi}{\partial\theta}\right\}_R = 0. \tag{IV.87}$$

We assume the following formulas for the longitudinal and transverse oscillations, respectively:

$$\chi_0 = j_1(\varkappa_\parallel r)\,Y_{10}(\theta\varphi) + A^0 j_3(\varkappa_\parallel r)\,Y_{30}(\theta\varphi) + B^0 j_5(\varkappa_\parallel r)\,Y_{50}(\theta\varphi), \tag{IV.88}$$

$$\chi_1 = j_1(\varkappa_\perp r)\,Y_{11}(\theta\varphi) + A^1 j_3(\varkappa_\perp r)\,Y_{31}(\theta\varphi) + B^1 j_5(\varkappa_\perp r)\,Y_{51}(\theta\varphi), \tag{IV.89}$$

where j_n denotes a spherical Bessel function. By substituting Eqs. (IV.88) and (IV.89), along with Eq. (IV.3), into Eq. (IV.87), we obtain after lengthy computations

$$\omega_\parallel = \varkappa_\parallel\sqrt{2k/M} = \omega_d\left(1 - 0.916\sqrt{\frac{5}{4\pi}}\beta + 0.033\beta^2\right), \tag{IV.90}$$

$$\omega_\perp = \varkappa_\perp\sqrt{2k/M} = \omega_d\left(1 + \frac{1}{2}0.916\sqrt{\frac{5}{4\pi}}\beta + 0.000\ldots\beta^2\right). \tag{IV.91}$$

Equations (IV.90) and (IV.91) result in

$$G_1 = -0.916\sqrt{15/(8\pi)}\,\hbar\omega_d, \tag{IV.92}$$

$$G_{20} = -0.043\hbar\omega_d, \tag{IV.93}$$

$$G_{22} = -0.050\hbar\omega_d. \tag{IV.94}$$

When the definition of Eq. (IV.73) is used, we have

$$\omega_\parallel = \omega_d\left(1 - 0.916\sqrt{\frac{5}{4\pi}}\beta + 0.112\beta^2\right), \tag{IV.95}$$

$$\omega_\perp = \omega_d\left(1 + \frac{1}{2}0.916\sqrt{\frac{5}{4\pi}}\beta + 0.079\beta^2\right), \tag{IV.96}$$

$$G_{20} = -0.34\hbar\omega_d. \tag{IV.97}$$

These calculations [Eqs. (IV.76) and (IV.77), (IV.81) and (IV.82), (IV.90), and (IV.91), and (IV.95) and (IV.96)] lead to the conclusion that the maximum distortion of the dipole spectrum due to terms which are quadratic in α_μ amounts to $0.01\hbar\omega_d \lessdot 0.1$-$0.15$ MeV, i.e., this distortion is immaterial.

Thus, when calculations are made with the schematic model, the accuracy of the approximation which is linear in α_μ is quite adequate.

Possible Consideration of Many Dipole-Type Degrees of Freedom. The greatest simplification of the model under consideration originates from the replacement of many dipole-type degrees of freedom by a single degree of freedom.

In principle, a formal generalization of the theory to the case of many "fast" degrees of freedom is possible when several "terms" of the type stated in (III.74) can be present in the region of a giant resonance.

This improvement is sensible if the number of terms is not very large because in the case of many terms, the spacing of the terms will not exceed the frequency of the quadrupole oscillations and, accordingly, the adiabaticity condition is violated. The probability of a transition from one term to another per unit time is large in this case and the "quasi-molecular" description is no longer applicable [87].

This situation is evidently encountered in heavy nuclei.

Attempts to use this description for light nuclei have been made [88]. But it is very doubtful that in this case macroscopic coordinates can be used for describing surface oscillations.

Conclusions

The following is a short list of the basic results.

1. Studies of the kinematics of the interaction of dipole radiation with optically anisotropic systems lead to the conclusion that scattering at oriented nuclei can provide a maximum of information on vector and tensor polarizabilities among the large number of experiments (scattering of polarized and unpolarized photons at nonaligned nuclei and absorption by and scattering of photons at aligned nuclei).

Photon scattering is the principal source of information on the parameters of optical anisotropy in the case of even–even nuclei. The determination of both tensor and vector polarizability from a single purely kinematic characteristic necessitates in this case the use of polarized photons.

2. We evaluated the effects of tensor polarizability of strongly deformed nuclei with a minimum of model assumptions.

In particular, no special assumptions concerning the binding of an odd nucleon to the surface were made.

The cross section of photon scattering at aligned deformed nuclei was estimated from currently available experimental results and expressed with the aid of parameters which are known from experiments. It follows from the resulting expression that the effect of tensor polarizability observed in scattering at aligned nuclei is approximately twice as large as the corresponding effect measured in the case of absorption at aligned nuclei.

3. Vector polarizability can help to determine rotational properties which manifest themselves in the dipole spectrum of a system.

The vector polarizability of heavy, strongly deformed nuclei is very small. On the other hand, considerable vector polarizability effects can be expected in the case of odd light nuclei.

4. Tensor polarizability of nuclei contributes to the quadrupole constant of the hyperfine structure of atomic and molecular spectra, with the magnitude of the constant being comparable to the second-order effects in hyperfine-structure interactions. It is therefore impossible to determine tensor polarizability from the hyperfine structure of atomic and molecular spectra.

5. Generally valid is the conclusion that the average energies of the longitudinal and transverse dipole excitation of spheroidal nuclei are inversely proportional to the lengths of the corresponding semiaxes.

We obtained this conclusion from summation rules and the "particle–hole" model. The relation obtained in the latter case reflects the fact that the independent particle model can correctly describe the shape of strongly deformed nuclei.*

The factors which in the particle–hole approximation result in broadening of the dipole maximum contribute to an increased energy spread of the transverse peak.

6. Our schematic model of the coupling between dipole and surface oscillations allows calculations of the optical anisotropy effects in any type of low-energy quadrupole excitations.

When applied to heavy spheroidal nuclei, the model leads to the conclusion that γ oscillations result in broadening of the transverse maximum by a value $\Delta\Gamma$ of less than 0.7 MeV. This result was obtained for the greatest possible amplitude ($\bar{\gamma} = 0.3$) and the lowest frequency ($\hbar\Omega = 0.75$ MeV) of γ oscillations and can be considered valid for the nuclei of the center of the rare-earth group. The effect of γ oscillations must be less pronounced in other cases.

The experimentally observed difference between the widths of the longitudinal and transverse maxima amounts to 1.5-2 MeV. The influences which we considered (an harmonicity, single-particle potential, other interactions between "particle–hole" states, and γ oscillations) can only to some extent account for this difference.

The model of a rotator with a slight deviation from axial symmetry, $\gamma_0 \simeq \bar{\gamma}$, and a rotation frequency $\hbar\Omega$, which differs from the γ oscillator model in the fine structure of the dipole spectrum, renders similar values for the broadening.

However, the value is much smaller than the broadening of about 3 MeV which stems from influences neglected in the schematic model. The agreement between experimental and theoretical photoabsorption cross sections is therefore not a proof of the validity of this model.

Raman scattering with excitation of γ oscillation terms is a very effective method of checking the schematic model. (The magnitude of the effect is proportional to $\Delta\Gamma/\Gamma_b$.)

We must bear in mind that, in general, the schematic model is hardly capable of describing a structure which is narrower than 2-3 MeV, because influences leading to broadening of a giant resonance by this amount are neglected in the schematic model.

We believe for this reason that also in all other cases (apart from the case considered in this paper), comparisons with Raman scattering rather than with the photoabsorption spectrum should be the principal method of checking the model.

Unfortunately, no experimental data on nuclear Raman scattering are currently available. A corresponding scattering experiment would be crucial for establishing the validity of the qualitative conclusions drawn from the schematic model.

Replacement of the set of "fast" degrees of freedom, which constitute a giant resonance, by a single degree of freedom is the main simplification of the model with dipole–quadrupole coupling.

When many fast degrees of freedom are included in the considerations, a quasi-molecular description of a giant resonance may be inadequate because the terms merge.

Like any other analogy, the quasi-molecular description is of limited applicability and will turn out to be an improperly crude approximation once more information on the details of giant resonances is available.

---

* This conclusion is not valid for moderately and weakly deformed nuclei.

The corresponding amount of information is not available at the present time.

The author thanks A. M. Baldin for proposing the subject and for continual assistance in the execution of the present work.

The author extends his gratitude to the group of the Laboratory of Photomesonic Processes for kindly providing him with experimental data.

The author is also indebted for many valuable comments to the participants of the seminars of the Laboratory of Photomesonic Processes, the Laboratory of the Atomic Nucleus, and the Optics Laboratory, and to the participants of the Seminar on Many-Body Problems.

References

1. J. Levinger, Photonuclear Reactions [Russian translation], Foreign Literature Press (1962).
2. L. E. Lazareva, in: Nuclear Reactions at Low and Medium Energies [in Russian], Izd. Akad. Nauk SSSR (1962).
3. E. J. Fuller and Evans Hayward, in: Nuclear Reactions [Russian translation], Atomizdat, Vol. 2 (1964).
4. B. I. Goryachev, Atom. Energy Rev., 2:71 (1964).
5. A. B. Migdahl, Zh. Eksp. Teor. Fiz., 15:81 (1945).
6. M. Goldhaber and E. Teller, Phys. Rev., 74:1046 (1948).
7. H. Steinvedel and J. H. D. Jensen, Z. Naturforsch., 5a:413 (1950).
8. G. E. Brown and M. Bolsterli, Phys. Rev. Lett., 3:472 (1959).
9. D. I. Thouless, Nucl. Phys., 22:78 (1961).
10. V. V. Balashov, V. G. Shevchenko, and N. P. Yudin, Zh. Eksp. Teor. Fiz., 41:1929 (1961).
11. V. V. Balashov, Izv. Akad. Nauk SSSR, Ser. Fiz., 26:1459 (1962).
12. V. G. Shevchenko and N. P. Yudin, Atom. Energy Rev., 3:3 (1965).
13. G. E. Brown, Unified Theory of Nuclear Models, North Holland Publ. Co. (1964).
14. A. B. Migdahl, Theory of Finite Fermi Systems and Properties of Atomic Nuclei [in Russian], Nauka Press (1965).
15. W. Brenig, Advances in Theoretical Physics, Academic Press, New York, Vol. 1, p. 1 (1965).
16. F. A. Zhivopistsev, V. M. Moskovkin, and N. P. Yudin, Izv. Akad. Nauk SSSR, Ser. Fiz., 30(2):306 (1966).
17. M. Danos and W. Greiner, Phys. Rev., 138:B876 (1963).
18. V. M. Mihailovic and M. Rosina, Nucl. Phys., 40:252 (1963).
19. B. I. Goryachev, L. Mailing, B. G. Neudachin, and B. A. Yur'ev, Yadernaya Fizika, 5:77 (1967).
20. A. M. Baldin, in: Nuclear Reactions at Low and Medium Energies [in Russian], Izd. Akad. Nauk SSSR (1958).
21. A. M. Baldin, Nucl. Phys., 9:237 (1958).
22. A. M. Baldin, Zh. Eksp. Teor. Fiz., 37:202 (1959).
23. A. M. Baldin, in: Nuclear Reactions at Low and Medium Energies [in Russian], Izd. Akad. Nauk SSSR (1962).
24. G. Placek, Rayleigh Scattering and Raman Effect [Russian translation], GNTIU (1935).
25. E. Ambler, E. G. Fuller, and H. Marshak, Phys. Rev., 138:B117 (1965).
26. K. Okamoto, Progr. Theor. Phys., 15:75 (1956).
27. K. Okamoto, Phys. Rev., 110:143 (1958).
28. M. Danos, Nucl. Phys., 5:23 (1958).
29. S. F. Semenko, Vestnik Mosk. Gosud. Univ., Ser. 3, No. 3, p. 75 (1961).
30. E. V. Inopin, Zh. Eksp. Teor. Fiz., 38:992 (1960).
31. S. F. Semenko, Optical Anisotropy of Atomic Nuclei [in Russian], FIAN SSSR, A-31 (1963).
32. P. A. Tipler, P. Axel, N. Stein, and D. C. Sutton, Phys. Rev., 129:2096 (1963).
33. A. I. Akhiezer and V. B. Berestetskii, Quantum Electrodynamics [in Russian], Fizmatgiz(1959).
34. A. M. Baldin, V. I. Gol'danskii, and I. A. Rozental', Kinematics of Nuclear Reactions [in Russian], Fizmatgiz (1959).
35. U. Fano, L. Spencer, and M. Berger, γ-Radiation Transfer [Russian translation], Gosatomizdat (1963).
36. G. R. Khutsishvili, Usp. Fiz. Nauk, 53:381 (1954).
37. A. P. Yutsis, I. B. Levinson, and V. V. Vanagas, Mathematics of the Theory of Momentum [in Russian], Gospolitizdat (1960).
38. A. Bussiere de Nercy, J. Phys. Radium, 22:535 (1961).
39. H. Langevin, J. M. Loiseaux, and J. M. Maison, Nucl. Phys., 54:114 (1964).
40. J. M. Daniels, Oriented Nuclei, Polarized Targets and Beams, Academic Press, New York (1965).
41. A. M. Baldin and S. F. Semenko, Zh. Eksp. Teor. Fiz., 39:434 (1960).
42. S. F. Semenko and B. A. Tulupov, Zh. Eksp. Teor. Fiz., 41:1996 (1961).

43. S. C. Fultz, R. L. Bramblett, J. T. Caldwell, and N. A. Kerr, Phys. Rev., 127:1273 (1962).
44. A. Bohr, Kgl. Danske Vid. Selskab. Mat.-Fys. Medd., Vol. 26, No. 14 (1952).
45. A. Bohr and B. Mottelson, Kgl. Danske Vid. Selskab. Mat.-Fys. Medd., Vol. 27, No. 16 (1953).
46. A. S. Davydov, Theory of the Atomic Nucleus [in Russian], Fizmatgiz (1958).
47. A. S. Davydov, Excited States of Atomic Nuclei [in Russian], Atomizdat (1967).
48. R. L. Bramblett, J. T. Caldwell, G. F. Auchampaugh, and S. G. Fultz, Phys. Rev., 129:2723 (1963).
49. E. G. Fuller and E. Hayward, Nucl. Phys., 30:613 (1962).
50. G. S. Landsberg, Optics[Russian translation], GITTL (1957).
51. S. Nilson, in: Deformation of Atomic Nuclei [Russian translation], Foreign Literature Press (1958).
52. B. S. Dolbilkin, Trudy Fiz. Inst. Akad. Nauk, 36:18 (1966) [English translation: Photodisintegration of Nuclei in the Giant Resonance Region, D. V. Skolel'tsyn, ed., p. 17, Consultants Bureau, New York (1967)].
53. F. A. Nikolaev, Trudy Fiz. Inst. Akad. Nauk, 36:84 (1966) [English translation: Photodisintegration of Nuclei in the Giant Resonance Region, D. V. Skolel'tsyn, ed., p. 77, Consultants Bureau, New York (1967)].
54. H. K. Howe, in: Structure of the Nucleus [Russian translation], Gosatomizdat (1962).
55. S. Geschwind, R. Gunther-Mohr, and C. H. Townes, Phys. Rev., 21:288 (1951).
56. N. Towns and A. Shavlov, Radiospectroscopy [Russian translation], Foreign Literature Press (1959).
57. G. Kopfferman, Nuclear Moments [Russian translation], Foreign Literature Press (1960).
58. C. Schwartz, Phys. Rev., 97:380 (1955); 105:173 (1957).
59. T. C. Wang, Phys. Rev., 99:566 (1955).
60. E. G. Fuller and M. S. Weiss, Phys. Rev., 122:560 (1958).
61. B. M. Spicer, H. H. Thies, J. E. Baglin, and F. R. Allum, Austral. J. Phys., 11:298 (1958).
62. R. W. Parsons and L. Katz, Canad. J. Phys., 37:809 (1959).
63. H. H. Thies and B. M. Spicer, Austral. J. Phys., 13:505 (1960).
64. O. V. Bogdankevich, B. I. Goryachev, and V. A. Zapevalov, Zh. Eksp. Teor. Fiz., 42:1501 (1962).
65. R. L. Bramblett, J. T. Caldwell, R. R. Harvey, and S. C. Fultz, Phys. Rev., 133:B869 (1964).
66. C. D. Bawman, G. F. Auchampaugh, and S. C. Fultz, Phys. Rev., 133:B876 (1964).
67. S. F. Semenko, Nucl. Phys., 37:486 (1962).
68. J. Goldstone and K. Gottfried, Nuovo Cim., 13:848 (1959).
69. S. Fallieros, Nucl. Phys., 26:594 (1961).
70. V. Gillet and E. A. Sanderson, Nucl. Phys., 54:472 (1964); V. Gillet, A. M. Green, and E. A. Sanderson, Phys. Lett., 11:44 (1964).
71. S. F. Semenko, Zh. Eksp. Teor. Fiz., 43:2188 (1962).
72. B. R. Mottelson and S. G. Nilsson, Nucl. Phys., 13:281 (1959).
73. B. R. Mottelson and S. G. Nilsson, Kgl. Danske Vid. Selskab. Mat.-Fys. Skr., Vol. 1, No. 8 (1959).
74. L. Schiff, Quantum Mechanics [Russian translation], Foreign Literature Press (1957).
75. V. V. Balashov and A. F. Tulinov, Zh. Eksp. Teor. Fiz., 43:702 (1962).
76. S. F. Semenko, Phys. Lett., 10:182 (1964).
77. M. Danos and W. Greiner, Phys. Rev., 134:B284 (1964).
78. S. F. Semenko, Phys. Lett., 13:157 (1964).
79. S. F. Semenko, Yadernaya Fizika, 1:414 (1965).
80. J. Le Tourneux, Kgl. Danske Vid. Selskab. Mat.-Fys. Medd., 34:11 (1965).
81. J. Le Tourneux, Phys. Lett., 1:325 (1964).
82. V. V. Balashov and V. M. Chernov, Zh. Eksp. Teor. Fiz., 43:227 (1962).
83. B. L. Birbrair, L. K. Peker, and L. Sliv, Zh. Eksp. Teor. Fiz., 36:803 (1959).
84. D. R. Bes, P. Federman, E. Maqueda, and A. Zuker, Nucl. Phys., 65:1 (1965).
85. Liu Yuan, V. G. Solov'ev, and A. A. Korneichuk, Zh. Eksp. Teor. Fiz., 47:252 (1964).
86. A. K. Kerman and H. K. Quang, Phys. Rev., 135:B883 (1964).
87. J. Hill and J. Wheeler, Usp. Fiz. Nauk, 52:83, 239 (1954).
88. E. Boeker, W. A. de Muynck, and C. C. Jonker, Compt. Rend. Congr. Internat. Phys. Nucl. Paris, Vol. 2, p. 405 (1964).

SPARK-CHAMBER SPECTROMETRY OF HIGH-ENERGY PHOTONS AND ELECTRONS

A.G. Gapotchenko, B.B. Govorkov, and S.P. Denisov

The article outlines the results of research performed on numerous spark chambers which were used as spectrometers for electrons and photons having energies of 50-200 MeV. The characteristics of the electromagnetic cascade in a chamber with lead plates were studied. It could be shown that a multi-layer spark chamber with Plexiglas plates can be used as a flight-path spectrometer for electrons and photons (energy resolution 35%).

Electron and photon detectors which are characterized by high resolution in space, time, and energy are required for research on electromagnetic interactions and radiative decays of particles. Spark chambers with a rapid response of about 10^{-6} sec and a spatial resolution of about 1 mm have been recently used as detectors of the above-described kind. We considered in our work possibilities of determining energies of photons and electrons passing through multilayer spark chambers in which layers or plates of a material are inserted in spark gaps. The electrons and photons had energies between 50 and 200 MeV. When material with a high atomic number was used, the relation between the particle energy and the characteristics of the electromagnetic shower produced by the particles, the total number of charged particles in the shower, and the dispersion of the particles was studied. When a material with a low atomic number was inserted, the possibility of determining the particle energy from the particle's range in the material was studied. The present article describes experiments which were made with shower chambers (with lead as intermediate material) and range chambers (with Plexiglas as intermediate material).

Geometry of the Experiment and Apparatus

The experiments were made with the 265-MeV synchrotron of the Physics Institute of the Academy of Sciences of the USSR. In order to generate monochromatic electrons and photons, a device described in the present collection on page 216 and in [1] was used. Figure 1 depicts the geometry of the experiment.

In experiments with γ quanta, the spark chamber for showers is inserted into bremsstrahlung radiation which is produced in a thin lead target K_1 by monochromatic electrons. The spark chamber for showers is controlled in the following fashion. When the energy of the electrons before the electrons intersect target K_1 is E_0 and scintillation counters C_1 and C_2 record simultaneously electrons with energy E_c, a photon with energy $E_\gamma = E_0 - E_e$ was incident on the spark chamber. Electrons with a certain energy E_0 and E_e were detected with magnetic spectrometers M_1 and M_2. When showers produced by electrons are recorded, spectrometer M_1 is used, the target is removed, and the counters are placed directly before the chamber.

When photons were recorded with a spark chamber for range measurements, the bremsstrahlung photons were converted in target K_2. When signals appeared simultaneously at the outputs of C_1, C_2, and C_4, a photon with an energy $E_\gamma = E_0 - E_e$ had been incident on the chamber. In experiments with electrons, converters K_1 and K_2 were removed, and the control circuit of the spark chamber was triggered by pulses derived from counters C_3 and C_4.

Figure 1 shows the principal components of a spark chamber for range measurements. The chamber consisted of 15 spark gaps which were formed by Plexiglas frames with internal dimensions of 150 × 150 × 12 mm, and of 15 Plexiglas blocks used as absorbers. For operation in the energy interval 60-150 MeV, the thickness of the Plexiglas blocks was rather great, 35 mm on the average, but without any excessive increase in the number of spark gaps. One-mm-thick duraluminum plates formed the electrodes which were glued directly upon the Plexiglas blocks. The polymer link of the Plexiglas had the chemical formula $C_5H_8O_2$.

The spark chamber of the shower type consisted of 14 discharge gaps with the dimensions 150 × 150 × 12 mm. The high-voltage electrodes were separated from the discharge spaces by 1.7-mm-thick glass plates. The first two electrodes were so thin that the probability of γ-quantum conversion in these electrodes were negligibly small. It became therefore possible to use a discharge in the first two gaps for discrimination between primary photons and electrons. These gaps played the role of the usually employed anticoincidence counters. The electrodes were made of steel plates (about 2 mm thick) and the converters proper, of lead plates (about 4 mm thick). The total amount of material in each electrode amounted to $0.95X_0$.

The control circuit of the spark chamber consisted of two coincidence circuits which are shown in Fig. 2, a generator delivering high-voltage pulses, and a control circuit for the photo camera [2].

Coincidence of pulses derived from FEU-36 photomultipliers used in the counters occurred at a 6V1P secondary-emission tube which was connected as slave relay with anode–cathode coupling and shut off via the control and screen grids. The shift voltage applied to the grids ($U_{c1} = -2$ V and $U_{c2} = -1$ V) was selected so that the relay was not actuated by single

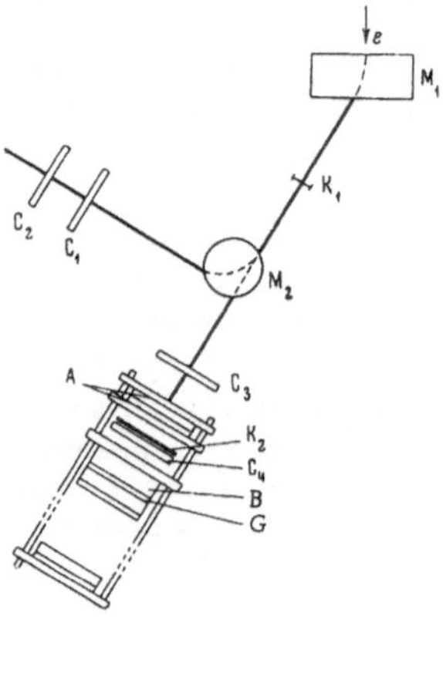

Fig. 1. Geometry of the experiments. M_1 and M_2 denote magnetic spectrometers; C_1, C_2, C_3, and C_4 denote scintillation counters; B denotes a Plexiglas block; G denotes the spark gap; and A denotes anticoincidence gaps.

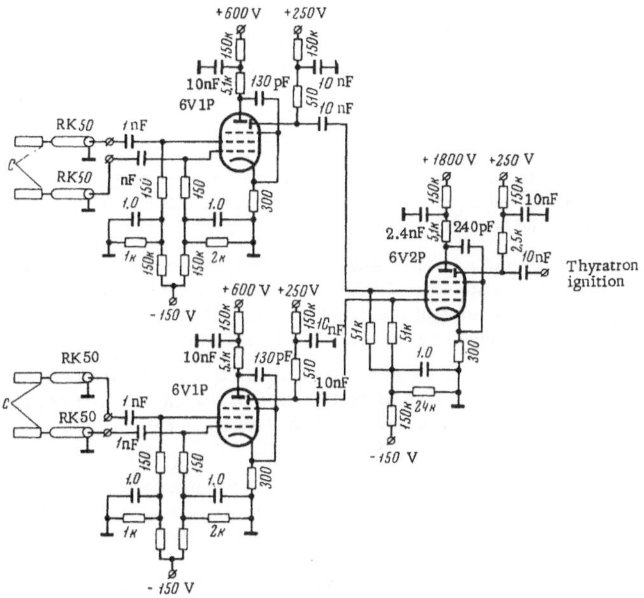

Fig. 2. Control circuit of the spark chamber.

signals having an amplitude of less than 50 V. When pulses which occur simultaneously and have an amplitude in excess of 5 V appear at the input of the system, avalanche processes in this circuit and a positive pulse with an amplitude of 70 V appears at the dynode load. The curvature of the rising front of the pulse is 3 V/nsec and the duration of the rising portion is 0.1 μsec. A negative pulse with a voltage of about 100 V is simultaneously shaped at the anode of the 6V1P tube. Measurements have shown that, without additional pulse shaping of the signals derived from the 6V1P photomultiplier, one can obtain a coincidence resolving time of about 3 nsec at 100% efficiency and a selection coefficient in excess of 30.

The positive pulse derived from a dynode of the 6V1P tube triggers a 6V2P tube. A positive pulse (of about 700 V) derived from a dynode of the 6V2P tube is used to trigger a TGI1 400/16 thyratron which applies a high-voltage pulse to the electrodes of the spark chamber or at the ignition electrode of a spark gap filled with nitrogen, with the spark chamber being supplied either from the thyratron or from the nitrogen-filled spark gap. The high-voltage pulse had the following parameters: amplitude 9-15 kV, duration of the rising front about 30 nsec, pulse duration about 150 nsec, and delay about 200 nsec.

The pulse derived from the anode of the 6V1P tube was used to trigger the control circuit of an RFK-5 photo camera in which a frame was exposed and pulsed illumination was switched on.

Two perpendicular projections of the chamber were photographed on a single frame with the aid of a mirror system. In order to take photographs of all the spark gaps of a chamber for range measurements over the total depth of the field, but from a rather short distance, a cylinder lens made of Plexiglas and filled with distilled water was employed.

The spark chambers were evacuated to 10^{-2} torr and filled with "particularly pure" neon, after preliminary cleaning. The chambers were refilled every 7-10 days. A decrease in the efficiency and other indicators pointing to aging of the gas were not detected within this time interval.

Shower Spectrometer

Measurements were made with primary photons having energies of 50, 100, and 150 MeV and with electrons having energies of 50, 100, 150, and 200 MeV (the experimental results are listed in Figs. 5 and 6 and in Table 1).

The background without a lead target was studied at clearing fields ranging from 0 to 200 V/cm in additional experiments. Conditions under which the number of sparks in a discharge gap is equal to the number of electrons passing through the gap are very favorable for recording showers. We discuss now the effects which can disturb this equality.

1. Shower particles going beyond the spark chamber. The total amount of matter in the chamber both along the axis of the shower and perpendicular to the direction of shower propagation was so large that the probability of a secondary electron traveling beyond the chamber was less than 1%.

2. Spatial resolution of the spark chamber. The projections of two sparks upon photographs could be reliably separated when the distance between the sparks was greater than 1.5 mm. The fraction of cases in which one spark is visible on one projection and two sparks, on the other is denoted by $\varkappa$. When $\varkappa \ll 1$, we use $\varkappa^2$ as a coarse estimate of the probability that two sparks cannot be resolved. Inspection of the photographs resulted in $\varkappa^2 \sim 0.01$.

3. Recording scattered electrons. Measurements which were made without the lead target have shown that the background produced by scattered electrons is negligibly small even when the clearing field is switched off.

4. Spurious sparks did not occur when high-voltage pulses of less than 10 kV were used (the operating voltage was 9 kV).

Surface discharges over the internal chamber walls occurred at higher voltages.

5. Efficiency for simultaneous recording of several electrons in a single discharge gap. In order to solve this problem, we made additional experiments with a multilayer spark chamber of special design. The chamber consisted of 8 discharge gaps with electrodes of 1-mm-thick aluminum plates. A lead layer with a thickness of about $2X_0$ was inserted between the second and third gaps. Thus, the primary electron was recorded with the first two gaps, whereas the secondary particles formed in the lead converter were observed in the ensuing 6 discharge gaps. The chamber was controlled by a telescope of scintillation counters. A nitrogen spark gap which generated a pulse with a pulse-front duration of 25 nsec was used for producing a high-voltage pulse. A storage capacitor of 3 nF was

TABLE 1

Primary particles	Photon			Electron			
Energy of the primary particle, MeV	50	100	150	50	100	150	200
Number of showers considered	274	538	398	358	575	675	656
Average number of sparks per shower	1.3	3.2	5.0	1.2	3.0	4.6	5.8
Dispersion of the number of sparks per shower	1.4	2.6	4.0	1.4	3.1	4.5	6.0

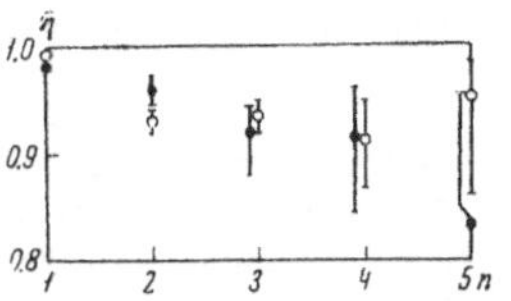

Fig. 3. Dependence of the average recording efficiency for shower particles in chambers with metal (circles) and insulator (squares) electrodes upon the number of particles in the spark gap.

charged with a voltage of 11.5-15 kV. The high-voltage pulse was applied with a delay of about 200 nsec to the electrodes which were connected in parallel. The tracks were photographed at a distance of 1.5 m. Two perpendicular projections were photographed on each frame. In order to improve the spatial resolution of the tracks, a blue SS-4 filter was inserted.

Six thousand 200 frames were processed. When the efficiency was determined, the 5th and 6th gap were checked. It was assumed that the secondary particle intersected the control spaces if in the 3rd, 4th, 7th, and 8th gap at least three sparks along a straight line were observed. The direction of the straight line was given by the direction in which the secondary electron moved (accuracy ±3°). The inspection of the photographs provided data on the recording efficiency for 1 to 5 particles. The number of photographs with 1 to 3 tracks sufficed for obtaining information on the angular dependence of the efficiency. The efficiency figures for the 5th and 6th gaps coincided within the statistical error limits. We list below the values which were obtained by averaging over these two gaps.

The principal results are shown in Figs. 3 and 4. The errors of the experimental points denote the limits of the confidence interval with the confidence coefficient 0.95. The circles of Fig. 3 denote the number of tracks which are recorded in a spark gap when 1 to 5 particles

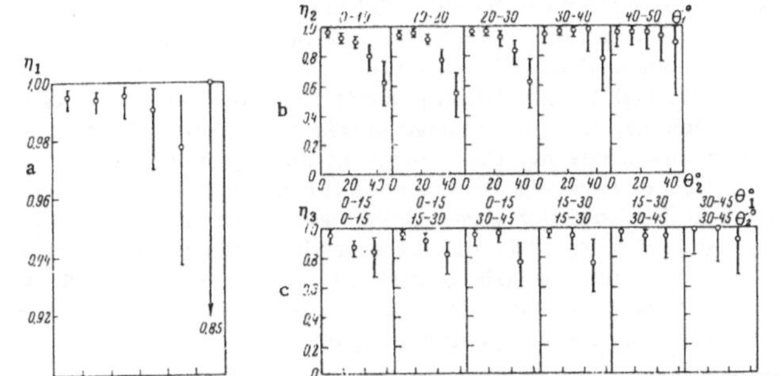

Fig. 4. Efficiency of recording a single particle in dependence of the angle θ between the particle trajectory and the direction of the electric field in the chamber (a); angular dependence of the probability of recording a single particle when another particle intersects the control gaps at a fixed angle θ_1 (b); dependence of the probability of recording a third particle upon the angle θ_3, when the other two particles pass through the chamber at angles θ_1 and θ_2.

pass through the gap. The data stem from averaging over the angular distribution of shower particles in the depth interval $2X_0$.

About 2000 additional photographs were inspected in order to establish a relation between the efficiency for recording two particles and the distance between the particles. It was found that, within the statistical error limits (which amount to several percent), the efficiency is constant when the distance between the particles ranges from 3 to 30 mm.

The efficiency of a spark chamber with insulated high-voltage electrodes, i.e., the chamber type used as a shower spectrometer, was examined in special tests. A nitrogen spark gap or a TGI1 400/16 thyratron were used as high-voltage discharge gaps for this chamber. The experiment established agreement between the results of the two tests. The results are denoted by squares in Fig. 3.

The following conclusions can be drawn from an analysis of the results: a) The average efficiency of recording secondary electrons amounted to 0.95 in our experiments; b) insulation of the electrodes of the spark chamber does not greatly improve the efficiency; c) the efficiency of the spark gap depends only slightly upon the number of tracks (for $2 \leq n \leq 5$) but decreases strongly when the difference of the track angles with respect to the direction of the field produced by the high-voltage pulse increases; and d) the efficiency of recording particles is, with an accuracy of several percent, indepndent of the distance between the particles (within the limits 3-30 mm).

Based on the conclusions of points 1-5, we can assume that the number of sparks measured in a shower corresponds with an error of at most several percent to the number of electrons intersecting the discharge space. Accordingly, the observed fluctuations of the number of sparks are entirely determined by the statistical nature of the processes leading to the development of showers.

The following remark is in place before we analyze the experimental results. Each electron which is incident on the spark chamber is recorded at least in two (anticoincidence) gaps. In the case of a photon, there exists a certain probability W(0) that not a single spark can be observed (all electrons of a shower can be absorbed in the chamber electrodes without passing into a discharge gap). The quantity $\eta = 1 - W(0)$ is one of the principal characteristics of γ detectors, i.e., the efficiency of detection. The W(0) value can be estimated when the number N_e of electrons passing through the scintillation counter and the number N_n of photographs on which no sparks are visible are known. We have in this case $W(0) = N_n/N_e$. In order to determine W(0), it is therefore necessary that a bremsstrahlung photon which is incident on the spark chamber correspond to each electron passing through C_1 and C_2. The geometry of the experiment was such that this condition was satisfied with an accuracy of about 10%. In calculating W(0), the background measured without the lead target was taken into account. Subtraction of the background resulted in large statistical errors which are indicated in Fig. 5. The error of W(0) was not considered in calculations of the mean-square deviation for the probability of cases having a spark number greater than zero.

The experimental data which are depicted by Figs. 5-7 and listed in Table 1 allow the following conclusions concerning the characteristics of chambers for showers.

1. The average number of sparks per showers, $\overline{N}$, is approximately proportional to the energy E_0 of the primary particle (this dependence is also observed at energies $E > 200$ MeV [5, 6]). This result agrees with the theoretical calculations if we take into account that the product $L = \overline{N}t_e$ (where t_e denotes the thickness of the electrodes in radiation units) is close to the sum of the lengths of flight projections of secondary electrons upon the shower axis (provided that t_e is much smaller than the dimensions of the shower). The quantity L represents the average number of sparks per shower in a spark chamber which has electrodes

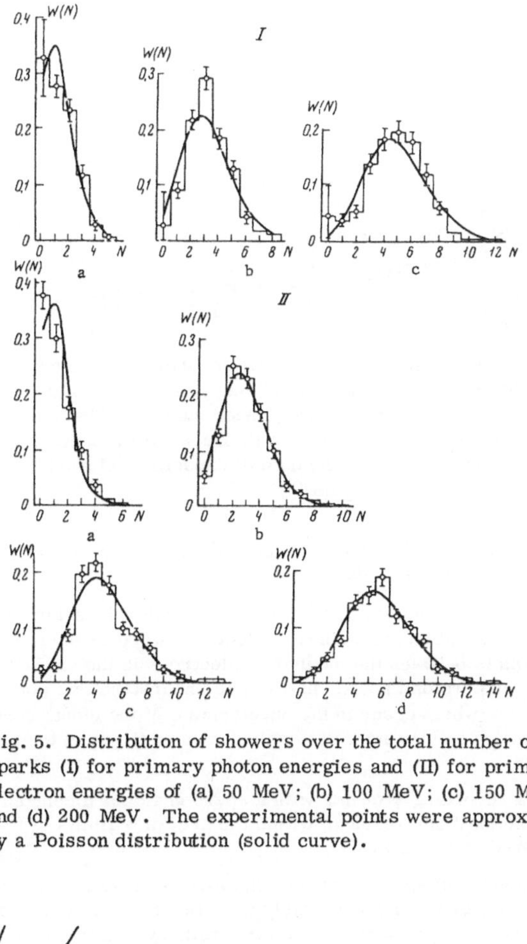

Fig. 5. Distribution of showers over the total number of sparks (I) for primary photon energies and (II) for primary electron energies of (a) 50 MeV; (b) 100 MeV; (c) 150 MeV; and (d) 200 MeV. The experimental points were approximated by a Poisson distribution (solid curve).

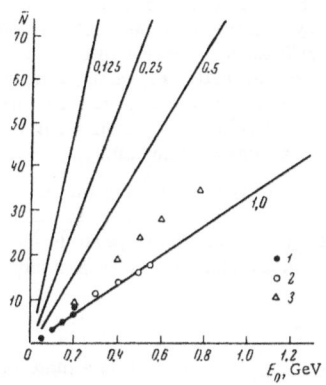

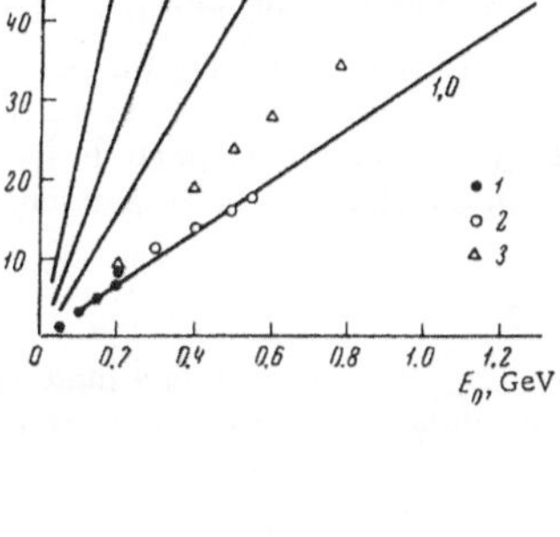

Fig. 6. Dependence of the average total number of electrons in a shower upon the energy of the primary electron. The solid curves denote the results of calculations. The numbers under the curves denote the thickness of the plates in units of radiation length. Experimental data (1) of [4]; (2) of [6]; and (3) of [7].

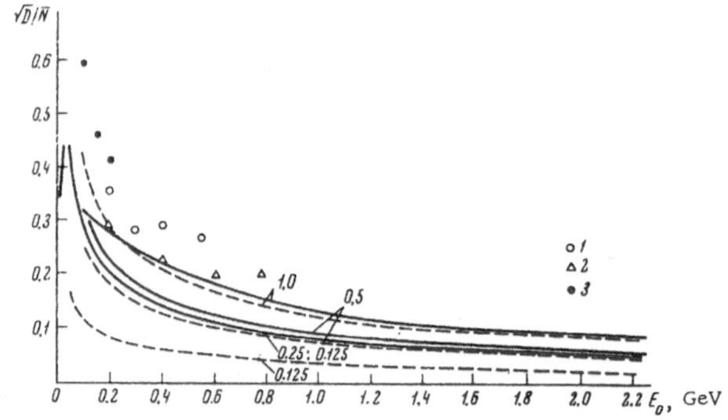

Fig. 7. Relation between $\sqrt{D}/\overline{N}$ and the energy of the primary electron; statistics of branching processes (solid curves) and Poisson statistics (dashed curves). The numbers under the curves denote the thickness of the plates; curves 1 and 2 represent the data of experimental work [5, 6]; curve 3 is taken from [3].

with a thickness $1X_0$. The experimental and calculated $\overline{N}$ values are indicated in Fig. 6 for various energies and thicknesses of the plates.

2. Fluctuations of the total number of sparks are described by a Poisson distribution. This result would be understandable, if the shower electrons were formed independently. But since a genetic relation exists between the number of electrons in the various populations, it is logical to expect a larger dispersion than $\overline{N}$. In order to understand this point, the dispersion of the total number of sparks which occur in the development of the electron–photon shower in a multiplate spark chamber was estimated. The calculation method and the results are given below.

3. The efficiency of recording γ quanta with a spark chamber amounted to 0.7 for γ quanta with an energy of 50 MeV and reached, within the accuracy limits of the experiments, unity at energies of 100 MeV and higher.

4. The measured number of sparks which in showers is close to the number of secondary electrons (as indicated above) exceeds considerably the value to be expected from calculations of electromagnetic cascades. These calculations were made by Crawford and Messel with the Monte-Carlo method for lead absorbers [7]. Since the main difference is observed beyond the maximum of the shower development at large thicknesses of the material, i.e., when particles with low energies begin to be important, whereas the calculations refer to electrons with energies above 10 MeV, it is logical to assume that the difference is related to the efficient recording of electrons with energies below 10 MeV by the particular spark chamber.

6. Calculation of the dispersion of the total number of sparks in the spark chamber. A simple model of an electron–photon shower was used to calculate the dispersion as a function of the plate thickness and the energy of the primary particle [8]. It was assumed for the calculations that the number of sparks in the chamber is proportional to the number of charged particles in the shower.

Let us outline the main assumptions of the model adopted.

1. When an electron with an energy below some energy E_{min} is generated in a plate of the chamber, no sparks appear. On the other hand, an electron which is generated with an en-

ergy in excess of E_{min} will be necessarily recorded by the chamber. The energy which the electron must have to pass through a plate of the chamber in a direction perpendicular to the plate is assumed as E_{min}:

$$E_{min} = \frac{800}{Z} (e^{\frac{R}{l_{rad}}} - 1) \qquad \text{(MeV)}. \tag{1}$$

The thickness of the plate (in cm) is denoted by R. Radiation and ionization losses are assumed in a very simple form of [9]:

$$\left(\frac{dE}{dx}\right)_{ion} = -\frac{800}{Zl_{rad}}, \quad \left(\frac{dE}{dx}\right)_{rad} = -\frac{E}{l_{rad}},$$

where l_{rad} denotes the radiation length for the material of the chamber plate.

2. When the electron energy E_0 is split by bremsstrahlung processes, only the emission of hard γ quanta which carry away a considerable part of the electron energy (between $E_0/3$ and E_0) is taken into account. In real radiative losses of the electron energy, rare processes as well as emission processes of soft γ quanta with energies $< E_0/3$ occur with equal frequency.

3. The probability of a single γ quantum being emitted by an electron in the energy interval between $E_0/3$ and E_0 is ascribed to the emission of a single photon with average energy $\frac{2}{3}E_0$. For paths on which the electron reduces its energy three times by losses due to ionization and emission of γ quanta with energies below $2E_{min}$ (secondary electrons from which no γ quanta can be detected), the probability is

$$p_{E_0} = 1 - 3^{-\frac{2}{3}\frac{E_0}{2E_{min}+800/Z}}. \tag{2}$$

For the probability of γ-quantum emission in the energy interval E, E + dE during passage of an electron over a path dx, we assume the dependence $(dx/l_{rad})(dE/E)$.

The hard γ quantum generated is assumed to have the probability 1 of forming a pair with equal electron and positron energies (the energy being E/3). Thus, the probability stated in Eq. (2) can be taken as probability of producing 3 charged particles of a new population with equal energies E/3. The probability $1 - p_{E_0}$ refers then to the transition of an electron into the next population with the energy $E_0/3$ without formation of new particles which could be detected.

These restrictions allow us to exclude from our considerations the γ quanta and to state the problem of electron multiplication in the generations. All particles of the r-th generation have the same energy $E_0/3^r$ (the initial particle was assumed to form the zeroth generation). Obviously, splitting the energy can be continued only as long as the energy has not been reduced to the value E_{min}. The probability of splitting the energy of a particle of the r-th generation into three equal portions of three particles of the (r + 1)-th population is

$$p_r = 1 - 3^{-\frac{2}{3}\frac{E_0/3^r}{2E_{min}+800/Z}}. \tag{3}$$

The probability of a particle of the r-th generation to pass into the (r + 1)-th generation without loss of energy (energy splitting) is $1 - p_r$.

The generating function of particle multiplication in the r-th generation has the form

$$G(x) = p_r x^3 + (1 - p_r)x \quad (x \leqslant 1). \tag{4}$$

The average number of particles generated from a single particle of the r-th generation is

$$m_r = \left(\frac{\partial G}{dx}\right)_{x=1} = 1 + 2p_r, \tag{5}$$

and the dispersion of the average number is

$$v_r = \left(\frac{\partial^2 G}{\partial x^2}\right)_{x=1} - m_r(m_r - 1) = 4p_r(1 - p_r). \tag{6}$$

We define $W_r(n, N)$ as the probability that n particles are present in the r-th generation and N particles exist in all generations from the zeroth to the r-th, inclusive. We state the corresponding generating function

$$\Pi_r(x, y) = \sum_{n=0}^{\infty} \sum_{N=n}^{\infty} x^n y^n w_r(n, N) \qquad (x \leqslant 1 \qquad y \leqslant 1). \tag{7}$$

When we consider in succession the generating functions for r = 0, 1, 2, ..., we can easily derive the following recurrence relation

$$\Pi_{r+1}(x, y) = \Pi_r\{G(x, y)\, y\}. \tag{8}$$

The first derivatives of Eq. (7) with respect to the arguments x and y at the point x = y = 1 determine the average values of the number of particles in the r-th generation and of the number of particles in all generations from the zeroth to r-th, inclusive. The dispersion of these quantities are defined as the second derivatives at the point x = y = 1.

Thus, by calculating the first and second derivatives of the two sides of Eq. (8) at the point x = y = 1, we obtain recurrence relations for the average number of particles in the r-th generation

$$\bar{n}_r = \bar{n}_{r-1}(2p_{r-1} + 1), \tag{9}$$

and we have for the dispersion of the average number

$$\overline{(n_r - \bar{n}_r)^2} \equiv d_r = (2p_{r-1} + 1)^2 d_{r-1} + 4p_{r-1}(1 - p_{r-1})\, n_{r-1}. \tag{10}$$

We obtain for the average number of particles in all generation from the zeroth to the r-th, inclusive:

$$\bar{N}_r = \bar{N}_{r-1} + \bar{n}_r, \tag{11}$$

and for the dispersion of this number:

$$\overline{(N_r - \bar{N}_r)^2} \equiv D(r) = D_{r-1} + d_r + 2(2p_{r-1} + 1) K_{r-1}, \tag{12}$$

where the correlation function K_r satisfies the recurrence relation

$$\overline{(n_r - \bar{n}_r)(N_r - \bar{N}_r)} \equiv K_r = (2p_{r-1} + 1) K_{r-1} + d_r. \tag{13}$$

The initial conditions are

$$\bar{n}_0 = \bar{N} = 1, \qquad d_0 = D_0 = K_0 = 0. \tag{14}$$

Equations (9)-(11) can be easily solved and we obtain

$$n_r = (2p_0 + 1)\ (2p_1 + 1) + \ldots, \tag{15}$$

$$\bar{N}_r = \sum_{i=0}^{r} \bar{n}_i, \tag{16}$$

$$\frac{d_r}{\bar{n}_r^2} = \frac{1}{\bar{n}_r} + 3\sum_{i=1}^{r} \frac{2p_i}{2p_i+1} \cdot \frac{1}{\bar{n}_i} - \frac{1}{\bar{n}_0}. \tag{17}$$

The influence of the genetic relation of the particles in the shower upon the relative dispersion of the number of particles in the r-th generation can be inferred from Eq. (17). In addition to the first term on the right side of Eq. (17), which defines the usual statistical deviations from the average value, there exists a second term which contains a sum over all preceding generations. The third term with the negative sign appears because, when no energy is absorbed, the number of particles in any generation is exactly known and the dispersion vanishes (all $p_i = 1$ and $n_i = 3$, $d_r = 0$).

Equations (10) and (11) were used to calculate $\bar{N}$ and $\sqrt{D}/\bar{N}$ for sets of lead plates having thickness of 1, 0.5, 0.25, and $0.125X_0$. The minimum energy which can be recorded is 17, 6.3, 2.8, and 1.33 MeV, respectively. In order to obtain smooth energy dependencies, the calculations were made for the limit energies of the particle generations (e.g., with the energies 17, 51, 153 MeV, etc., for $1X_0$) and the results were referred to the average energy between generations. Figure 7 depicts the results of $\sqrt{D}/\bar{N}$ calculations. Figure 6 shows the average value of the particle number in the shower as a function of the energy E_0 of the primary electron. This figure includes the results of experimental work. It follows from Fig. 6 that the results of the $\bar{N}$ calculations can be considered to be in agreement with the experimental results, i.e., agreement exists between the linear dependence of $\bar{N}$ upon E_0 and the absolute values of $\bar{N}$.

Figure 7 depicts the dependence of the square root of the relative dispersion of the average total number of shower particles, $\sqrt{D}/N$, upon the energy of the primary electron; this dependence was calculated with the model considered in place of the experimental results on the spectral properties of chambers [3, 5, 6]. The calculated curves and the experimental data have about the same energy dependence $\sim 1/\sqrt{E}$. A certain difference between the experimental $\sqrt{D}/\bar{N}$ values of other work [3, 5] may originate from differences in the composition of the chamber plates.

It follows from the calculations that the $\sqrt{D}/N$ values for the thicknesses 0.25 and $0.125X_0$ are almost in agreement (they are represented by a single curve on Fig. 7). This curve is the "upper estimate" of the energy resolution which can be anticipated for a spark chamber when the particular electron–photon shower model is used. "Upper estimate" means that the results for $\sqrt{D}/N$ can be too large because the development of the shower is assumed to be a branching process with considerable energy splitting between the particles (assumption 2). However, under real conditions, a considerable part of the particle energy is spent in the emission of soft γ quanta and the generation of low-energy pairs. We therefore compare our results with the result of the other limit case, in which particle multiplication in a shower is assumed to obey the Poisson statistics. The total number of shower particles is in this case defined as $\bar{N} = E_0/E_{min}$. We therefore have $\sqrt{D}/\bar{N} = 1/\sqrt{\bar{N}} \sim 1/\sqrt{E_0}$. The corresponding functions are represented by dashed lines in Fig. 6. It follows from Fig. 7 that at large thicknesses (about $1X_0$), the dispersion of the total number of particles is independent of the nature of the process and that even when a genetic relation exists between the particles, the particle statistics of the shower resembles a Poisson statistics. This conclusion has been confirmed in experiments [3, 5, 6]. When the thickness decreases, the relative dispersion of the number of shower particles begins to depend upon the specific assumptions which are made for the process. Since the actual process is neither a pure branching process nor a pure Poisson process, the real limit of the energy resolution which can be obtained in multiplate spark chambers with lead is situated somewhere in the middle between the lower solid line and the dashed line of Fig. 7.

Spark-Chamber Spectrometer for Range Measurements

Measurements were made with electrons of the three energies 60, 95, and 140 MeV and with photons of the two energies 95 and 140 MeV.

TABLE 2

E, MeV	$E_c = 60$	$E_e = 95$	$E_e = 145$	$E_\gamma = 95$	$E_e = 145$
Number of tracks considered	930	986	916	321	778
Fraction of tracks remaining after application of a criterion (efficiency of the chamber)					
Criterion 1	0.80	0.85	0.65		
Criterion 2	0.50	0.55	0.55	0.44	0.50
Criterion 3	0.28	0.33	0.33	0.31	0.23
Criterion 4	0.25	0.23	0.24	0.15	0.17
Energy resolution (%)	30	35	32	37	38
t_{max}, cm	20	34	44	24	37

We determined the electron-range straggling and estimated the energy resolution of the spectrometer from the curve representing the differential range distribution. We determined for photons the distribution of the total range of electrons formed by primary γ quanta in lead converter K_2 which was situated before the chamber.

In order to reduce range straggling due to fluctuations of the bremsstrahlung of electrons passing through layers of the chamber material, certain conditions were imposed upon the tracks. We rejected tracks which did not satisfy the following criteria: 1) passage beyond the chamber and 2) entry angle less than 5°. These two criteria characterized basically the chamber design and the quality of the beam. The importance of these criteria can be minimized and they will not have great influence upon the efficiency of the spectrometer; 3) electrons emitting photons which in the following blocks render electrons recorded by the chamber, i.e., interrupted tracks were rejected; 4) tracks with sharp direction changes were rejected since they could be related to large energy losses by radiation. Changes in the direction of motion of the electron relative to the direction of motion in the preceding gap must not exceed 5° in the first and second gaps, 10° in the third and fourth gaps, and 30° in the other gaps.

We note that the angle of multiple scattering of electrons having an energy of 100 MeV amounts to about 2° in a 40-mm-thick Plexiglas block.

Table 2 lists the results of the experiments. The efficiency η_i of the spectrometer was defined as the ratio of the number of cases remaining after application of the i-th criterion to the total number of tracks evaluated. Thus, by rejecting some of the tracks, i.e., by reducing the efficiency, we can reduce the range straggling and, hence, improve the energy resolution of the spectrometer. Figure 8 shows the distribution of the electron ranges after application of criteria 1 and 2 (curve a) for E_0 = 95 MeV and the change in the form of the distribution when, in addition, criteria 3 and 4 are applied.

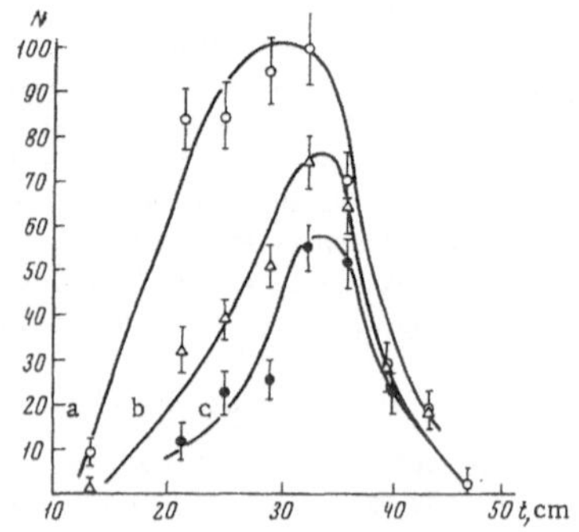

Fig. 8. Effect of the sampling criterion upon the form of the distribution curve of electron ranges. Range distribution after application of criteria 1 and 2 (curve a); distribution after application of criterion 3 (curve b); distribution after application of criterion 4 (curve c).

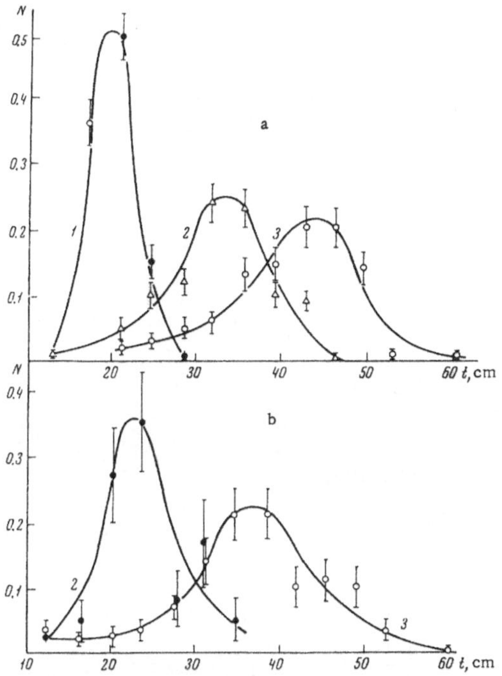

Fig. 9. Differential distribution of (a) electron ranges and (b) photon ranges at the energies (1) E = 60 MeV; (2) E = 95 MeV; and (3) E = 145 MeV.

The energy resolution was estimated with the formula $\sigma = \Delta t / t_{max}$, where Δt denotes the half-width of the range-distribution curve, and t_{max}, the position of the maximum of the distribution. The resolutions and t_{max} values are listed in Table 2, and the range distribution is shown in Fig. 9. The function $t_{max}(E_0)$ is linear within the accuracy limits.

Equipment with Spark Chambers for Recording Two-Photon Decays of Neutral Mesons [10]

Since spark chambers have a poorer time resolution than scintillation counters, it was of interest to investigate whether it is possible to make measurements with spark chambers when a strong background originating from the operation of an electron-accelerating synchrotron is present.

The equipment is schematically shown in Fig. 10. Photons which originate from the decay of pions produced in the target by γ quanta of the bremsstrahlung beam of the synchrotron were incident upon multilayer spark chambers. The first two gaps were in anticoincidence and allowed measurements of γ quanta on a strong background of charged particles. A steel plate with a thickness of $0.5X_0$ was inserted between each two successive spark gaps (the chamber comprised 8 spark gaps) so that the total thickness of the converters before the scintillation counters amounted to about $1.5X_0$.

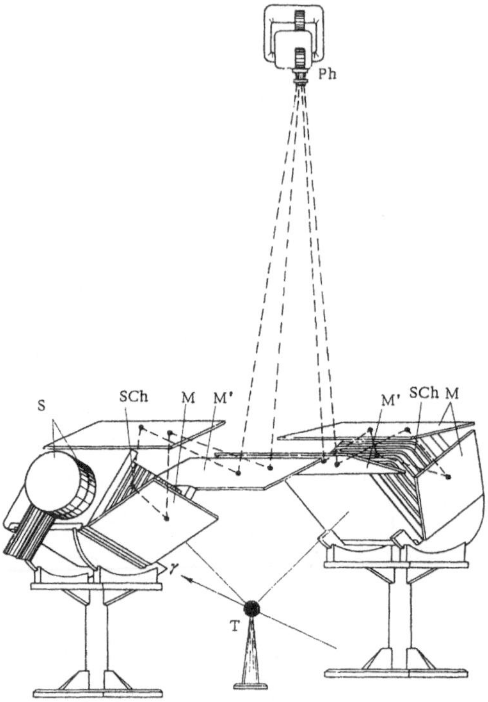

Fig. 10. Equipment with spark chambers for recording γ quanta originating from the decay of π^0 mesons. T) Target; M and M') mirrors; S) scintillation counters; SCh) spark chambers; Ph) photo camera.

The electron–positron pairs which were formed by γ quanta in one of the steel electrodes were recorded by two scintillation counters. The pulses of all four counters were fed to the four-channel coincidence circuit described above.

The sensitivity of the spark chambers to the background of scattered particles was examined in special tests under normal operation conditions. The control counters were in these tests situated so that the particles recorded by these counters were not incident upon the spark chamber. The average number of background particles in the chambers depends greatly upon the duration of the bremsstrahlung pulse of the synchrotron (the total radiation flux per operation cycle of the accelerator was constant). The number of background sparks increased sharply when the clearing field was reduced to $U_0 = 0$ V/cm and amounted to about 0.5 when the field was of the order of 100 V/cm (at a radiation-pulse duration of T = 2 msec). The formation of neutral pions was studied at T = 2 msec and $U_0 = 100$ V/cm. In about nine out of ten cases, the tracks produced by decay γ quanta can be reliably discriminated from the background. It was therefore shown that when certain conditions are satisfied, proper selection of the pulse duration of the accelerator and of the clearing field in the spark chamber facilitate measurements with the main beam of a ring accelerator for electrons.

Conclusions

The following is a list of the conclusions of our research work.

1. The characteristics of a multilayer spark chamber having a layer thickness of $1X_0$ were investigated, the chamber being operated as a shower detector of γ quanta and electrons in the energy range 50-200 MeV. It was shown that the average total number $\overline{N}$ of sparks per shower is proportional to the energy of the primary particle. Fluctuations of the total number of sparks around the average value obey the Poisson distribution law. The efficiency of recording γ quanta with a spark chamber is 0.7 at γ-quantum energies of 50 MeV, and unity at γ-quantum energies above 100 MeV.

2. Additional investigations of the efficiency with which a spark chamber records several electrons in a single gap revealed that the efficiency is practically independent of the number of tracks (for $2 \leq n \leq 5$) and the distance between the tracks (in the range 3-30 mm), but that the efficiency decreases strongly with increasing difference of the angle between the particle trajectories and the direction of the electric field in the chamber.

3. The energy resolution of the chamber for photons and electrons with energies of 50 to 200 MeV was worse than the resolution which can nowadays be obtained with total-absorption Cherenkov spectrometers.

4. A simple model of an electron—photon shower was used to show that there exists a limit of the energy resolution of shower-type spark chambers. Methods were outlined for obtaining an energy resolution which is typical for total-absorption Cherenkov spectrometers.

5. The characteristics of a multilayer spark chamber used as a spectrometer for range measurements of γ quanta and electrons in the energy range 60-150 MeV were experimentally investigated. It was shown that the energy resolution amounts to about 35% at these energies.

6. The results of our experiments prove that it is possible to build and to use highly efficient devices with a rather high energy resolution for research of electromagnetic interactions.

We express our gratitude to D. A. Stoyanova, S. S. Starostin, N. G. Kotel'nikov, and A. I. Shlyukov for their assistance in the experiments and in the evaluation of the experimental results, and to A. B. Govorkov for useful discussions and advice.

References

1. V. P. Agafonov, B. B. Govorkov, S. P. Denisov, and E. V. Minarik, Pribory Tekhn. Eksp., No. 5, 47 (1962).
2. S. P. Denisov and D. A. Stoyanova, FIAN Preprint No. 32 (1966).
3. A. G. Gapotchenko, B. B. Govorkov, S. P. Denisov, N. G. Kotel'nikov, and D. A. Stoyanova, Pribory Tekhn. Eksp., No. 5, 60 (1966).
4. A. G. Gapotchenko, B. B. Govorkov, S. P. Denisov, and S. S. Starostin, FIAN Preprint No. 146 (1967).
5. R. Kajikava, J. Phys. Soc. Japan, 18:1365 (1963).
6. J. E. Augustin, P. Marin, and F. Rumpf, Preprint LaL-1114, September (1964), Orsay, France.
7. B. F. Crawford and H. Messel, Nucl. Phys., 61:145 (1965); Phys. Rev., 128:2352 (1962).
8. A. B. Govorkov and B. B. Govorkov, Pribory Tekhn. Eksp., No. 1, 54 (1967).
9. E. Fermi, Nuclear Physics, University of Chicago Press (1950).
10. A. G. Gapotchenko, B. B. Govorkov, S. P. Denisov, N. G. Kotel'nikov, and D. A. Stoyanova, FIAN Preprint No. 56 (1966).

TECHNIQUE FOR MEASURING THE CHARACTERISTICS OF γ DETECTORS

B.B. Govorkov and S.P. Denisov

A method of producing monochromatic γ quanta is described so that efficiency, energy resolution, and other characteristics of photon detectors can be determined.

Introduction

In several experiments with accelerators (photoproduction of π^0 mesons, Compton effect, etc.), γ quanta with energies of several dozen or several hundred MeV must be recorded. It is important to know the energy dependence of the efficiency and the energy resolution of the γ detectors when these experiments are analyzed. Techniques of determining these characteristics of γ counters have been described in numerous papers [1-10].

The energy dependence $\varepsilon(E_\gamma)$ of the efficiency of telescopes consisting of scintillation counters was calculated with a statistical trial technique (Monte-Carlo method) in [1-3], which used the equations of cascade theory. But these calculations render only estimates because it is not possible to take fully into account ionization losses and multiple scattering of shower particles. The properties of shower particles were therefore experimentally introduced in a large number of papers on the use of γ detectors. Beams of monoenergetic electrons were used for this purpose in [4-6]. The measurements resulted in this case in a correction which accounts for differences in the development of the electromagnetic cascades generated by γ quanta and electrons.

Simple measurements of the energy dependence of the efficiency of γ detectors must be made when a technique is practiced, in which a photon telescope is directly inserted into the beam of the synchrotron and the dependence of the counting rate upon the maximum energy K_{max} of the bremsstrahlung spectrum is measured. However, in order to avoid overload of the counters, the beam intensity must be greatly reduced (about 10^6 times below the usual intensity) for the measurements, and it is therefore difficult to monitor the beam [5]. These difficulties were very intelligently eliminated in [7]. A γ telescope consisting of scintillation counters was mounted at an angle of 3° with respect to the primary bremsstrahlung beam of the synchrotron and recorded γ quanta which had undergone Compton scattering at the electrons of the target made of a material with a low Z value (carbon). The photon spectrum of Compton scattering could be easily taken into account by theoretical considerations. The usual operation of the accelerator need not be modified during the measurements. The following are the usual shortcomings of techniques making use of the continuous bremsstrahlung spectrum and comprising measurements of a certain effect in dependence of K_{max}: need for absolute measurements of the bremsstrahlung flux (this refers also to measurements involving electrons) and poor energy resolution.

A technique which is based on kinematic relations of the reaction $\gamma + p \rightarrow p + \pi^0$ was used in [8] to measure the efficiency of a γ telescope. By recording protons in a certain interval of emission angles and energies, photons originating from the decay of the pion are discriminated in a certain energy interval at the particular angle. Observations of $\gamma - p$ coincidences allow in this case measurements of the energy dependence of the efficiency without introduction of the energy of the primary photon flux. Low energy resolution (about 20%) in the energy range of 20-200 MeV of the photons is the main shortcoming of this technique.

Monochromatic photons are used in the most straightforward technique of measuring the characteristics of γ detectors. First attempts to generate monochromatic photons in the energy range of several hundred MeV were made in [9]. The photons which were produced in the target of the synchrotron and the electrons emitted through these photons were in that work recorded by a circuit for time-dependent coincidences. The magnetic field of the accelerator was used to determine the energy of the electrons. The energy E_0 of an accelerated electron and the energy E_e of an emitted electron define the energy of the photon: $E_\gamma = E_0 - E_e$. However, the energy resolution of the γ quanta amounted in this technique to at most ± 20%.

Thus, the methods considered had a poor resolution of the energy E_γ which resulted in inaccuracies in the determination of $\varepsilon(E_\gamma)$. These techniques were therefore not used to measure the energy resolution of γ spectrometers. The energy resolution was usually estimated from data obtained with monoenergetic electrons. In order to study the properties of γ detectors, we used in our work since 1961 a simple technique [10] which is based on the generation of monochromatic γ quanta from the external bremsstrahlung beam of the synchrotron. We based our work on the simple idea that the spectral properties and the efficiency of γ detectors can be investigated when a flux of monochromatic photons with an intensity of 1-10 photons/cm^2 · sec is incident [because $\varepsilon(E_\gamma)$ of modern detectors is 0.1-1]. But the energy resolution should be improved, as much as possible (to 1-2%).

Generation of Monochromatic Photons

Figure 1 shows the scheme of the setup. A lead target (1) which formed an intensive source of electron–positron pairs emitted in the direction of the primary beam of γ quanta of the synchrotron was placed in the path of the collimated bremsstrahlung beam (3 cm diameter) of the 264-MeV synchrotron of the Physics Institute of the Academy of Sciences of the USSR.

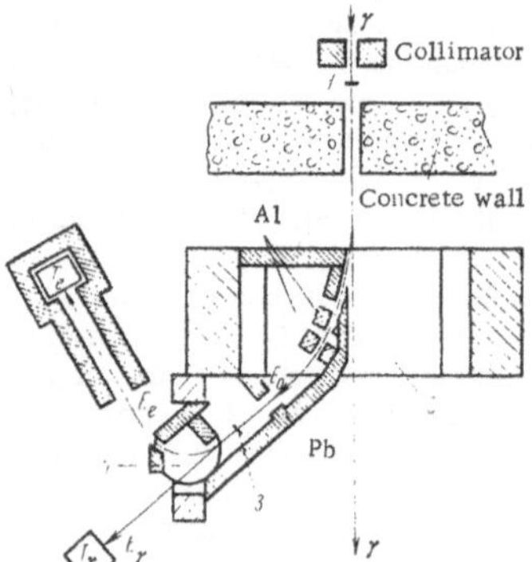

Fig. 1. Setup for generating a monochromatic photon beam from the accelerator. 1) First target; 2) magnet I; 3) second target; 4) magnet II.

The magnetic field of a first spectrometer (2) was used to select from the beam electrons with a certain energy E_0 (primary electrons). A second lead target (3) was placed in the path of the primary electrons. When an electron passed through the lead target, it could emit a bremsstrahlung γ quantum of the energy E_γ.

It is important that the emission has a sharp direction dependence in forward direction. The γ detector T_γ was placed perpendicular to the direction of the bremsstrahlung emitted.

The electron spectrum behind the second target ("discharge" electrons) was analyzed once more with the magnetic field of spectrometer (4). Electrons with a certain energy E_e were therefore incident on counter T_e. T_γ and T_e were connected in coincidence, and the coincidence counts indicated that γ quanta with a certain energy $E_\gamma = E_0 - E_e$ were detected. Moreover, the geometry of the setup was chosen so that each recording of an electron with energy E_e corresponded to the incidence of a γ quantum with the energy E_γ upon T_γ. The counting rate of the electronic counting equipment determined therefore the number of γ quanta having the energy E_γ and being incident on T_γ. The ratio of γ–e coincidences to electrons recorded during the same time rendered the efficiency $\varepsilon(E_\gamma)$.

Equipment

Electromagnets (the magnets of spectrometers 2 and 4 in Fig. 1) with high stabilization of the magnetization current were used to produce the magnetic fields. The trajectories of the electrons in the magnetic fields were determined with a flexible wire through which a current was sent [11]. The entire path of an electron before and after emission of the photon was in our case determined as a whole. To this end, a shunt was connected parallel to the wire portion corresponding to the electron trajectory before the emission of radiation. By varying the resistance of the shunt, the required ratio of the currents in the various wire sections was chosen so that the current discontinuity corresponded to the change in the electron's momentum upon emission of a bremsstrahlung γ quantum of a certain energy.

The shunt was connected to the wire via a thin copper hair thread (0.08 mm diameter). The tension of the wire was determined with an accuracy of 0.2% with the aid of a pendulum suspension.

The accuracy of the electron-path determination was estimated at 0.3% after repeatedly tracing the trajectory of electrons with a fixed energy E_0 and by tracing the trajectories of electrons with an energy differing from E_0 by ± 2.5%.

The first spectrometer (Fig. 1) was operated in the range in which the current dependence of the magnetic field is linear (1500–4500 G). It was therefore possible to restrict the tracing of the electron paths to a single magnetic field value H_0 and to a single energy of the primary electrons. At a different field H, electrons with an energy

$$E = \frac{HE_0}{H_0}$$

must move along this trajectory.

Since the field of the second spectrometer was kept constant during the measurements and, hence, since electrons with a fixed energy E_0 were incident upon the electron counter, the energy of photons recorded at the field H of the first magnet amounted to

$$E_\gamma = \frac{H(E_0 - E_e)}{H_0}.$$

A telescope comprising two scintillation counters (mounted one behind the other) was used to record the electrons. In some cases, the counters were moved apart in order to reduce the background counts which resulted from the passage of cosmic ray particles through telescope T_e. A current-free diode coincidence circuit with a resolution $\tau = 10^{-8}$ sec was used to record coincidences between the counters in the telescopes; γe coincidences were discriminated with a Rossi circuit ($\tau = 0.3 \cdot 10^{-6}$ sec).

Radial Intensity Distribution and Energy Spectrum of the Monochromatic γ Quanta

The radial intensity distribution of the monochromatic bremsstrahlung beam depends upon both the size and the angular divergence of the electron beam incident on the lead target (target 3 in Fig. 1), upon multiple electron scattering in the target before emission of radiation, and upon the angular dependence of the differential cross section of bremsstrahlung.

The radial intensity distribution was experimentally investigated [12] with the aid of a spark chamber for showers. The spark chamber was placed at the position of T_γ at the distance d = 2.6 m from the target. The projections of tracks upon vertical and horizontal planes were photographed. The accuracy in the determination of the coordinates of e^+e^- pairs formed by γ quanta in the lead plates of the chamber amounted to 0.5 cm. The cross section of the electron beam at the point at which the target was located formed a square with the lateral length $l = 3.8$ cm. The angular divergence of the electron beam was given by the rotation of the electrons in the magnetic field of spectrometer 1 and amounted to $\theta = 0.013$ r in a plane which was perpendicular to the direction of the magnetic field (horizontal plane). The divergence of the electrons in the vertical plane could be ignored. The radial distribution was investigated for γ-quantum energies of 50, 100, and 150 MeV (the electron energy after emission amounted to 60 MeV for all energies of the photons generated). The experimental results are depicted in Fig. 2.

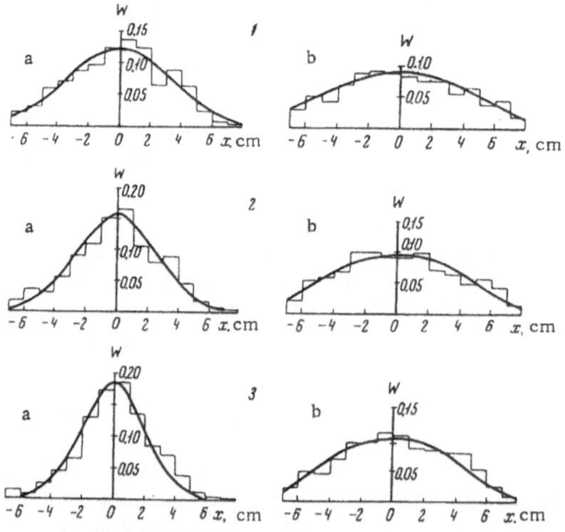

Fig. 2. Intensity distribution of monochromatic radiation (a) in the vertical and (b) in the horizontal plane for E_γ = 50 MeV (curves 1), 100 MeV (curves 2), and 150 MeV (curves 3). The histograms represent experimental data, and the curves were obtained from calculations.

If we assume that the radiating electrons are uniformly distributed over the cross section of the beam and that the spread of the projections of emission angles of bremsstrahlung photons upon a plane which is perpendicular to the electron flux is given by a Poisson law with the dispersion α, the projection of the radial intensity distribution of a monochromatic photon beam can be described by the following formula:

$$W(x) = \frac{1}{2l'}\left\{I\left[\frac{1}{\sigma}\left(x + \frac{l'}{2}\right)\right] - I\left[\frac{1}{\sigma}\left(x - \frac{l'}{2}\right)\right]\right\}, \tag{1}$$

where

$$I(x) = \sqrt{\frac{2}{\pi}}\int_0^x \exp\left(-\frac{t^2}{2}\right)dt$$

denotes the probability integral; $\sigma^2 = \alpha d^2$; $l' = l$ for the projection upon the vertical plane, and $l' = l + 2d\theta$ for the horizontal projection. It was assumed in the derivation of Eq. (1) that multiple electron scattering in the target has no influence upon the probability of electrons being recorded by the equipment. Figure 2 is a comparison of the results of W(x) calculations and experimental data. The dispersion of the projections of the angles of multiple electron scattering in the target [13] was used as α in the calculations, along with a correction for the range straggling of the electrons before the emission of radiation. The dispersion of the angular dependence of the differential cross section of the bremsstrahlung was by one order of magnitude smaller than the dispersion of multiple scattering and was disregarded in the calculations. It follows from Fig. 2 that the agreement between the calculations and the experimental results is quite satisfactory. The measured radial intensity distribution of the γ radiation of background events (without target 3) was close to the distribution obtained with the target. This means that the background originates mainly from bremsstrahlung processes of electrons passing through the air column between spectrometers 2 and 4 (see Fig. 1).

The energy E_γ which is spent by an electron in the emission process in the lead target can be distributed among various γ quanta (multiple bremsstrahlung event). The target can be made very thin (0.04 radiation units) so that the probability of a bremsstrahlung photon having an energy greater than some energy E_{min} is very small:

$$t\int_{E_{min}}^{E_\gamma} \Phi(E_0, E_{\gamma 1})\, dE_{\gamma 1} \ll 1, \tag{2}$$

where $\Phi(E_0, E_{\gamma 1})$ denotes the effective cross section of bremsstrahlung, and t denotes the target thickness in radiation units. The quantity E_{min} was introduced so that, when the emission of photons having an energy of less than E_{min} is ignored, no unbounded quantities appear in calculations of the spectrum [$\Phi(E_0, E_{\gamma 1})$ for $E_{\gamma 1} \to 0$]. The numerical value of E_{min} is chosen so that the ambiguity in the energy (which is proportional to E_{min}) does not introduce a noticeable error in measurements of this characteristic of the γ detector (it is hardly sensible to use an E_{min} value below 1-2 MeV with the currently available equipment). As will be shown below, a certain ambiguity in the selection of E_{min} is immaterial. When inequality (2) is satisfied, the probability of n-fold bremsstrahlung processes is proportional to $t^n/n!$ and, since $t \ll 1$, it suffices to consider in a first approximation the contribution of single emission processes and the emission of two photons in calculations of the spectrum of the monochromatic photon beam.

The single-emission spectrum is a narrow peak with its maximum near E_γ. The width δ of the peak depends upon the energy resolution of the magnetic spectrometers and upon fluctuations of the ionization losses of the electrons and is less than several MeV. The emission of photons having an energy below $E_{min} = 2$ MeV has practically no influence upon both the

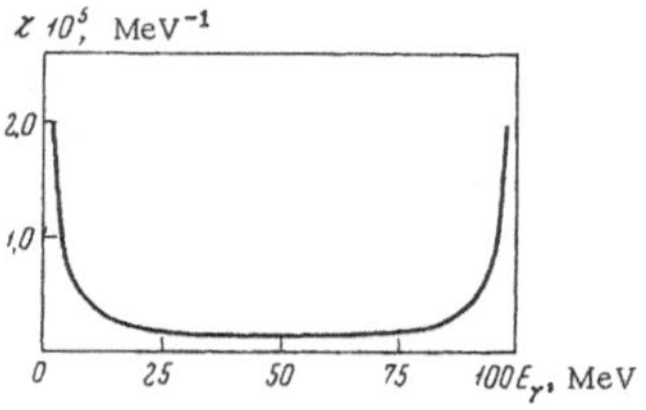

Fig. 3. Spectrum $\chi(E_{\gamma 1})$ of double bremsstrahlung processes for $E_0 = 180$ MeV, $E_\gamma = 100$ MeV, $\delta = 1$ MeV, and $t = 0.04$, radiation units.

position and the width of the peak so that the average energy which is carried away by these photons amounts only to 0.1 MeV. The form of the peak depends upon the geometry of the setup, the properties of the spectrometers, and fluctuations in the interaction between the electrons and the target. It is therefore difficult to determine the details of the form of the peak. The form of the spectrum of double bremsstrahlung emission, on the other hand, can be rather accurately calculated. It is easy to show with Eq. (2) that the probability of one of the quanta having an energy within the interval $E_{\gamma 1}$ and $E_{\gamma 1} + dE_{\gamma 1}$ under the condition $E_{\gamma 1} + E_{\gamma 2} = E_\gamma + \delta/2$ is given by

$$\chi(E_{\gamma 1})\,dE_{\gamma 1} = \frac{\delta t^2}{2}\left[\Phi(E_0, E_{\gamma 1})\,\Phi(E_0 - E_{\gamma 1}, E_\gamma - E_{\gamma 1}) + \Phi(E_0, E_\gamma - E_{\gamma 1})\,\Phi(E_0 - E_\gamma + E_{\gamma 1}, E_{\gamma 1})\right] dE_{\gamma 1}. \quad (3)$$

The spectrum $\chi(E_{\gamma 1})$ which is shown in Fig. 3, is superimposed upon the narrow peak of the single-event bremsstrahlung processes and leads to a long "tail" in the spectrum of the monochromatic γ-quantum beam at low energies.

We must bear in mind that multiple bremsstrahlung does not lead to an additional error if the detector characteristic under investigation is proportional to the energy. Errors can be made in other cases. For example, when the probability of detection with the scintillator telescope for photons is measured near the threshold multiple emission results in reduced values if the target has not been made thin enough. This effect is illustrated by the experimental results shown in Fig. 4.

The spectrum $\chi(E_{\gamma 1})$ must not be calculated for a preliminary estimate of the importance of multiple radiation processes. It may suffice to know the ratio of the probability of two-photon emission (W_2) to the probability of single-photon emission (W_1). When we assume $\Phi(E_0, E_{\gamma 1}) = 1/E_{\gamma 1}$, we obtain

$$W_{21} = \frac{W_2}{W_1} = t \ln \frac{E_\gamma}{E_{\min}}. \quad (4)$$

It follows from the above formula that the integral contribution of double bremsstrahlung events to the spectrum of the monochromatic photon beam depends only logarithmically upon $E_{\min}$. The ambiguity in the selection of $E_{\min}$ is therefore only important in the form of this function.

A small correction (about 2%), which results from the conversion of bremsstrahlung photons in the lead target, must be introduced in the absolute normalization of the spectrum.

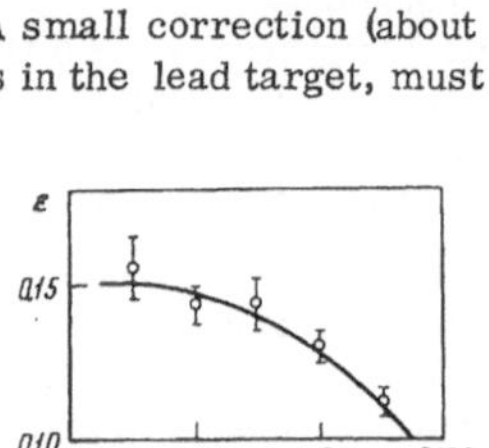

Fig. 4. Dependence of the efficiency of a scintillation γ telescope consisting of three counters upon the thickness of the lead target (expressed in radiation units) at $E_\gamma = 60$ MeV.

We have discussed in detail the basic characteristics of the apparatus because knowledge of these characteristics is required for the correct calibration of γ detectors and for estimating the capabilities of apparatus of this type.

Measurements of the Characteristics of γ Detectors

The Laboratory of Photomesonic Processes of the Physics Institute of the Academy of Sciences of the USSR has now developed a stationary setup in which monochromatic γ quanta are generated and which is used for measurements of the characteristics of various types of γ detectors. The setup was particularly used to measure the energy dependence of the efficiency of γ telescopes consisting of scintillation counters [10, 14]. In order to determine the efficiency, the dependence of the counting rate of γ−e coincidences and the counting rate of electron counters upon the magnetic field in the first spectrometer was recorded. The counting rate with the second target, or without the second target, was determined for each field strength value. Disregarding the area close to the threshold of photon recording, the ratio of counts with target to counts without target was 3 in the case of the electron telescope and 10 in the case of γ−e coincidences. The counting rate without the second target was assumed to be the background and was subtracted from the counting rate obtained with the target. Multiple measurements of each point of the efficiency curve rendered results which agree within the statistical accuracy limits. A thin-wall ionization chamber was used for relative measurements of the intensity of the bremsstrahlung.

Since the results of the measurements refer to a certain design of the γ telescopes, we return to the description of the equipment. The γ telescopes consisted of two or three identical scintillation counters. Plastic scintillators (terphenyl and polystyrene) with a diameter of 15 cm and a thickness of 2-2.5 cm were used. The distance between the central planes of the counters of the telescope was 6 cm. A 6-mm-thick lead converter was placed before the first counter of the γ telescope. The results of the efficiency measurements made with γ telescopes consisting of two or three counters are shown in Fig. 5. The errors σ indicated in Fig. 5 were calculated for the efficiency with the aid of a formula for the binomial distribution of the γ−e coincidence counts

$$\sigma = \sqrt{\frac{\varepsilon(1-\varepsilon)}{N_e}},$$

where N_e denotes the total number of electrons counted with telescope T_e. The measured functions $\varepsilon(E_\gamma)$ were approximated by the expression

$$\varepsilon(E_\gamma) = \varepsilon_\infty\left[1 - \exp\left(-\frac{E_\gamma - E_n}{E_1}\right)\right]. \qquad (5)$$

The coefficient ε_∞ characterizes the limit efficiency for $E_\gamma \to \infty$. The parameter E_1 determines the form of the efficiency curve, whereas the absolute value of the parameter E_n in-

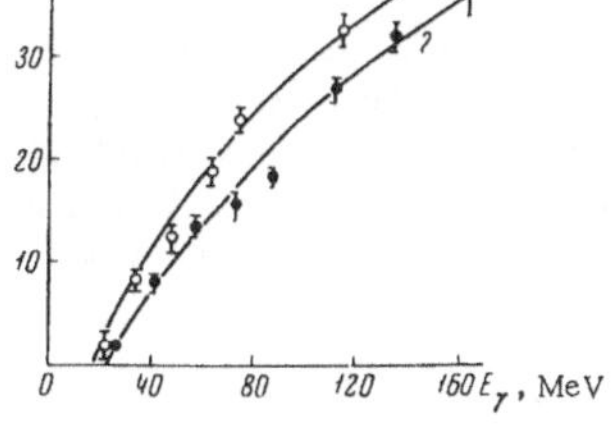

Fig. 5. Relation between the efficiency of the γ telescopes and the energy of the γ quanta recorded. 1) Telescope consisting of two counters; 2) telescope consisting of three counters; the solid curves correspond to Eq. (5).

dicates the minimum energy of the γ quanta which are still detected with the telescope. We obtain the following parameter values for the γ telescope consisting of two counters: $\varepsilon_\infty = 0.58$, $E_n = 16$ MeV, and $E_1 = 119$ MeV; the corresponding values for a telescope comprising three counters were $\varepsilon_\infty = 0.57$, $E_n = 22$ MeV, and $E_1 = 140$ MeV.

The parameters ε_∞ and E_n can be theoretically calculated. ε_∞ can be easily calculated when we assume that, at very high energies of the γ quanta, the efficiency depends entirely upon the probability of pair formation in the converter. Thus, $\varepsilon_\infty = 1 - \exp(-\sigma_p x)$, where σ_p denotes the cross section of pair formation in lead at very high energies; x denotes the thickness of the converter; and E_n can be set equal to the minimum energy which an electron needs to pass through the first scintillator and the filter of the γ telescope. The calculations result in the values $\varepsilon_\infty = 0.55$ and $E_n = 15$ MeV for the telescope consisting of two counters and $E_n = 23$ MeV for the telescope consisting of three counters. The calculations are in good agreement with the experimentally obtained values. This means that the efficiency of recording γ quanta is entirely determined by the design of the telescope and the thickness of the converter. In calculating ε_∞ we used the experimental value $\sigma_p = 39.3$ barn which was obtained in [15] for $E_\gamma = 1$ GeV. Similar measurements were afterwards made in [16] on the same setup. Direct measurements of the efficiency of the γ telescopes made it possible to greatly improve the results obtained in measurements on electromagnetic processes with the aid of these telescopes.

The setup which extracts γ quanta with a certain energy from the bremsstrahlung beam was successfully employed for measurements of the energy resolution of spectrometers, namely of a spark chamber for showers and of a scintillation counter with NaI(Tl) crystals having a size of 10 × 10 cm [18]. Measurements were made in addition to calibration tests on this setup. Thus, a multiplate spark chamber was used to study the development of an electromagnetic cascade produced in lead by monochromatic γ quanta having the energies 50, 100, and 150 MeV [17].

Conclusions

The proposed technique makes it possible to generate monochromatic γ fluxes with an energy resolution of about 1%. The fluxes of extracted γ quanta incident upon the detector can be greatly increased once the trajectories for several energies of the "discharged" electrons have been established and, hence, the number of electron counters has been increased accordingly. The measurements will then provide a simultaneous determination of the entire energy dependence of a certain parameter of a γ detector.

The principal advantages of our method of measuring the characteristics of γ detectors are: a) high energy resolution; b) no need for absolute measurements; and c) simple evaluation of the results and possibility of investigating, during the measurement process, effects which affect the efficiency of a γ telescope or other characteristics of a γ detector. It is therefore possible to establish the optimum operation conditions of γ detectors and of the related electronic equipment.

Recently, there appeared a publication which mentioned that a similar setup has been inserted in other electron accelerators [19] in order to calibrate γ detectors. Since the direction dependence of the bremsstrahlung increases with increasing energies, the technique of external monochromatization of the photon beam is very useful at higher energies for both methodological work and for research work in which monochromatic photon beams of relatively low intensities are required.

References

1. J. L. Steinberger, W. K. H. Panofsky, and J. S. Steller, Phys. Rev., 78:802 (1950).
2. L. J. Koester and F. E. Mills, Phys. Rev., 105:1900 (1957).
3. G. E. Modesitt, Ph. D. Thesis, MIT (1958).
4. C. L. Oxley, Phys. Rev., 110:733 (1957).
5. L. L. Higgins, Thesis UCRL-3688 (1957).
6. A. S. Belousov, S. V. Rusakov, and E. I. Tamm, Zh. Eksp. Teor. Fiz., 35:355 (1958).
7. V. I. Gol'danskii, O. A. Karpukhin, and V. V. Pavlovskaya, Pribory Tekhn. Eksp., No. 3, 3 (1960).
8. P. S. Baranov, L. I. Slovokhotov, G. A. Sokol, and L. N. Shtarkov, Pribory Tekhn. Eksp., No. 3, 63 (1961).
9. J. W. Weil and B. P. McDaniel, Phys. Rev., 86:582 (1952).
10. V. P. Agafonov, B. B. Govorkov, S. P. Denisov, and E. V. Minarik, Pribory Tekhn. Eksp., No. 5, 47 (1962).
11. M. S. Kozodaev and A. A. Tyapkin, No. 1, 21 (1956).
12. B. B. Govorkov and S. P. Denisov, FIAN Preprint No. 97 (1966).
13. G. Moliere, Z. Naturforsch., 3a:78 (1948).
14. B. B. Govorkov, S. P. Denisov, and E. V. Minarik, Yadernaya Fizika, 5:190 (1967).
15. E. Malamud, Phys. Rev., 115:687 (1959).
16. A. S. Belousov, S. V. Rusakov, E. I. Tamm, and L. S. Tatarinskaya, Pribory Tekhn. Eksp., No. 6, 125 (1962).
17. A. G. Gapotchenko, B. B. Govorkov, S. P. Denisov, N. G. Kotel'nikov, and D. A. Stoyanova, Pribory Tekhn. Eksp., No. 5, 60 (1966).
18. A. S. Belousov, S. V. Rusakov, E. I. Tamm, and L. S. Tatarinskaya, FIAN Preprint No. A-42 (1965).
19. W. K. H. Panofsky, Proc. Intern. Symposium on Electron and Photon Interactions at High Energies, Hamburg, 1, 138 (1965).

ONCE MORE THE TWO-STAGE CATHODE FOLLOWER

V.A. Zapevalov

The article describes a new method of calculating the amplification coefficient of cathode followers and states expressions for calculating the rise time of cathode followers. Experimental results are given.

The cathode follower, the circuit of which is shown in Fig. 1, has been frequently discussed in the literature. The stage is a resistance transformer with high input and low output resistances and exhibits a poor transmission characteristic at large input signal amplitudes [1].

The two-stage cathode follower of Fig. 2 is a circuit of greater complexity than the conventional cathode follower, but is characterized by a better transmission characteristic and a reduced output resistance.

Calculation of the Transmission Coefficients

The two-stage cathode follower was analyzed for the first time in 1946 [2]. Equivalent circuits were thereafter used to derive the transmission coefficient of the cathode follower [3, 4]. Since both the conventional and the two-stage cathode followers are systems with negative feedback, their transmission coefficient can be expressed by a single formula

$$K_{fb} = \frac{K_0}{1+K_0} \qquad (\beta = 1), \tag{1}$$

where K_0 denotes the amplification coefficient of the system with open feedback loop and K_{fb} the amplification (transmission) coefficient with feedback.

Let us determine the transmission coefficient of the conventional cathode follower with Eq. (1). We use the circuit of Fig. 1b for the calculation (feedback circuit open):

$$K_0 = \frac{\mu R_\kappa'}{R_i + R_\kappa'}; \qquad R_\kappa' = \frac{R_\kappa R_H}{R_\kappa + R_H}; \qquad K_{fb} = \frac{\mu R_\kappa'}{R_i + (1+\mu) R_\kappa'}.$$

Let us now determine the transmission coefficient of the two-stage cathode follower with the method described above.

In order to determine K_0, we open the feedback circuit and, hence, connect the input-signal generator as shown in Fig. 2b. We obtain in this case $K_0 = U_{out}'/U_{in}$. We determine U_{out}' from the circuit of Fig. 3a and from the corresponding equivalent circuit of Fig. 3b. The tube with the resistor R_H connected in parallel is shown in the circuit of Fig. 3a as an equivalent

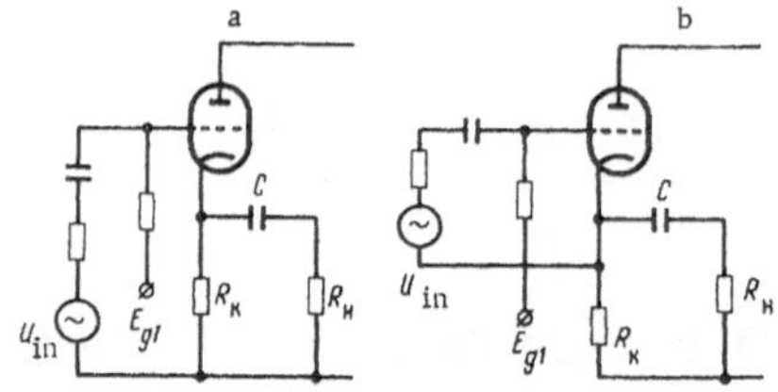

Fig. 1. Cathode follower circuit (a); open feedback (b).

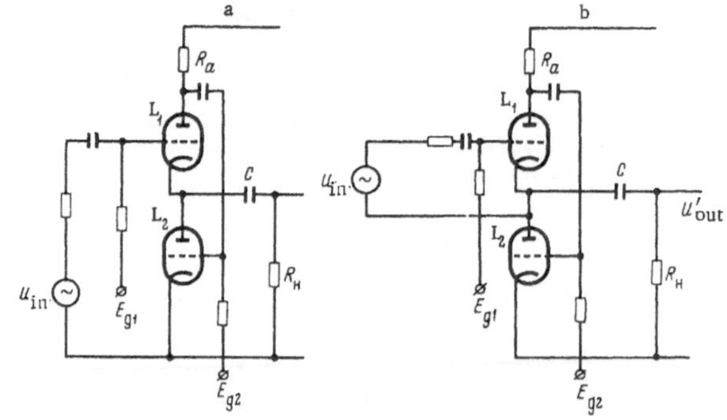

Fig. 2. Two-stage cathode follower circuit (a); open feedback (b).

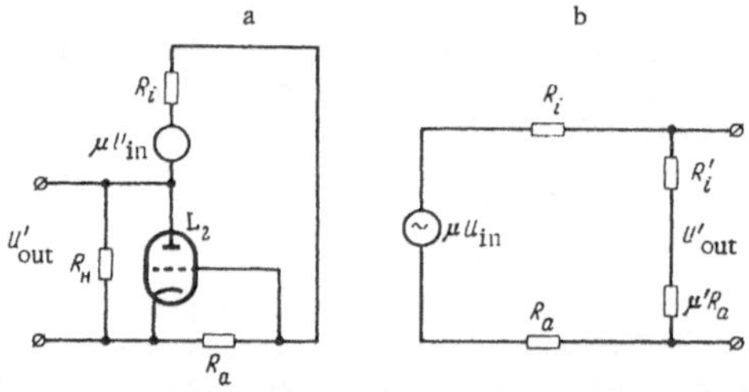

Fig. 3. Equivalent circuit of the two-stage cathode follower; a) tube L_1 replaced; b) total equivalent circuit.

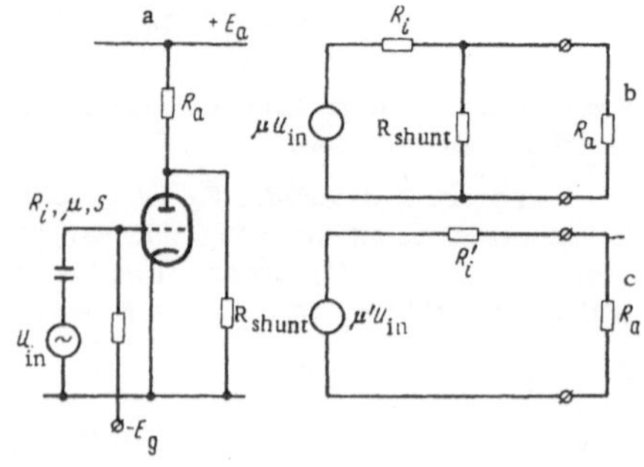

Fig. 4. To the computation of the parameters of the circuit shown in Fig. 3.

tube with parameters $\mu' = \mu \frac{R_H}{R_i + R_H}$ and $R_i' = \frac{R_i \cdot R_H}{R_i + R_H}$. This follows from Fig. 4. The transition from the circuit of Fig. 4b to the circuit of Fig. 4c is made with the theorem of the equivalent generator.

We have for the circuit of Fig. 3b:

$$K_0 = \frac{U'_{out}}{U_{in}} = \frac{\mu R_H (R_i + \mu R_a)}{[(R_i + R_a)(R_i + R_H) + (R_i + \mu R_a) R_H]}. \tag{2}$$

By substituting Eq. (2) into Eq. (1), we obtain

$$K = K_{fb} = \frac{\mu R_H (R_i + \mu R_a)}{[(R_i + R_a)(R_i + R_H) + (1+\mu)(R_i + \mu R_a) R_H]} = \frac{S(1 + SR_a) R_H}{[(1+\alpha)(1+\varkappa) + S(1 + SR_a) R_H]}. \tag{3}$$

where $\alpha = R_a/R_i$, $\varkappa = R_H/R_i$, and $\mu \gg 1$.

If tubes L_1 and L_2 are pentodes, we have

$$K = \frac{S(1 + SR_a) R_H}{1 + S(1 + SR_a) R_H} = \frac{S^* R_H}{1 + S^* R_H}. \tag{3'}$$

The output resistance of the two-stage cathode follower is

$$R_{out} = \frac{1 + \alpha}{S(1 + SR\alpha)}. \tag{4}$$

The transmission coefficient was calculated for two identical tubes.

Transient Characteristic of the Two-Stage Cathode Follower

The transient characteristics of the two-stage cathode follower have not been thoroughly investigated. A rather complete discussion of these characteristics has been outlined in [4].

A large amount of useful information is given in [4], though the paper does not include a simple calculation formula for a rapid evaluation of the rise time of the two-stage cathode follower. The transient characteristics were not considered in [4] for cathode follower operation with large signals (signals of the order of several tenths of a volt were considered).

In electronic equipment for the purposes of nuclear physics, the two-stage cathode follower is frequently working on a capacitive load, with large input signal amplitudes being applied. The operation conditions of the cathode follower are in this case such that the upper tube may be shut off and the leading edge of a negative, as well as the rear edge of a positive output pulse, are defined by the discharge of the cathode circuit of the upper tube.

In order to determine the rise time of the pulse in the cathode circuit, we use a negative input signal which satisfies the inequality $U_{in} \gg U_{cutoff1}$ with rise time zero (U_{in}/p in the graphical representation).

When these restrictions are imposed on the input signal, tube L_1 is shut off at $U_{in} = U_{cutoff1} - U_{gk01}$ (where U_{gk01} denotes the potential at the control grid of tube L_1 relative to the cathode potential without input signal, and $U_{cutoff1}$ denotes the cutoff voltage of tube L_1).

When tube L_1 is shut off, the voltage in the cathode circuit will, in the case of a capacitive load, depend upon the discharge of the cathode circuit, i.e., upon the load resistance, the capacitor C_H, and both the type and the operation conditions of tube L_2.

Let us consider the two cases: 1) tube L_2 is a triode and 2) tube L_2 is a pentode or tetrode.

Tube L_2 Is a Triode. When tube L_1 is shut off, a positive voltage pulse appears at the anode of this tube. This pulse appears also at the control grid of tube L_2. This pulse has the amplitude $I_{a01} \cdot R_a = U_{g2}$, where U_{g2} denotes the pulse applied to the grid of the tube L_2 and I_{a01}, the current through tube L_1 at $U_{in} = 0$.

When tube L_1 is shut off, the voltage appearing between the grid and the cathode of tube L_2 is $U'_{g2} = U_{gk\,02} + U_{g2}$ (see Fig. 5). The circuit with tube L_2 can be represented as a battery E_b connected in series with the internal resistance R_2 of the tube (Fig. 5b, circuit section 1).

When we transform the circuit of Fig. 5b into the circuit of Fig. 5c, we can use the theorem of the equivalent generator to determine $U_C(t)$, with the initial voltage U_{CO} applied to the capacitor. The output voltage can be stated in the form

$$U_{out}(t) = U_{CO} - U_C(t), \tag{5}$$

$$U_C(t) = \left(U_{CO} - \frac{E_b R_H}{R_{i2} + R_H}\right) e^{-\frac{t}{\frac{R_{i2} R_H}{R_{i2} + R_H} C_H}} + \frac{E_b R_H}{R_{i2} + R_H}, \tag{6}$$

and, hence,

$$U_{out}(t) = U_{CO} - \left[\left(U_{CO} - \frac{E_b R_H}{R_{i2} + R_H}\right) e^{-\frac{t}{\frac{R_{2i} R_H}{R_{2i} + R_H} C_H}} + \frac{E_b R_H}{R_{i2} + R_H}\right]. \tag{7}$$

The rise time of the output voltage can be easily derived from Eq. (7).

According to Fig. 6, the rise time t_H can be defined as the time required for changing the output voltage by the amount

$$U_{in} + U_{g\kappa 01} - U_{cutoff\,1}. \tag{8}$$

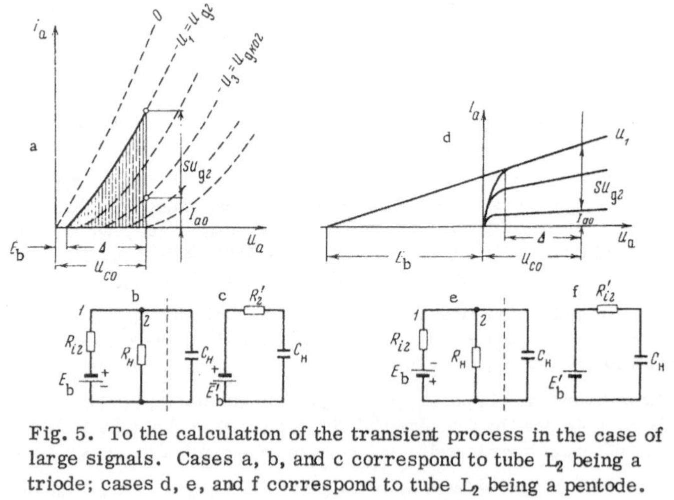

Fig. 5. To the calculation of the transient process in the case of large signals. Cases a, b, and c correspond to tube L_2 being a triode; cases d, e, and f correspond to tube L_2 being a pentode.

$$\left(R'_{i2} = \frac{R_{i2} R_H}{R_{i2} + R_H}; \qquad E'_b = \frac{E_b R_H}{R_{i2} + R_H}\right)$$

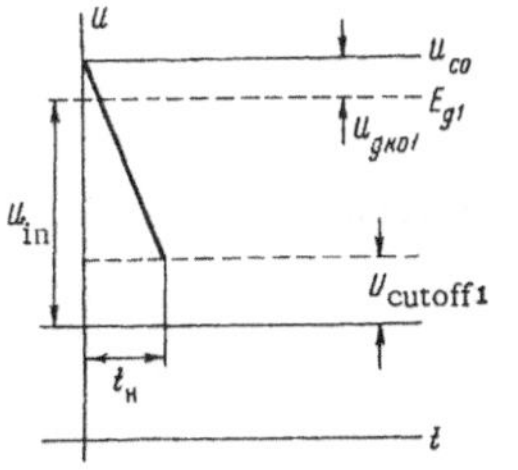

Fig. 6. Determination of the rise time of the two-stage cathode follower in the case of large negative input signals.

By substituting Eq. (8) into the left side of Eq. (7) and solving Eq. (7) with respect to t_H, we obtain

$$t_H = \tau \ln \frac{U_{co} - \frac{E_b R_H}{R_{i2} + R_H}}{U_{co} - U_{in} - U_{gк01} + U_{cutoff1} - \frac{E_b R_H}{R_{i2} + R_H}},$$

$$\tau = \frac{R_{i2} R_H}{R_{i2} + R_H} C_H. \tag{9}$$

Tube L_2 Is a Pentode or Tetrode. When tube L_2 is a pentode or tetrode, the rise time of the output pulse can be derived from the equivalent circuit of Fig. 5b and, as can be easily verified, can be obtained by changing the sign of the terms with E_b in Eq. (9)

$$t_H = \tau \ln \frac{U_{co} + \frac{E_b R_H}{R_{i2} + R_H}}{U_{co} - U_{in} - U_{gк01} + U_{cutoff1} + \frac{E_b R_H}{R_{i2} + R_H}}. \tag{10}$$

Absolute values were used for all potentials in Eqs. (8)–(10).

Experimental Results. The rise times which were calculated for the two-stage cathode follower with Eqs. (9) and (10) and the experimentally obtained rise times of the two-stage cathode follower circuit are listed in Tables 1 and 2, respectively.

TABLE 1

C_H, pF	210	350	320
t_H^*, sec	$5.1 \cdot 10^{-7}$	$8.5 \cdot 10^{-7}$	$13 \cdot 10^{-7}$
$t_H^\dagger$, sec	$6 \cdot 10^{-7}$	$10 \cdot 10^{-7}$	$16 \cdot 10^{-7}$

TABLE 2

C_H, pF	200	250	610
t_H^*, sec	$4.2 \cdot 10^{-7}$	$5.2 \cdot 10^{-7}$	$1.1 \cdot 10^{-7}$
$t_H^\dagger$, sec	$4.7 \cdot 10^{-7}$	$5.2 \cdot 10^{-7}$	$1.4 \cdot 10^{-7}$

* t_H denotes calculated values.
† t_H denotes experimental values.

References

1. K. É. Érglis and V. Levin, Radiotekhnika, 3:1963 (1963).
2. Calwin M. Hammack, Electronics, 19:206 (1946).
3. A. E. Voronkov, L. N. Korablev, I. D. Murin, and I. V. Shtranikh, VINITI [in Russian] (1957).
4. M. Brown, Rev. Sci. Instr., 31:403 (1960).

SELECTION OF A TWO-STAGE CATHODE FOLLOWER CIRCUIT WORKING ON A LOW-RESISTANCE LOAD

V.A. Zapevalov

The article evaluates the factors which determine the maximum input signal of a two-stage cathode follower and describes a modified two-stage cathode follower circuit which accomplishes a very satisfactory transfer of the maximum signal in operation on a low-resistance load.

In nuclear physics experiments, pulses must be frequently transmitted over large distances with the aid of high-frequency cables having wave resistances between 50 and 150 Ω.

Cathode followers or two-stage cathode followers with a low output resistance are often used as electronic equipment for establishing a link between a low-resistance load and a sensor with a rather high internal resistance.

The present article brings an expression for determining the maximum input signal of a two-stage cathode follower and gives recommendations for the design of two-stage cathode followers used for transmission of input signals with large amplitudes.

Let us consider the two-stage cathode follower circuit of Fig. 1. According to another article,* the voltage between the grid and the cathode of tube L_1 is

$$U_{g\kappa 1} = U_{in}(1-K) = U_{in}\left[1 - \frac{S(1+SR_{a1})R_н}{(1+\alpha)(1+\varkappa)+S(1+SR_{a1})R_н}\right] = \frac{(1+\alpha)(1+\varkappa)U_{in}}{(1+\alpha)(1+\varkappa)+S(1+SR_{a1})R_н}. \tag{1}$$

The voltage in the anode circuit of tube L_1 is

$$U_{a1} = U_{g2} = \frac{U_{in}SR_{a1}(1+\varkappa)}{(1+\alpha)(1+\varkappa)+S(1+SR_{a1})R_н} \tag{2}$$

and, hence,

$$\frac{U_{g2}}{U_{g\kappa 1}} = \frac{SR_{a1}}{1+\alpha}, \tag{3}$$

where K denotes the transmission coefficient of the two-stage cathode follower; $\alpha = R_{a1}/R_i$, $\varkappa = R_a/R_i$; and U_{g2} denotes the signal applied to the control grid of tube L_2. It follows from

*V. A. Zapevalov, "Once More the Two-Stage Cathode Follower," this volume, p. 225.

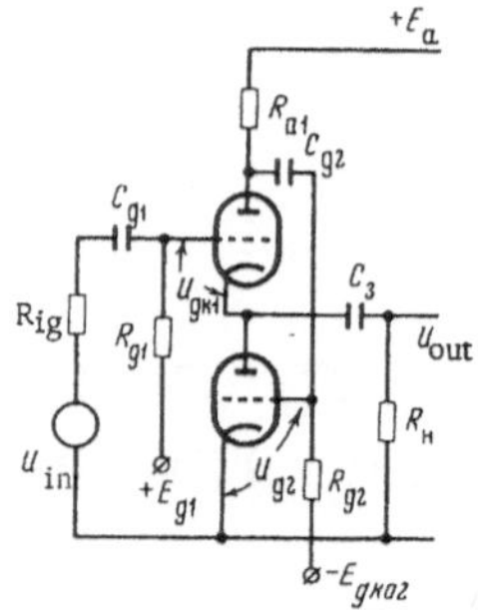

Fig. 1. Circuit of the two-stage cathode follower.

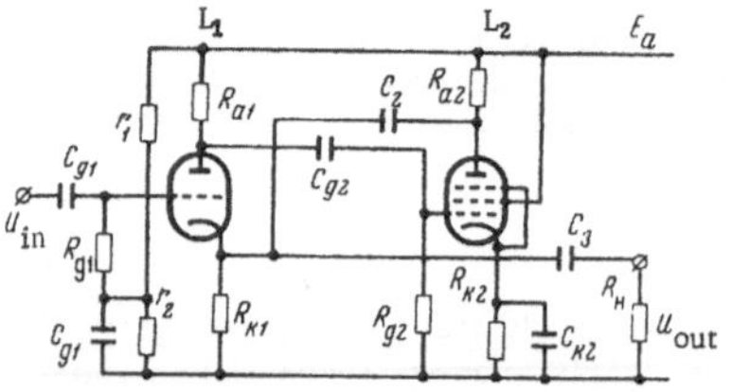

Fig. 2. Modified two-stage cathode follower circuit recommended for operation with a low-resistance load.

Eq. (3) that the signal applied to the control grid of tube L_2 is approximately SR_{a1} times greater than U_{gk1}.

Obviously, in order to increase the range of input signal amplitudes, tube L_2 must have a rather large cutoff voltage which must be greater that of tube L_1.

The conventional, extensively described two-stage cathode follower circuits (e.g., Fig. 1) are not suitable when a very large input signal must be handled by a particular tube and with a given supply voltage. This situation is encountered when tubes L_1 and L_2 are used in various operation conditions as current amplifiers (a unipolar input signal is considered). It is recommended that the basic two-stage cathode follower circuit shown in Fig. 2 be used in this case.

The circuit of Fig. 2 is particularly useful in the case of small load resistances because this circuit is free of any d.c. connection between tubes L_1 and L_2. The resistance R_H of expressions (1)-(3) is in this circuit equal to resistances R_{k1}, R_{a2}, and R_H connected in parallel.

The maximum amplitude of a unipolar input signal of this two-stage cathode follower can be obtained with the expression

$$U_{\text{in max}} \simeq \frac{U_{\text{cutoff}\,2}(1+\alpha)}{(1-K)\,SR_{a1}}, \tag{4}$$

where $U_{\text{cutoff}\,2}$ denotes the cutoff voltage of tube L_2.

THRESHOLD AMPLIFIER

V.A. Zapevalov

The amplifier is extensively used in equipment for research work in nuclear physics. The amplifier is designed mainly for amplification of a part of a spectrum of pulses with a statistical amplitude distribution, the amplification beginning from a certain amplitude threshold. The article outlines the calculation of the amplification coefficient of the amplifier, the maximum input signal, and the transient characteristics and discusses the improved linearity of the initial portion of the amplifier's transfer characteristic. The improvement was obtained with the "shifting threshold" and by nonlinear feedback.

Figure 1 shows the circuit of the threshold amplifier. An amplifier of this type is frequently used in electronic equipment for nuclear physics, e.g., as an amplifier with the amplification coefficient one in linear circuits of multichannel amplitude analyzers, in various amplitude discriminator circuits, etc.

We believe that the present article supplements the information which can be found in the literature on calculations of an amplifier of the above type.

Calculation of the Application Coefficient

The amplification coefficient of the threshold amplifier can be represented as a product of the amplification coefficient K_1 of a cathode follower (with tube L_1) with a cathode load equal to the input resistance of the circuit portion (tube L_2, R_{k12}) from the cathode times the amplification coefficient of an amplifier with cathode input and feedback through the cathode follower (Fig. 2).

A tube is occasionally inserted in these circuits in place of resistor R_{k12}. A pentode or a triode stage with current feedback can be used. The resistance from the anode side of these stages is $R_{i3} + (1 + \mu_3)R_{k3}$ and can be made large enough to avoid any effect on the transmission coefficient K_1.

Let us determine the amplification coefficient of the circuit shown in Fig. 2.

We assume that the generator of this circuit is a voltage generator, i.e., $R_{ig} = 0$. Figure 3 is the equivalent circuit to the circuit of Fig. 2. We have $\beta' = K_3 \frac{R_1}{R_1 + R_2}$; K_3 denotes the transmission coefficient of the cathode follower. The current derived from the generator is in the latter circuit given by

$$I = \frac{(\mu_2 + 1)\, U_g}{R_{i2} + R_{a2} + \mu_2\beta' R_{a2}} \simeq \frac{U_g}{\frac{1}{S_2} + \frac{R_{a2}}{1 + \mu_2} + \beta' R_{a2}}. \qquad (1)$$

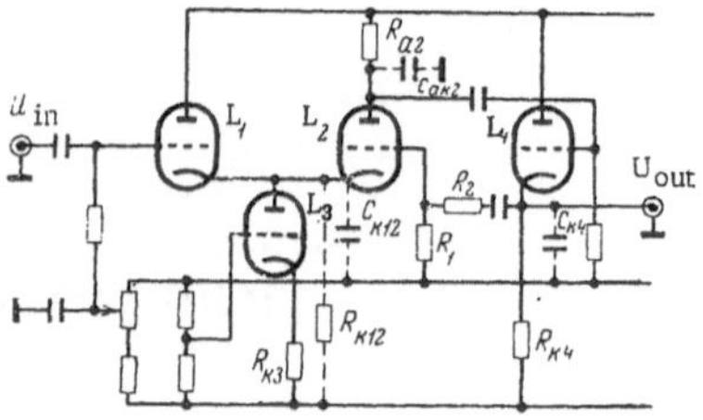

Fig. 1. Threshold amplifier circuit.

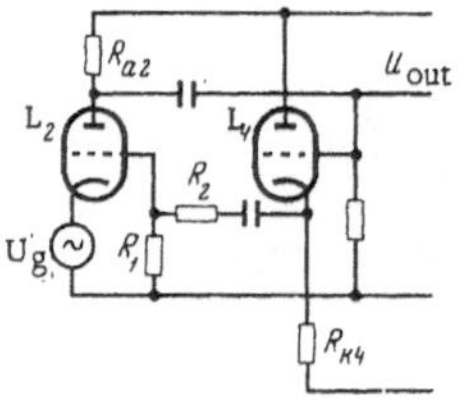

Fig. 2. To the determination of the amplification of a portion of the amplifier circuit.

It follows from Eq. (1) that the input resistance of the circuit of Fig. 2 is at the generator side equal to

$$R_{\text{in}} = \frac{1}{S_2} + \frac{R_{a2}}{1+\mu_2} + \beta' R_{a2}, \tag{2}$$

and in the case of a pentode, for which $\frac{R_{a2}}{1+\mu_2} \ll \frac{1}{S_2} + \beta' R_{a2}$, the input resistance is

$$R_{\text{in}} = \frac{1}{S_2} + \beta' R_{a2}. \tag{2'}$$

As can be inferred from Fig. 3, the amplification coefficient of the circuit of Fig. 2 is

$$K_2 = \frac{(1+\mu_2) R_{a2}}{R_{i2} + (1+\mu_2\beta') R_{a2}}. \tag{3}$$

Let us derive the transmission coefficient of tube L_1. It is assumed that $R_{k12} \gg R_{\text{in}}$:

$$K_1 = \frac{\mu_1 R_{\text{in}}}{R_{i1} + (1+\mu_1) R_{\text{in}}}. \tag{4}$$

The total amplification coefficient of the amplifier can be written in the form

$$K'_{\text{tot}} = K_1 K_2, \tag{5}$$

if the output voltage is taken from the anode of tube L_2; and

$$K''_{\text{tot}} = K_1 K_2 K_3, \tag{6}$$

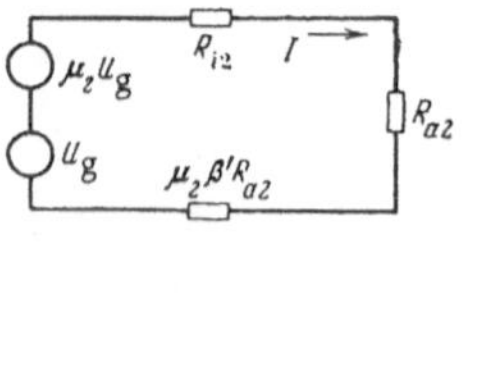

Fig. 3. Equivalent circuit of the amplifier circuit shown in Fig. 2.

if the output voltage is taken from the cathode of tube L_4. Thus, we have

$$K'_{tot} = \frac{\mu_1 R_{in}}{R_{i1} + (1+\mu_1) R_{in}} \frac{(1+\mu_2) R_{a2}}{R_{i2} + (1+\mu_2\beta') R_{a2}}, \quad (7)$$

$$K''_{tot} = \frac{\mu_1 R_{in}}{R_{i1} + (1+\mu_1) R_{in}} \frac{(1+\mu_2) R_{a2}}{R_{i2} + (1+\mu_2\beta'') R_{a2}} \frac{\mu_4 R'_{k4}}{R_{i4} + (1+\mu_4) R'_{k4}}, \quad (8)$$

where $\beta'' = \frac{R_1}{R_1 + R_2}$ and $R'_{k4} = \frac{(R_1 + R_2) R_{k4}}{R_1 + R_2 + R_{k4}}$.

The output resistance of the amplifier using tube L_2 as the output tube is

$$R'_{out} = \frac{\frac{1}{S_2} + R_{k12}}{\beta'}$$

if tube L_1 is shut off, and

$$R'_{out} = \frac{\frac{1}{S_2} + \frac{1}{S_1}}{\beta'}$$

if tube L_1 is conducting.

Maximum Input Voltage of the Amplifier

Let us calculate the maximum input voltage for the circuit of Fig. 2, i.e., we have to determine the U_g value at which tube L_2 is shut off.

We consider a variable component and, hence, the subscript 0 refers to constant voltages:

$$U_{gk2} = U_g - U_{gk2} K_{02} \beta'. \quad (9)$$

where K_{02} denotes the amplification coefficient of the stage with tube L_2 without feedback

$$U_{gk2\,max} = U_{cutoff\,2} - U_{gk02}. \quad (10)$$

When we assume in Eq. (9) $U_{gk2} = U_{gk2\,max}$, we obtain

$$U_{g\,max} = U_{gk2\,max}(1 + \beta' K_{02}). \quad (11)$$

The maximum input voltage of the first stage (with the maximum input voltage stated in Eq. (11) to be considered) can be graphically determined, as shown in Fig. 4. We plot the section OA = $U_{g\,max}$ to the right side of the abscissa. We determine $I'_{a0} = U_{g\,max}/R_{in}$ and plot this section to the ordinate (section OB). After that, we connect points A and B by a straight line. A straight line which is parallel to the abscissa axis is drawn from point B and intersects the characteristic of tube L_1 at point B'. A straight line is drawn parallel to AB from intersection point B'

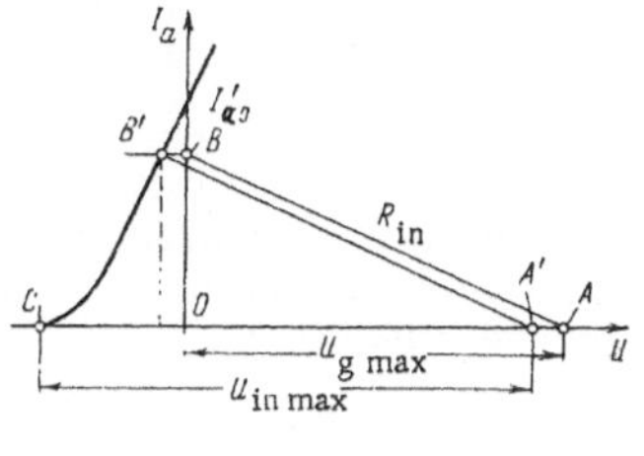

Fig. 4. To the determination of the maximum input voltage of the amplifier.

and intersects the abscissa axis. Section A'C is equal to the maximum input voltage of the entire amplifier.

It was assumed that the maximum input signal derived from the anode of tube L_2 is transmitted by the cathode follower with tube L_4. In calculating the maximum output signals, it was assumed that the pulses have positive polarity.

Transient Characteristic of the Amplifier

In calculating the transient characteristic $h'(t)$ (which corresponds to K'_{tot}), the following assumptions were made: 1) the influence of capacitor C_{k12} was disregarded (see Fig. 1); 2) the transient characteristic of the cathode follower with tube L_4 was disregarded; 3) β'' is independent of the frequency; and 4) the calculations are made under the assumption that tubes L_1 and L_2 are pentodes:

$$Z_{in}(p) = \frac{1}{S_2} + \frac{\beta' R_{a2}}{1 + p\tau_{a2}},$$

$$SZ_{in}(p) = \frac{1 + p\tau_{a2} + \beta' K_{02}}{1 + p\tau_{a2}},$$

$$K_1(p) = \frac{1 + p\tau_{a2} + \beta' K_{02}}{2(1 + p\tau_{a2}) + \beta' K_{02}}, \qquad K_2(p) = \frac{K_{02}}{1 + p\tau_{a2} + \beta' K_{02}},$$

$$K'_{tot}(p) = \frac{K_{02}}{2(1 + p\tau_{a2}) + \beta' K_{02}}.$$

According to the formula for switching on, we have

$$h(t) = \frac{M(0)}{N(0)} + \sum_{i=1}^{n} \frac{M(p_i)\, e^{p_i t}}{\left.\frac{dN}{dp}\right|_{p=p_i} p_i},$$

$$h'(t) = \frac{K_{02}}{2 + \beta' K_{02}} \left(1 - e^{-\frac{t}{\frac{2\tau_{a2}}{2+\beta' K_{02}}}}\right). \tag{12}$$

Thus, the rise time can be estimated with the formula

$$t'_r = 2.2 \frac{2\tau_{a2}}{2 + \beta' K_{02}}. \tag{13}$$

The transient process in this amplifier could be considered taking into account all capacitors and the frequency-dependent feedback. In this case we would obtain very complicated final expressions which are hard to use in actual calculations.

The exponential form of the transient characteristic of Eq. (12) makes it possible to use expressions for approximate calculations of the rise time of the entire amplifier

$$t'_{r\,tot} = \sqrt{t_{r1}^2 + t_r'^2} \tag{13'}$$

if the signal is derived from the anode of tube L_2, and

$$t''_{r\,tot} = \sqrt{t_{r1}^2 + t_r'^2 + t_{r4}^2} \tag{13''}$$

if the cathode of tube L_4 is used as the output terminal. The notation is interpreted as follows: t_{r1} denotes the rise time of the cathode follower with tube L_1 (calculated under the assumption $1/S_1 \ll R_{in}$); and t_{r4} denotes the rise time of the cathode follower with tube L_4.

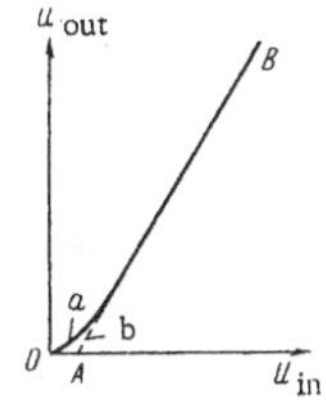

Fig. 5. Characteristic of voltage transmission by the amplifier. a) Characteristic of the uncompensated amplifier; b) characteristic which can be obtained by correcting the nonlinearity with the aid of the threshold-bias technique.

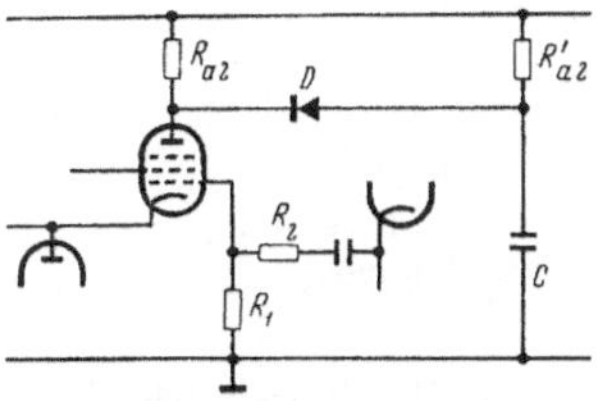

Fig. 6. Compensation for amplifier nonlinearity with the threshold-bias technique.

Obviously, nonlinearity of an amplifier of this type can be eliminated by a relationship $U_{out} = f(U_{in})$ represented by section OAB in Fig. 5. The amplification on section OA vanishes, but assumes the nominal value for all input pulses having amplitudes greater than the voltage represented by section OA. A threshold amplifier with this characteristic is an amplifier with an amplification threshold shifted by $\Delta U_{thresh} = UO_A$.

A characteristic resembling section OAB can be obtained by modifying the anode circuit of tube L_2 as shown in Fig. 6.

With a diode D and proper selection of the resistor ratio R_{a2}/R'_{a2}, we can obtain a voltage–current characteristic similar to that shown in Fig. 7.

The signal which appears in the cathode circuit of tubes L_1 and L_2 and at which the diode in the anode circuit of tube L_2 is completely shut off is denoted by U_{k12}. We can therefore design the amplifier so that once the diode has been shut off completely, the input signal biases tube L_1 to the very beginning of the linear section of the grid characteristic (point A in Fig. 8).

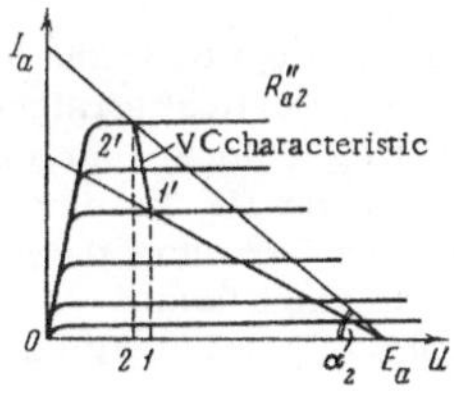

Fig. 7. Construction of the voltage–current characteristic for the circuit shown in Fig. 6.

$$\left(R''_{a2} = \frac{R_{a2} \cdot R'_{a2}}{R_{a2} + R'_{a2}}\right)$$

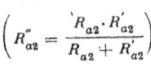

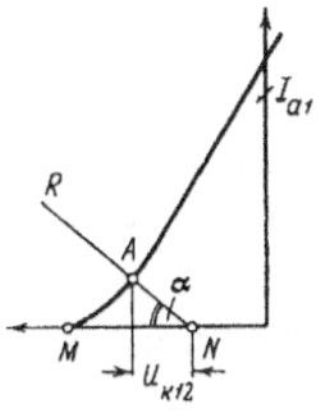

Fig. 8. To the determination of the threshold-shift voltage.

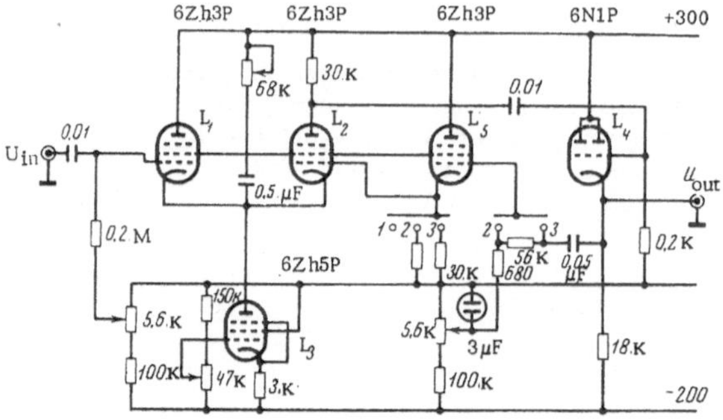

Fig. 9. Compensation for the nonlinearity of the threshold amplifier with the nonlinear feedback technique.

Let us assume that our threshold amplifier has some amplification and that the nonlinearity must be eliminated by shifting the threshold. Furthermore, we assume that we use some diode with forward resistance R_d (the diode characteristic is assumed to be linear).

The point which corresponds to the beginning of the linear section is found on the characteristic $I_{a1} = f(U_{gk1})$ (Fig. 8). We draw a straight line through this point in dependence upon the effective resistance in the cathode circuit of tube L_1, the resistance being $R = (1/S_2) + \beta' R_d$.

Voltage U_{k12} (Fig. 8) is defined as the voltage which in the anode circuit of tube L_2 causes complete shutoff of the diode, i.e., $U_d = U_{k12}K_2'$, where K_2' denotes the amplification coefficient of tube L_2 with feedback under the condition that diode D is not yet shut off, and U_d denotes the voltage appearing at the open diode.

In order to determine (in the anode circuit of tube L_2) the resistor R_{a2}' which with a particular diode produces a voltage drop at the forward resistance of the diode, so that with a voltage U_{in} producing in the cathode circuit of tubes L_1 and L_2 a signal U_{k12}, the diode is completely shut off, we must determine $R_{a2}'' = \dfrac{R_{a2}R_{a2}'}{R_{a2} + R_{a2}'}$ and, to do this under the assumption that R_{a2} is given:

1. We plot voltage U_d to the left side of point 1 (Fig. 7).
2. We draw a vertical straight line which is parallel to the ordinate from point 2.
3. We plot the voltage–current characteristic of resistors R_d and R_{a2} connected in parallel from point 1'. The intersection point of this characteristic with the line extending from point 2 renders the new quiescent point 2', i.e., the stationary point of the compensated amplifier.
4. We connect point 2' with point E_a on the abscissa so that

$$\alpha_2 = \tan^{-1} \frac{1}{\dfrac{R_{a2}R_{a2}'}{R_{a2} + R_{a2}'}}.$$

After that, we determine resistor R_{a2}'. Capacitor C is chosen so that during a pulse passing through the amplifier, the potential remains constant at the capacitor. The threshold shift ΔU_{thresh} is equal to section MN (Fig. 8).

Nonlinear feedback, which was proposed and tested by the author, is a second technique for improving the linearity of the amplifier.

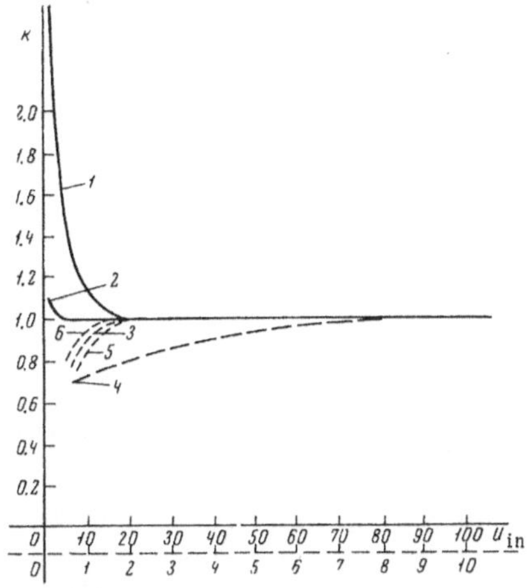

Fig. 10. Normalized transmission characteristics of a threshold amplifier with compensation by means of nonlinear feedback.

It has been proposed to use, as a nonlinear element in the feedback circuit, another tube L_1 which must be operated in a mode similar to that of tube L_1,* i.e., the tube should be shut off and the bias of tube L_5 should be variable to attain adjustment of optimum linearity (Fig. 9). Figure 10 shows the normalized transmission coefficient k for two amplification coefficients of the amplifier ($K''_{tot} = 10$, curves 3 and 4, and $K''_{tot} = 1$, curves 1 and 2).

The curves reveal that the nonlinearity is reduced by the introduction of a nonlinear element in the feedback circuit. For example, an overall nonlinearity of 4% is reached in a circuit without compensating tube, when $K''_{tot} = 10$ and the input signal is 6.3 V, whereas the same result is obtained with a compensating element, when the input signal amounts to 1.5 V. The curves show, in addition, that overcompensation can occur for $K''_{tot} = 1$ and that linearity can be adjusted by proper selection of the bias of tube L_5 (curves 2, 5, and 6).

* The cathode resistor of tube L_5 must be $(1/S_2)(1 + \beta' K_{02})$.

ONCE MORE THE METHOD OF INCREASING THE LINEARITY OF AMPLIFIERS WITH NEGATIVE FEEDBACK

V.A. Zapevalov

It is shown that the linearity of amplifiers with negative feedback can be improved by distributing the amplification among various amplifier stages so that the output stage renders the greatest contribution to the total amplification. The article outlines the calculation of the amplification coefficient of a stage whose anode load is a compound circuit, i.e., essentially a bipole with high resistance for alternating current.

Two terms which describe the nonlinearity of amplifiers, namely, overall nonlinearity and differential nonlinearity, have been introduced for amplifiers used in nuclear spectrometers.

Overall nonlinearity is defined as deviation of the actual transmission curve from the ideal transmission curve (which is a straight line). Differential nonlinearity is usually defined as a variation in the slope of the voltage-transmission curve, i.e., as a change in the amplification coefficient.

One or the other of these amplifier characteristics must be known when certain forms of a spectrum must be measured, e.g., when exact knowledge of the form of the measured spectrum is important, both integral and differential nonlinearity must be known; when only the energy of the peaks in the spectrum is of interest, it suffices to know the overall nonlinearity of the system. Distortions of the form of the spectrum due to differential nonlinearity result in a second-order effect.

Finally, if we wish to know the intensities of various spectral sections, linearity is not important because the area of the spectrum is independent of the transmission curve. Second-order errors due to nonlinearity, which arise in the determination of the area limits, are disregarded.

Figure 1 shows changes which result from amplifier nonlinearities in a pulse spectrum. We assume that we use an ideal, linear amplifier with a transmission characteristic $U_{out} = f(U_{in})$ shown in Fig. 1a (straight line 1). The form of the spectrum recorded with a multichannel differential amplitude analyzer is shown in Fig. 1b for this case (straight line 1). It is assumed that the input spectrum is such that the counting rate is everywhere the same in the equally spaced analyzer channels. But when the amplifier has the transmission curve of Fig. 1a (curve 2), the differential spectrum of the output pulses of the amplifier has the form shown in Fig. 1b (curve 2). It follows from Fig. 1 that various intervals of the input pulses correspond to equal intervals of the output pulses when the transmission curve is not linear.

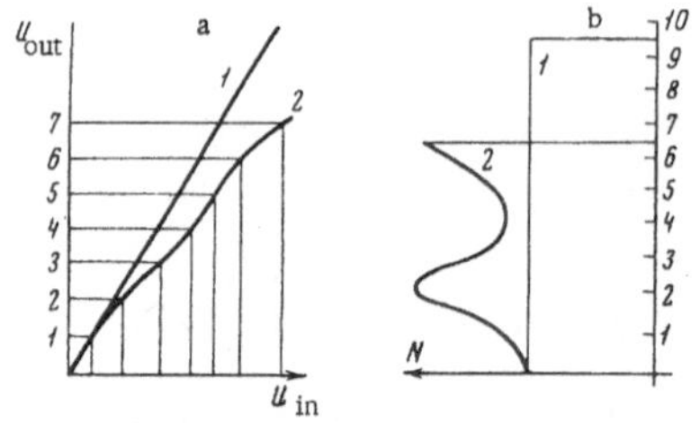

Fig. 1. Changes which amplifier nonlinearity produces in the form of a spectrum [1]. N denotes the counting rate.

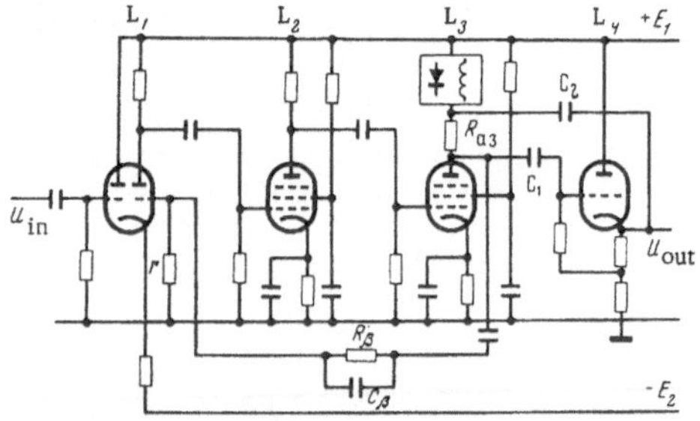

Fig. 2. Amplifier circuit in which our technique of improving linearity is used.

Strong negative feedback is an effective means of eliminating the nonlinearity of an amplifier system.

The circuit which is shown in Fig. 2 and which was used in, say, [2, 3] is one of the generally accepted circuits for reducing nonlinearity. The present article considers in detail the operation of this circuit and derives the amplification coefficient of tube L_3 with proper consideration of the effect of tube L_4.

When we consider the functioning of amplifiers with negative feedback (Fig. 2), we can easily show that for any particular value $K_0 = K_1K_2K_3K_4$, which corresponds to a certain amplification K_{fb} and to a certain stability of the amplification, the linearity increases with in-

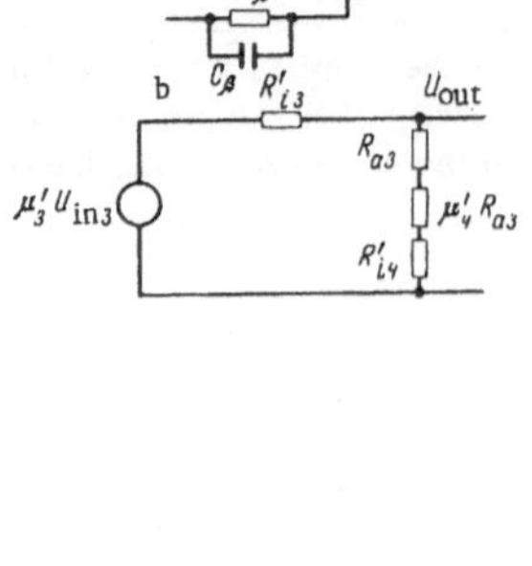

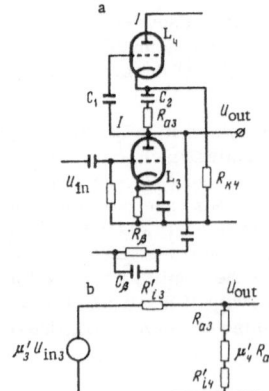

Fig. 3. Equivalent circuit of the last amplifier stages.

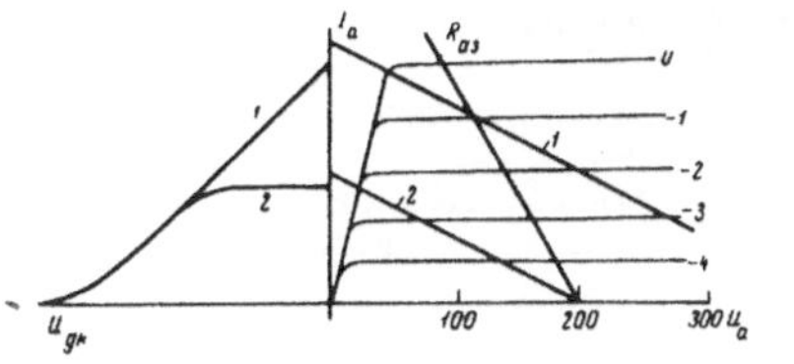

Fig. 4. Load lines. Curve 1 corresponds to the resistance of the two-terminal element I–I of Fig. 3 for a.c.; curve 2) corresponds to a conventional resistor whose resistance is equal to the resistance of the two-terminal element.

creasing amplification coefficient K_3 of the stage with tube L_3. We have

$$U_{g3} = U_{in} K'_{fb} = \frac{U_{in} K_{12}}{1 + \beta^* K_{12}} \simeq \frac{U_{in}}{\beta K_3},$$

where $\beta^* K_{12} \gg 1$; $K_4 = 1$; K'_{fb} denotes the amplification coefficient when the voltage applied to the grid of tube L_3 is considered output voltage; U_{g3} denotes the signal appearing at the grid of tube L_3; K_{12} is the product of the amplification coefficient of the first two tubes; $\beta^* = \beta K_3$. An increase in K_3 reduces the signal at the control grid of tube L_3 and, hence, improves the linearity of the amplifier.

An increase in the amplification of tube L_3 can be obtained either by a simple increase in the anode load or by modification of the circuit in the fashion shown in Fig. 2. An effective increase in the amplification of any stage can be obtained by using a two-terminal d.c. element as the load in the anode circuit (Fig. 3a, circuit section between points marked I). Figure 4 includes the load lines which correspond to a conventional resistor 2 and to a two-terminal resistor element 1 (R_d) for a.c. It follows from the figure that when a two-terminal element is used in the anode circuit, a greatly increased linearity and a higher amplification is obtained than in the circuit in which a simple resistor with resistance R_d is used as the load, because the load line corresponding to R_d is located within an area of increased curvature of the tube characteristic. This, in turn, corresponds to the linear section of the tube characteristic. These advantages are obtained with the low resistance of the two-terminal element for d.c., or, in the circuit shown in Fig. 2, with the small resistance R_{a3}, because the resistance in the cathode circuit of the tube of the two-terminal element and the tube itself are coupled through the variable component. We use the circuit shown in Figs. 3a and 3b for calculating the amplification coefficient of the circuit portion shown in Fig. 2 (tubes L_3 and L_4).

Let us determine the parameters of the tubes which are equivalent to tubes L_3 and L_4:

$$\mu'_3 = \mu_3 R/(R_{i3} + R), \tag{1}$$

$$R'_{i3} = R_{i3} \cdot R/(R_{i3} + R), \tag{2}$$

where

$$R = r + R_\beta,$$

$$\mu'_4 = \mu_4 \cdot R_{K4}/(R_{i4} + R_{K4}), \tag{1'}$$

$$R'_{i4} = R_{i4} \cdot R_{K4}/(R_{i4} + R_{K4}). \tag{2'}$$

we obtain from Fig. 3b

$$K_3 = \frac{U_{out}}{U_{in\,3}} = \frac{\mu'_3 [R'_{i4} + (1 + \mu'_4) R_{a3}]}{R'_{i3} + R'_{i4} + (1 + \mu'_4) R_{a3}}, \tag{3}$$

or, when we substitute Eqs. (1), (2), (1'), and (2') into Eq. (3); we find

$$K_3 = \frac{\dfrac{\mu_3 R}{R_{i3} + R}\left[\dfrac{R_{i4} \cdot R_{K4}}{R_{i4} + R_{K4}} + R_{a3}\left(1 + \dfrac{\mu_4 R_{K4}}{R_{i4} + R_{K4}}\right)\right]}{\dfrac{R_{i3} \cdot R}{R_{i3} + R} + \dfrac{R_{i4} \cdot R_{K4}}{R_{i4} + R_{K4}} + \left(1 + \dfrac{\mu_4 R_{K4}}{R_{i4} + R_{K4}}\right) R_{a3}}. \tag{4}$$

References

1. H. W. Koch and R. W. Johnston (eds.), Multichannel Pulse Height Analyzers, Proceedings of an Informal Meeting (1957).
2. E. Feirstein, Rev. Sci. Instr., 27:475 (1956).
3. A. A. Sanin, Electronic Equipment of Nuclear Physics [in Russian], Fizmatgiz (1961).

ACTIVE MAGNETIC MEMORY WITH FERRITE CORES HAVING A SMALL COERCIVE FORCE

V.A. Zapevalov

An active magnetic memory device with a capacity of 32 numbers of 16 binary digit bits each is described; the memory consists of 0.16 VT cores with low coercive force. A system with direct number access, provided with two cores for read-out and shifting, is used.

Two systems, namely, the system using the principle of matching currents and the system with direct number access [1, 2], are most widely used among the various magnetic memory systems with ferrite cores which have a rectangular hysteresis loop.

Figure 1 shows the ferrite core configuration in the magnetic memory block of a matching-current system. The magnetic memories of the majority of modern computers are designed according to the principle of matching currents.

Figure 2 shows the ferrite core configuration in a magnetic memory block with direct number access.

The functioning of these memory devices has been described in detail in the literature and need not be recited here.

Magnetic memories of national production, which employ the principle of direct number access, are usually designed in the form of a master system (z system with boosted read current).

Some modifications of the master system have been described in the literature and are found listed in the table of [3].

A magnetic memory device with 0.16 VT ferrite cores (3 × 2.2 × 1.4) having a low coercive force has been developed in the laboratory of photonuclear reactions of the Physics Institute of the Academy of Sciences of the USSR. The magnetic memory has a capacity of 32 numbers, with each number comprising 16 binary digit bits. The control current set and shift-current unit correspond to system B of [3].

Each bit of the numeric row consists of two cores, i.e., of one operating (main) core and one compensating core (Fig. 3). The compensating core improves the rectangular shape of the hysteresis loop of the main core and therefore increases the ratio of the unity read-out signal to the zero read-out signal. The shift current of the memory matrix is 60–70 mA, and the write current amounts to about 130 mA.

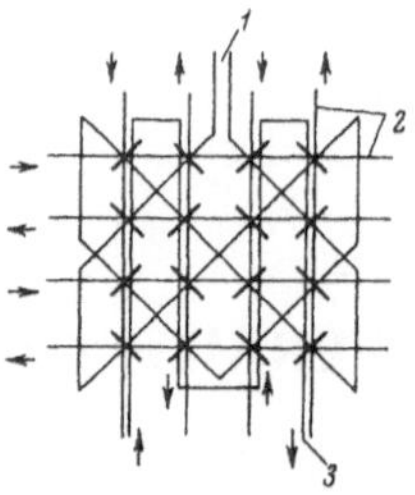

Fig. 1. Configuration of the cores in a magnetic memory device working as a system with matching currents. 1) Read driver; 2) coordinate wires; 3) write driver.

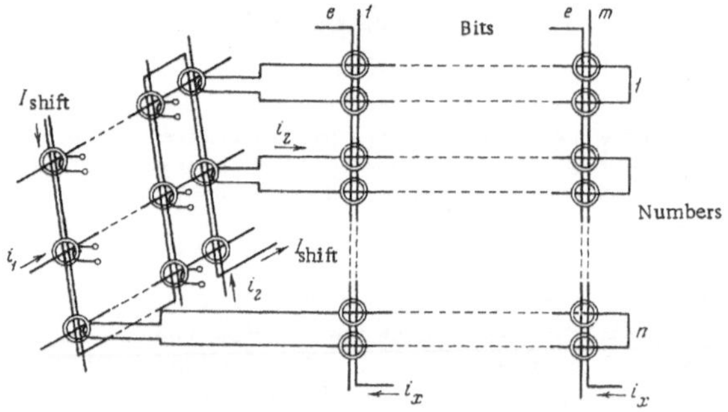

Fig. 2. Configuration of the cores in a magnetic memory device with direct number access.

When the magnetic memory is referenced, only one core of the magnetic decoder (Fig. 4a) is actuated. This is equivalent to a potential applied at one of the horizontal bus bars and to a current flowing through one of the vertical bus bars.

Differentiating transformers which transform unipolar current pulses into bipolar pulses in the output drive wire (numeric driver) w_3 are used as transformers in the decoder.

One bit of the number register of the magnetic memory device is shown in Fig. 5.

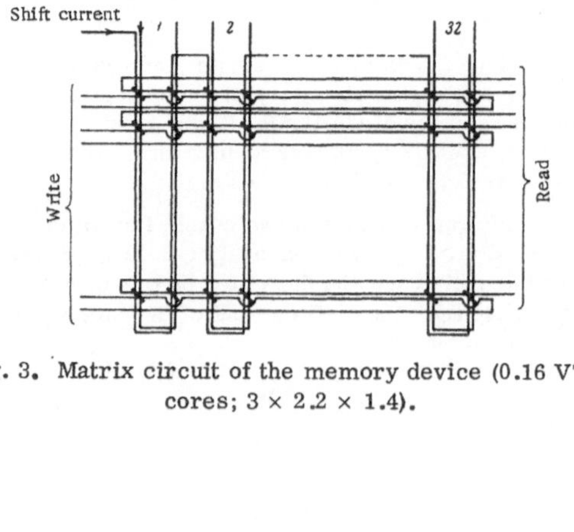

Fig. 3. Matrix circuit of the memory device (0.16 VT cores; 3 × 2.2 × 1.4).

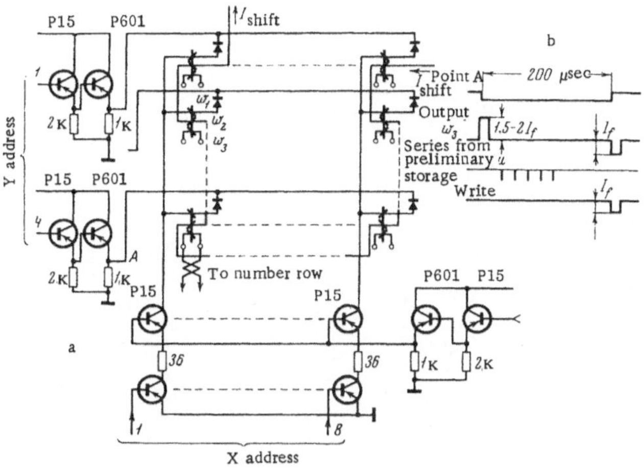

Fig. 4. Magnetic decoder (TR-VT-5, 7 × 5 × 2, w_1 = 50 V, w_2 = 15 V, and w_3 = 12 V).

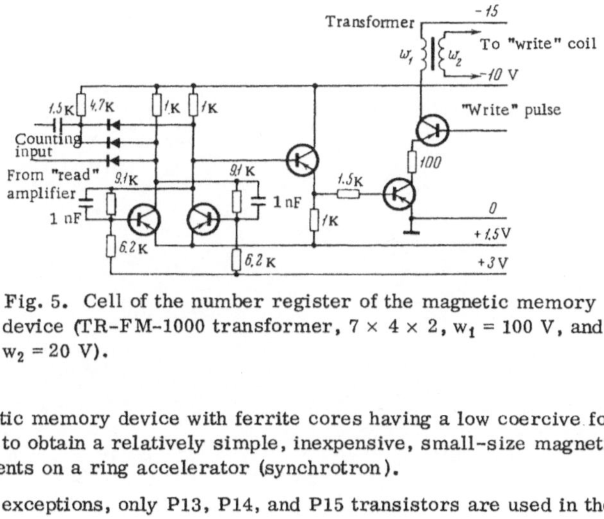

Fig. 5. Cell of the number register of the magnetic memory device (TR-FM-1000 transformer, 7 × 4 × 2, w_1 = 100 V, and w_2 = 20 V).

The magnetic memory device with ferrite cores having a low coercive force was developed in order to obtain a relatively simple, inexpensive, small-size magnetic memory for physics experiments on a ring accelerator (synchrotron).

With a few exceptions, only P13, P14, and P15 transistors are used in the magnetic memory device.

References

1. I. A. Rajchman, Proc. IRE, January (1961).
2. V. V. Bardish, "Digital Magnetic Elements" [Russian translation], Nauka Press (1965).
3. D. B. G. Edwards, M. I. Lenigan, and T. Kilburn, Inst. Electr. Eng., 107(Part B):36 (1960).
4. O. V. Bogdankevich, Atomnaya Énergiya, 3:12 (1962).

INTERMEDIATE MEMORY DEVICE

A.M. Goryachev and V.A. Zapevalov

The logic circuits distributing read and write pulses over the rows of an intermediate core are described. The incoming five-digit binary code is written into the first unfilled row, whereas writing is effected from the first filled row.

In experiments with accelerators for charged particles which are delivered in a pulsed beam, the useful information arrives generally in the form of pulse groups so that the average interval between the events in a group can be much shorter than the resolving time of the currently available recording equipment.

The response of multichannel recording systems, which is accordingly limited, results in counting losses and errors in the results.

Counting losses can be greatly reduced when an intermediate memory device with a short dead time is inserted before the main recording equipment [1-3].

When batch reading is used, the information at the output of the intermediate memory is distributed so that the minimum interval between two groups of code pulses is defined by the reference cycle for the next main recording system for which a ferrite memory is usually employed.

The present article is a short description of an intermediate memory device with three lines. Static transistorized triggers are the information carriers in the lines.

Each line of the intermediate memory device (Fig. 1) consists of six triggers, i.e., five code triggers and one trigger of the code sign.

The unity state of the trigger of the code sign indicates that a code has been written in that line; the null state means that no code is present. The triggers of the code sign are connected to logic circuits which distribute the write and read pulses over the lines of the intermediate memory device. The principle of writing a five-mark code, which appears at inputs 1-5 (Fig. 1), into the first unfilled line is used in the operation of the logic circuit for writing. The logic circuit for reading the stored information implies return to the first filled line after delivery of information.

These read and write principles are characterized by the advantage that when information appears in statistical sequence for writing and when information is read out in batches, the recorded codes are grouped in small number cells which increase, to some extent, the fast response during the write operation, because the transfer time is reduced in the logic circuit. When a line of the intermediate memory device has been completely filled, the next write pulse

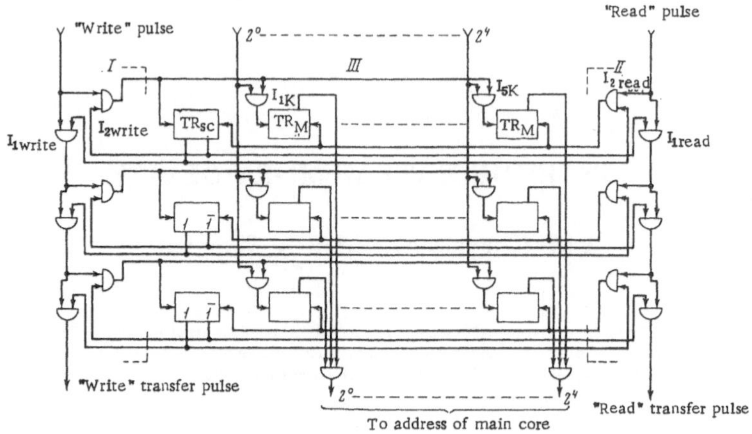

Fig. 1. Simplified block diagram of the intermediate memory device. I) "Write" logic block; II) "read" logic block; III) memory block.

passes to the output of the logic circuit for writing and, hence, the number of overflows can be easily recorded.

Therefore, an additional blocking circuit is not required for the input of the write logic since no information can be written into any of the lines in the case of an overflow. When information has been read out from all lines, the read-in pulse appears at the output of the logic circuit for reading and blocks the trigger of the control circuit of the main core.

In order to describe the system in read and write operation (Fig. 1), logic expressions can be used.

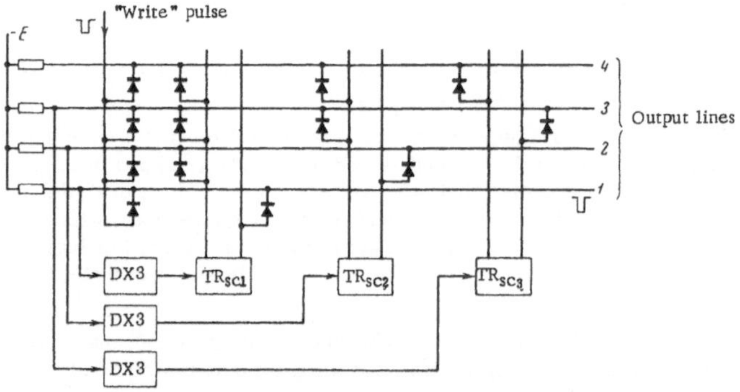

Fig. 2. Write circuit of the intermediate memory device. TR_{sc} denotes a trigger of the code sign; in the null state, the right triode is closed, the left open. D denotes a trigger-actuating delay circuit; the delay time is equal to the length of the write pulse.

TABLE 1

State of the triggers of the coil size			Output lines			
1	2	3	1	2	3	4
0	0	0	1	0	0	0
1	0	0	0	1	0	0
0	1	0	1	0	0	0
1	1	0	0	0	1	0
0	0	1	1	0	0	0
1	0	1	0	1	0	0
0	1	1	1	0	0	0
1	1	1	0	0	0	1

The condition that a code be transferred into a free line can be written in the following form:

$$s_i = p_{i-1}\bar{y}_i \quad (i = 1, 2, \ldots, n). \tag{1}$$

where s_i denotes the write signal in line i; p_{i-1} denotes the signal for the write transfer into line i, after filling the preceding lines; and y_i denotes the state of the trigger of the code sign of the intermediate memory device.

The condition that a write pulse is transferred into line i + 1 has the following form:

$$p_i = p_{i-1}y_i. \tag{2}$$

Equation (1) can be written in the form

$$s_i = p_0\bar{y}_i y_{i-1} \ldots y_1, \tag{3}$$

where p_0 denotes the write signal at the input of the circuit.

The condition for reading a code from the i-th filled line is given by the expression

$$s_i' = p_{i-1}' y_i \quad (i = 1, 2, \ldots n), \tag{4}$$

$$p_i' = p_{i-1}'\bar{y}_i, \tag{5}$$

where s_i' denotes the signal for read-out from line i; p_{i-1}' denotes the signal for read-out transfer into line i if the preceding lines are free; and p_0' denotes the signal at the input of the read circuit. Figure 1 shows the circuit in which this principle is employed.

In the actual operation of the basic circuit of Fig. 1, the write time depends upon the number of the line, due to the delays in the transfer circuits.

The logic expression of Eq. (3) makes it possible to build a circuit for writing the codes into the lines of the intermediate memory device (Fig. 2) so that the write time is independent of the number of the line.

The triggers of the code sign are actuated by delayed write pulses which appear at the bus bars of the lines.

The operation of the circuit is explained by Table 1.

References

1. R. E. Bell, Canad. J. Phys., 34:563 (1956).
2. I. V. Shtranikh, Reports of the 5th Scientific-Technological Conference on Nuclear Electronics, Vol. 2, Part 1, p. 47 [in Russian], Atomizdat (1963).
3. B. E. Zhuravlev, T. Shetet, and V. D. Shibaev, Preprint OIYaI 10-3120 [in Russian], Dubna (1967).